TEXTBOOK OF QUANTUM PHYSICS

TEXTBOOK OF QUANTUM PHYSICS

By

Dr. Ajit Kumar Sharma

Deptt. of Physics
D.A.V. (P.G.) College
Bulandshahr
(U.P.)

DISCOVERY PUBLISHING HOUSE PVT. LTD.
NEW DELHI-110 002

Published by:

Tilak Wasan

DISCOVERY PUBLISHING HOUSE PVT. LTD.
4383/4B, Ansari Road, Darya Ganj
New Delhi-110 002 (India)
Phone : +91-11-23279245, 23253475, 43596065
E-mail : discoverypublishinghouse@gmail.com
sales@discoverypublishinggroup.com
web : www.discoverypublishinggroup.com

Edition: **2020**

ISBN: 978-81-8356-849-4

Textbook of Quantum Physics

Printed at:
Infinity Imaging Systems
Delhi

Preface

The present book deals with " Text Book of Quantum Physics" in general properties of matter and sound. Reason fore this change is that paper division in the examination is different Universities. One of salient course syllabi features of the present edition of the book is the inclusion of a large number of typical worked out problems which will elucidate various abstract principles and theories discussed in the text. It covers the complete syllabus of physics prescribed by Technical Universities. The treatment given is simple, lucid and comprehensive. The language used is simple.

The classification of the questions, specially the short-answer type questions, will, I hope, help the students to prepare for the final examination. I have spared no pains in applying my long experience of degree class teaching to make the book useful to the students.

I shall, therefore, consider, my labours amply rewarded if the book receives warm reception and appreciation from the teachers and the taught. Though nothing can be claimed as original but the subject matter has been arranged in my own style.

Suggestion for improvement of the book will be gratefully acknowledged.

— *Author*

Contents

1

Impulsive Forces and Quadratic Function of Velocities

INTRODUCTION

When the forces are finite,

$$\frac{d}{dt}\left(\frac{\partial T}{\partial \dot{q}_\alpha}\right)-\frac{\partial T}{\partial q_\alpha}=\phi_\alpha \text{ where } \phi_\alpha=\sum_v V_v\cdot\frac{\partial r_v}{\partial q_\alpha}$$

Integrating both sides of (1) w.r.t. "t" form $t = 0$ to $t = \tau$,

we get

$$\left[\left(\frac{\partial T}{\partial \dot{q}_\alpha}\right)\right]_{t=0}^{t=\tau}-\int_0^\tau \frac{\partial T}{\partial q_\alpha}dt=\int_{t=0}^{\tau}\phi_\alpha dt=\sum_v\left\{\left(\int_{t=0}^{\tau}F_v dt\right).\frac{\partial r_v}{\partial q_\alpha}\right\} \qquad ...(1)$$

Taking the limit as $\tau\to 0$, we get

$$\lim_{\tau\to 0}\left\{\left(\frac{\partial T}{\partial \dot{q}_\alpha}\right)_{t=\tau}-\left(\frac{\partial T}{\partial \dot{q}_\alpha}\right)_{t=0}\right\}-\lim_{\tau\to 0}\int_0^\tau\frac{\partial T}{\partial q_\alpha}dt$$

$$=\sum_v\left\{\left(\lim_{\tau\to 0}\int_0^\tau F_v dt\right)\cdot\frac{\partial r_v}{\partial q_\alpha}\right\}$$

$$\Rightarrow\left(\frac{\partial T}{\partial \dot{q}_\alpha}\right)_1-\left(\frac{\partial T}{\partial \dot{q}_\alpha}\right)_0=\Sigma^* 1_v\cdot\frac{\partial r_v}{\partial q_\alpha}=**P_\alpha\text{(say)} \qquad ...(2)$$

where subscripts 0 and 1 denote respectively quantities before and after the application of the impulsive force.

These equations are known as Lagrange's equations under impulsive rces.

LAGRANGIAN FUNCTION

If the forces are derivable from a potential function V, then

$$\phi_\alpha = \frac{\partial W}{\partial q_\alpha} = -\frac{\partial V}{\partial q_\alpha},$$

since the potential, or potential energy is a function of q's only (and possibly the name t) then, we have

$$\frac{d}{dt}\left(\frac{\partial T}{\partial \dot{q}_\alpha}\right) - \frac{\partial T}{\partial q_\alpha} = -\frac{\partial V}{\partial q_\alpha} \Rightarrow \frac{d}{dt}\left[\frac{\partial}{\partial \dot{q}_\alpha}(T-V)\right] - \left(\frac{\partial T}{\partial q_\alpha} - \frac{\partial V}{\partial q_\alpha}\right) = 0$$

$$\Rightarrow \frac{d}{dt}\left(\frac{\partial L}{\partial \dot{q}_\alpha}\right) - \frac{\partial L}{\partial q_\alpha} = 0, \text{ where } L = T - V \qquad ...(1)$$

The function L defined by L = T – V is said to be Lagrangian function.

QUADRATIC FUNCTION OF VELOCITIES

If at time t, the position of the vth particle (mass m), of a holonomic system is defined by r_v then K.E. is given by

$$T = \frac{1}{2}\Sigma_{v=1}^{N} m_v \dot{r}_v^2,$$

where $r_v = r_v\,(q_1, ...q_n; t)$ ...(1)

So that $\dot{r}_v = \dot{q}_1 \frac{\partial r_v}{\partial q_1} + \dot{q}_2 \frac{\partial r_v}{\partial q_2} + ... + \dot{q}_n \frac{\partial r_v}{\partial q_n} + \frac{\partial r_v}{dt}$

$$\Rightarrow T = \frac{1}{2}\Sigma_{v=1}^{N} m_v \left(\dot{q}_1 \frac{\partial r_v}{\partial q_1} + \dot{q}_2 \frac{\partial r_v}{\partial q_2} + ... + \dot{q}_n \frac{\partial r_v}{\partial q_n} + \frac{\partial r_v}{\partial t}\right)^2$$

$$= \frac{1}{2}[(a_{11}\dot{q}_1^2 + a_{22}\dot{q}_2^2 + ... + a_{nn}\dot{q}_n^2 + 2a_{12}\dot{q}_1\dot{q}_2 + 2a_{1n}\dot{q}_1\dot{q}_n + ...)$$

$$+ 2(a_1\dot{q}_1 + a_2\dot{q}_2 + . + a_n\dot{q}_n) + a] \qquad ...(2)$$

$$\left.\begin{array}{l} \text{where } 5a_{rs} = \Sigma_{v=1}^{N} m_v(\partial r_v/\partial q_r).(\partial r_v/\partial q_s)(s \geq r) \\ a_{rr} = \Sigma_{v=1}^{N} m_v(\partial r_v/\partial q_r)^2, a = \Sigma_{v=1}^{N} m_v(\partial r_v/\partial t)^2, = \Sigma_{v=1}^{N} m_v\left(\frac{\partial r_v}{\partial q_r}\right).\left(\frac{\partial r_1}{\partial t}\right) \end{array}\right\}$$

From (2), we see that T is a quadratic function of the of the generalised velocities.

The case t is not explicity involved, is of considerable importance.

Hence we have $\frac{\partial r_v}{\partial t} = 0$ and therefore (2) implies that

$$T=\frac{1}{2}(a_{11}\dot{q}_1^2+a_{22}\dot{q}_2^2+\ldots+a_{nn}\dot{q}_n^2+2a_{12}\dot{q}_1\dot{q}_2+\ldots) \quad \ldots(3)$$

$$=\frac{1}{2}\Sigma_{s=1}^{n}\Sigma_{r=1}^{n}a_{rs}\dot{q}_r\dot{q}_s \text{ where } a_{rs}=a_{sr.} \quad \ldots(4)$$

Now using Euler's theorem for homogeneous functions, we get

$$\dot{q}_1\frac{\partial T}{\partial \dot{q}_1}+\dot{q}_2\frac{\partial T}{\partial \dot{q}_2}+\ldots+\dot{q}_n\frac{\partial T}{\partial \dot{q}_n}=2T$$

$$\Rightarrow 2T=\Sigma_{\alpha=1}^{n}\dot{q}_\alpha\frac{\partial T}{\partial \dot{q}_\alpha}p_\alpha\dot{q}_\alpha$$

i.e. $2T=p_1\dot{q}_1+p_2\dot{q}_2+\ldots+p_n\dot{q}_n$

TO DEDUCE THE PRINCIPLE OF ENERGY FROM THE LAGRANGE'S EQUATIONS (CONSERVATIVE FIELD)

Lagrange's equations are:

$$\frac{d}{dt}\left(\frac{\partial T}{\partial \dot{q}_\alpha}\right)-\frac{\partial T}{\partial q_\alpha}=-\frac{\partial V}{\partial q_\alpha}; \qquad (\alpha=1,2,\ldots.n) \quad \ldots(1)$$

Also by 7.11 equation we know that

$$T=\frac{1}{2}(a_{11}\dot{q}_1{}^2+a_{22}\dot{q}^2+\ldots.+a_{m}\dot{q}_n{}^2+2a_{12}\dot{q}_1\dot{q}_2+\ldots)$$

that is, T can be expressed as a quadratic expression in generalised velocities. Hence applying Euler's theorem. We get

$$\Sigma_{\alpha=1}^{n}\dot{q}_\alpha\frac{\partial T}{\partial \dot{q}_\alpha}=2T \quad \ldots(2)$$

$$\text{Also } \frac{dT}{dt}=\Sigma_{\alpha=1}^{n}\frac{\partial T}{\partial q_\alpha}\dot{q}_\alpha+\Sigma_{\alpha=1}^{n}\frac{\partial T}{\partial \dot{q}_\alpha}\ddot{q}_\alpha \quad \ldots(3)$$

Now multiplying the n equations of (1) by $\dot{q}_1,\dot{q}_2,\ldots\dot{q}_n$, respectively and then adding. we get

$$\left\{\dot{q}_1\frac{d}{dt}\left[\frac{\partial T}{\partial \dot{q}_1}\right]+\ldots+\dot{q}_n\frac{d}{dt}\left[\frac{\partial T}{\partial \dot{q}_1}\right]\right\}-\left\{\dot{q}_1\frac{\partial T}{\partial q_1}+\ldots+\dot{q}_n\frac{\partial T}{\partial \dot{q}_n}\right\}$$

$$=-\left\{\dot{q}_1\frac{\partial V}{\partial q_1}+\ldots+\dot{q}_n\frac{\partial V}{\partial q_n}\right\}$$

$$\Rightarrow\frac{d}{dt}\left\{\Sigma_{\alpha=1}^{n}\dot{q}_\alpha\frac{\partial T}{\partial \dot{q}_\alpha}\right\}-\left(\Sigma_{\alpha=1}^{n}\ddot{q}_\alpha\frac{\partial T}{\partial \dot{q}_\alpha}\right)-\left\{\Sigma_{\alpha=1}^{n}\dot{q}_n\frac{\partial T}{\partial q_\alpha}\right\}$$

$$= -\left(\Sigma_{\alpha=1}^{n} \dot{q}_{\alpha} \frac{\partial V}{\partial q_{n}}\right)$$

$$\Rightarrow \frac{d}{dt}(2T) - \frac{dT}{dt} = -\frac{dV}{dt} \Rightarrow \frac{dT}{dt} + \frac{dV}{dt} = 0 \Rightarrow \frac{d}{dt}(T+V) = 0 \Rightarrow T+V = \text{constant.}$$

SMALL OSCILLATIONS

To explain how Lagrange's equations are used in case of small oscillations. (when there exist three generalised co-ordinates),

In order to investigate the theory of small oscillations by the use of Lagrange's equation about the position of equilibrium, the generalised co-ordinates q_1, q_2, q_3 must be chosen in such a way that they vanish in the position of equilibrium.

But the system makes small oscillations about the position of equilibrium, so q_1, q_2, q_3 and their differentials $\dot{q}_2$, $\dot{q}_2$, $\dot{q}_3$ will remain small during the whole motion. Hence we shall reject all the powers of small quantities except the lowest one.

Now, let T be kinetic energy, and W the work-function of the system; then we have

$$T = A_{11}\dot{q}_1{}^2 + A_{22}\dot{q}_2{}^2 + A_{33}\dot{q}_3{}^2 + 2A_{12}\dot{q}_2\dot{q}_1 + 2A_{23}\dot{q}_2\dot{q}_3 + 2A_{13}\dot{q}_1\dot{q}_3 \quad ...(1)$$

$$W = C + B_1q_1 + B_2\, q_2 + B_3q_3 + B_{11}q_1{}^2 + B22\, q_2{}^2 + q_3{}^2 \quad ..(2)$$

Now assume that q_1, q_2, q_3 can be expressed in terms of X, Y, Z by the equations of the form

$$q_1 = \lambda_1 X + \lambda_2 Y + \lambda_3 Z; q_2 = \mu_1 X + \mu_2 Y + \mu_3 Z; \quad q_3 = v_1 X + v_2 Y + v_3 Z;$$

and further assume $\lambda_1, \lambda_2, \lambda 3, \mu_1, \mu_2, \mu_3, v_1, v_2, v_3$ in such a way that if we put value of q_1, q_2, q_3 in (1) and (2), then T does not involve $\dot{Y}\dot{Z}$, $\dot{Z}\dot{X}$, $\dot{X}\dot{Y}$ and W does not involve YZ, ZX, XY. Then X,T, Z, are called the *normal or Principal Co-ordinates.*

Thus equation (1) and (2) $\Rightarrow$

$$T = A'_{11}\dot{X}^2 + A'_{22}\dot{Y}_{22}\dot{Y}^2 + A'_{33}\dot{Z}^2$$

$$W = C' + B_1X + B'_2Y + B'_3Z + B'_{11}X^2 + B_{22}Y^2 + B'_{33}Z^2$$

Now, using Lagrange's equations,

viz. $\frac{d}{dt}\left(\frac{dT}{d\dot{X}}\right) - \left(\frac{dT}{dX}\right) = \left(\frac{dW}{dX}\right)$ etc., we obtain

$$2A'_{11}\ddot{X} = B'_1 + 2B'_{11}X; \quad 2A'_{22}\ddot{Y} = B'_2 + 2B'_{22}Y; \quad 2A'_{33}\ddot{Z} = B'_3 + 2B'_{33}Z$$

which can be put in the forms

$$\ddot{X} = -\omega^2{}_1 X, \ddot{Y} = -\omega^2{}_3 Y, \ddot{Z} = -\omega^2{}_3 Z \text{ etc.}$$

which represents simple Harmonic Motions, Or in other words they give small oscillations about the position equilibrium

DEGREES OF FREEDOM

The number of independent co-ordinates required to specify the position of a system of one or more particles is called the number of degrees of freedom of the system.

Example 1:

A particle moving freely in space require 3 co-ordinates, e.g. (x, y, z), to specify its position. Thus the number of degrees of freedom is 3.

Example 2:

A system containing of N-particles moving freely in space require 3N co-ordinates to specify the position. The number of degrees of freedom is 3N.

A rigid body which can move freely in space has 6 degrees of freedom i.e. 6. co-ordinates are required to specify the position.

Let 3 non-collinear points of a rigid body be fixed in space, then the rigid body also fixed in space. Let these points have co-ordinates (x_1,y_1,z_1); (x_2,y_2,z_2); (x_3,y_3,z_3) respectively, a total of 9. Since the body is rigid, we must have

$$(x_1 - x_2)^2 + (y_1 - y_2)^2 + (z_1 - z_2)^2 = \text{constant.}$$

$$(x_2 - x_3)^2 + (y_2 - y_3)^2 + (z_2 - z_3)^2 = \text{constant.}$$

$$(x_3 - x_1)^2 + (y_3 - y_1)^2 + (z_3 - z_1)^2 = \text{constant.}$$

Hence 3 co-ordinates can be expressed in terms of the remaining six. Thus six independent co- ordinates are needed to describe the motion *i.e.* there exit six degrees of freedom.

LAGRANGE'S EQUATION

Let F be the net external force acting on the Vth particle of a system, then bt Newton's second law applied to Vth particle, we have

$$m_v \ddot{r}_v = F_v$$

$$\Rightarrow m_v \ddot{r}_v \bullet \frac{\partial r_v}{\partial q_\alpha} = F_v \bullet \frac{\partial r_v}{\partial q_\alpha} \qquad ...(1)$$

$$\Rightarrow \Sigma_{v=1}^{N} m_v \ddot{r}_v \bullet \frac{\partial r_v}{\partial q\alpha} = \Sigma_v^N F_v \bullet \frac{\partial r_v}{\partial q\alpha}$$

$$\Rightarrow \frac{d}{dt}\left(\Sigma_{v=1}^{N} m_v \dot{r}_v \bullet \frac{\partial r_v}{\partial q_\alpha}\right) - \Sigma_{V=1}^{N} m_v \dot{r}_v \bullet \frac{d}{dt}\left(\frac{\partial r_v}{\partial q\alpha}\right) = \Sigma_{v=1}^{N} F_v \bullet \frac{\partial r_v}{\partial q_\alpha}$$

But $r_v = r_v(q_1, q_2, ..., q_n; t)$ ...(2)

$$\therefore \dot{r}_v = \frac{\partial r_v}{\partial q_1}\dot{q}_1 + \frac{\partial r_v}{\partial q_2}\dot{q}_2 + ... + \frac{\partial r_v}{\partial q_n}\dot{q}_n + \frac{\partial r_v}{\partial q\alpha} \quad ...(3)$$

$$\Rightarrow \frac{\partial \dot{r}_v}{\partial \dot{q}\alpha} = \frac{\partial r_v}{\partial q\alpha} \quad \text{[Cancellation law of the dots]} \quad ...(4)$$

$$\text{Also, } \frac{\partial \dot{r}_v}{\partial q\alpha}(\dot{r}_v) = \frac{\partial}{\partial q\alpha}\left(\frac{\partial r_v}{\partial q_1}\dot{q}_1 + \frac{\partial r_v}{\partial q_2}\dot{q}_2 + ... + \frac{\partial r_v}{\partial q_n}q_{\dot{n}} + \frac{\partial r_v}{\partial t}\right)$$

$$= \frac{\partial^2 r_v}{\partial q_\alpha \partial q_1}\dot{q}_1 + \frac{\partial^2 r_v}{\partial q_\alpha \partial q_1}\dot{q}_2 + ... + \frac{\partial^2 r_v}{\partial q_\alpha \partial q_n}\dot{q}_n + \frac{\partial}{\partial q_\alpha}\left(\frac{\partial r_v}{\partial t}\right)$$

$$= \frac{\partial}{\partial q_1}\left(\frac{\partial r_v}{\partial q_\alpha}\right)\dot{q}_1 + \frac{\partial}{\partial q_2}\left(\frac{\partial r_v}{\partial q_\alpha}\right)\dot{q}_2 + ... + \frac{\partial}{\partial q_\alpha}\left(\frac{\partial r_v}{\partial q_n}\right)\dot{q}_n + \frac{\partial}{\partial t}\left(\frac{\partial r_v}{\partial q_\alpha}\right)$$

$$\text{or } = \frac{\partial}{\partial q_\alpha}\left(\frac{\partial r_v}{\partial t}\right) = \frac{d}{dt}\left(\frac{\partial r_v}{\partial q_\alpha}\right) \Rightarrow \frac{d}{dt}\left(\frac{\partial}{\partial q_\alpha}\right) \equiv \frac{\partial}{\partial q\alpha}\left(\frac{d}{dt}\right) \quad ...(5)$$

[interchange law of the order of operators]

$$\text{Now } \frac{d}{dt}\left\{\Sigma_{v=1}^{N} m_v \dot{r}_v \bullet \frac{\partial r_v}{\partial q_\alpha}\right\} - \Sigma_{v=1}^{N} m_v \dot{r}_v \bullet \frac{\partial \dot{r}_v}{\partial q_\alpha} = \Sigma_{v=1}^{N} F_v \bullet \frac{\partial r_v}{\partial q_\alpha} \quad ...(6)$$

$$\text{and } T = \frac{1}{2}\sum_v m_v \dot{r}_v^2 = \frac{1}{2}\sum_v m_v(\dot{v}_v \cdot \dot{r}_v) \quad ...(7)$$

$$\Rightarrow \frac{\partial T}{\partial q_\alpha} = \sum_v m_v \dot{r}_v \cdot \frac{\partial \dot{r}_v}{\partial q_\alpha} \quad ...(8)$$

$$\text{and } \Rightarrow \frac{\partial T}{\partial \dot{q}_\alpha} = \sum_v m_v \dot{r}_v \cdot \frac{\partial \dot{r}_v}{\partial \dot{q}_\alpha} = \Sigma m_v \dot{r}_v \cdot \frac{\partial v_v}{\partial q_\alpha} \quad \text{[using (12)]} \quad ...(9)$$

$$\therefore (14) \Rightarrow \frac{d}{dt}\left(\frac{\partial T}{\partial \dot{q}_\alpha}\right) - \frac{\partial T}{\partial q_\alpha} = \phi_\alpha = \frac{\partial W}{\partial q_\alpha} \quad \text{using (7)} \quad ...(10)$$

Note. The quantity $p_\alpha = \frac{\partial T}{\partial \dot{q}_\alpha}$ is called the generalised momentum associated with the general co-ordinates q_α.

TRANSFORMATION EQUATION

Let $rv = x_v i + y_v j + z_v k$ be the position vector of v-th particle with respect to xyz co-ordinates system. The relationships of th generalised co-ordinates q_1

$q_{2,.....,}q_n$ the position co-ordinates are given by the transformation equations.

$$\left.\begin{aligned} x_v &= x_v(q_1, q_2, ..., q_n; t) \\ y_v &= y_v(q_1, q_2, ..., q_n; t) \\ z_v &= z_v(q_1, q_2, ..., q_n; t) \end{aligned}\right\} \quad ...(1)$$

where t denotes the time. In vector (1) can be written as

$$r_v = r_v(q_1, q_2, ..., q_n; t) \quad ...(2)$$

where the functions in (1) or (2) are continuous and have continuous derivatives.

CLASSIFICATION OF MECHANICAL SYSTEMS

(a) Scleronomic System

The mechanical system in which t, the time, does not enter explicity in equations (1) or (2) is called a scleronomic system.

(b) Rheonomic System

The mechanical system in which the moving constraints are involved and the time t does enter explicity is called a Rheonomic system.

(c) Holonomic System and Non Holonomic System

Let q_1 , q_2 ;.....,q_n denote the generalised co-ordinates describing a system and let t denote the time. If all the constraints of the system can be expressed as equations having the form $(q_1,q_1,....q_n;t) = 0$ or their equivalent, then the system is said to be Holonomic otherwise it is be Non- Holonomic system.

(d) Conservation and Non-conservative System

If the forces acting on the system are derivable from a potential function [or potential energy] V, then the system is called conservative, otherwise it is non-conservative.

ILLUSTRATIVE EXAMPLES

Example:

(i) *Set up the Lagrangian for a simple pendulum, and*

(ii) *Obtain an equation describing its motion.*

Solution:

(i) Choose as generalised coordinates, the angle θ made by the string OB of the pendulum and the vertical OA. Let l be the length of OA, then K.E. is given by

$$T = \frac{1}{2}mv^2 = \frac{1}{2}m(l\dot{\theta})^2 = \frac{1}{2}ml^2\dot{\theta}^2$$

where m is the mass of the bob.

The potential energy of mass m is given by

$V = mg\ (OA - OC) = mg\ (l - l\cos\theta)$

$= mgl\ (1 - \cos\theta)$

$$\therefore L = T - V = \frac{1}{2}ml^2\dot{\theta}^2 - mgl(1-\cos\theta)$$

(ii) Hence Lagrange's θ equation gives

$$\frac{d}{dt}\left(\frac{\partial L}{\partial \dot{\theta}}\right) - \frac{\partial L}{\partial \theta} = 0 \Rightarrow \frac{d}{dt}(ml^2\dot{\theta}) - (-mgl\sin\theta) = 0$$

$$\frac{d}{dt}\left(\frac{\partial L}{\partial \dot{\theta}}\right) - \frac{\partial L}{\partial \theta} = 0 \Rightarrow l\dot{\theta} = -g\sin\theta \Rightarrow \ddot{\theta} = -\frac{g}{l}\sin\theta,$$

which is the required equation of motion.

KINETIC ENERGY AND GENERALISED VELOCITIES

The K.E of the system is $T = \frac{1}{2}\Sigma_{v=1}^{n} mv\, \dot{r}_v$ ...(1)

The K.E of the system can be written as a quadratic form in the generalised **co-ordinates** q_α If the system is independent of time explicitly *i.e.* scleronomic quadratic form has only terms of the type $q_{\alpha\beta}\ q_\alpha\ q_\beta$ In case the system is Rhenomic, linear terms in q_α are also present.

GENERALISED FORCES

If W is the total work done on a system of particles by forces Fv acting on the v-th particle, then $dW = \Sigma_{\alpha=1}^{n} \phi_\alpha dq_\alpha$ where $\phi_\alpha = \Sigma_{v=1}^{N} F_v \bullet \frac{\partial r_v}{\partial q_\alpha}$ is called the *generalised force* associated with generalised co-ordinator q_α

Suppose that a system undergoes increments $dq_1, dq_2, \ldots dq_n$; then the v-th particle undergoes a displacement.

$$dr_v = \Sigma_{\alpha=1}^{n} \frac{\partial r_v}{\partial q_\alpha} dq_\alpha \qquad ...(1)$$

∴ Total work done is given by

$$dW = \Sigma_{v=1}^{N} F \bullet dr_v = \Sigma_{v=1}^{N} \left\{ \Sigma_{\alpha=1}^{n} F_v \bullet \frac{\partial r_v}{\partial q_\alpha} \right\} dq_\alpha \qquad ...(2)$$

Now, let $\phi_\alpha = \Sigma_{v=1}^{N} \left(\Sigma_{v=1}^{N} F_v \bullet \frac{\partial r_v}{\partial q_\alpha} \right)$

Now, let $\phi_\alpha = \Sigma_{v=1}^{N} F_v \bullet \dfrac{\partial r_v}{\partial q_\alpha}$

then (5) $dW = \Sigma_{\alpha=1}^{N}\left(\Sigma_{v=1}^{N} F_v \bullet \dfrac{\partial r_v}{\partial q_\alpha}\right) d_{a\alpha} = \Sigma_{\alpha=1}^{n} \phi_\alpha d_{a\alpha}$ ...(3)

We have $dW = \Sigma_{\alpha=1}^{N} \dfrac{\partial W}{\partial q_\alpha}, \therefore \dfrac{\partial W}{\partial q_\alpha} = \phi_\alpha$ ...(4)

Note:

(i) varies from (1) to n, the number of degrees of freedom.

(ii) varies from 1 to N, the number of particles in the system.

GENERALISED MOMENTUM

We defined $p_\alpha = \dfrac{\partial T}{\partial \dot{q}_\alpha}$ *to be the generalised momentum associated with generalised co-ordinates* q_α *or the conjugate momentum.*

In case the system is conservative, we have

$$T = L + V \Rightarrow (\partial T/\partial \dot{q}_\alpha) = (\partial L/\partial \dot{q}_\alpha) + (\partial V/\partial \dot{q}_\alpha) = (\partial L/\partial \dot{q}_\alpha)$$

because V, the P.E. of the system does not depend upon $\dot{q}_\alpha$

$$\therefore P_\alpha = (\partial L/\partial q_\alpha).$$

REVISION AT A GLANCE

Generalised Co-ordinates

Suppose that a particle or a system of N-particles moves subject to possible constraints, as for example a particle moving along a circular wire or a rigid body moving along an inclined plane, then there will be necessarily a minimum number of independent co-ordinates then needed to specify the motion.

These co-ordinates denoted by $q_1, q_2, \ldots. q_n$ are called generalised co-ordinates. These may be distances, angles or quantities to them.

Degrees of Freedom

The number of co-ordinates required to specify the position of a system of one or more particles is called the number of degrees of freedom of the system.

Transformation Equations

Let $r_v = x_v i + y_v j + z_v k$ be the position vector of vth particle with respect to xyz co-ordinate system. The relationships of the generalised co-ordinates

$q_1, q_2, \ldots q_n$ to the position co-ordinates are given by the transformation equations.

$$\left.\begin{aligned} x_v &= x_v(q_1, q_2, \ldots, q_n, t) \\ y_v &= y_v(q_1, q_2, \ldots, q_n, t) \\ z_v &= z_v(q_1, q_2, \ldots, q_n, t) \end{aligned}\right\} \quad \ldots(1)$$

where t denotes the time in vector (1) can be written as

$$r_v = r_v(q_1, q_2, \ldots, t)$$

where the function in (1) or (2) are continuous and have continuous derivatives.

Classification of Mechanical Systems

(a) *Scleronmic system* : The mechanical system in which the moving constraints are involved and the time t does enter explicitly is called a Rhenomic system.

(b) *Rhenomic systm* : The mechanical system in which t, the time, does not enter explicitly in equations (1) to (2) is called a sceleronimic system.

(c) *Holonomic system and Non Holonomic system* : Let $q_1, q_2, \ldots q_n$ denote the generalised co-ordinates describing a system and let t denote the time. If all the constraints of the system m can be expressed as equations having the form $f(q_1, q_2, \ldots q_n\, t) = 0$. or their equivalent, then the system is said to be Holonomic otherwise it is be Non-Holonomic system.

(d) *Conservative and non conservative system* : If the forces acting on the system are derivable from a potential function [or potential energy V], then the system is called conservative, otherwise it is non-conservative.

Kinetic Energy and Generalised Velocities

The K.E. of the system is $T = \frac{1}{2}\sum_{v=1}^{n} m_v \dot{r}_v^{\,2}$

The K.E. of the system can be written as a quadratic form in the generalised $\dot{q}_\alpha$ If the system is independent of time explicitly *i.e.* Sclemomic then the system is Rheonomic, linear terms in q_α are also present.

Generalised Forces

If W is the total work done on a system of particles by forces F acting on the Vth particle then

$$dW = \sum_{\alpha=1}^{n} \phi_\alpha dq_\alpha \text{ where } \phi_\alpha = \sum_{v=1}^{n} F_v \bullet \frac{\partial r_v}{\partial q_\alpha}$$

is called the generalised force associated with generalised coordinates q_α

Lagrange's Equations

$$\frac{d}{dt}\left(\frac{\partial T}{\partial \dot{q}_\alpha}\right)-\frac{\partial T}{\partial q_\alpha}=\phi_1=\frac{\partial W}{\partial q_\alpha}$$

For a = 1, 2, ..., n, we have n different equations which are called Lagrange's equations.

Note: The quantity $q_\alpha=\left(\frac{\partial T}{\partial \dot{q}_\alpha}\right)$ is called the generalised momentum associated with the general co-ordinate q_α

If the Forces are Derivable from a Potential V, then

$$\phi_\alpha\left(\frac{\partial W}{\partial q_\alpha}\right)=-\frac{\partial V}{\partial q_\alpha},$$

since the potential, or potential energy is a function of q's only (and possible the time t) then, we have

$$\frac{d}{dt}\left(\frac{\partial L}{\partial \dot{q}_\alpha}\right)-\frac{\partial T}{\partial q_\alpha}=-\frac{\partial V}{\partial q_\alpha}\Rightarrow\frac{d}{dt}\left[\frac{\partial}{\partial \dot{q}_\alpha}(T-V)\right]-\left(\frac{\partial T}{\partial q_\alpha}-\frac{\partial V}{\partial q_\alpha}\right)=0$$

$$\Rightarrow\frac{d}{dt}\left(\frac{\partial L}{\partial \dot{q}_\alpha}\right)-\frac{\partial L}{\partial q_\alpha}=0 \text{ where } L=T-V \qquad ...(2)$$

The function L defined by L = T V is said to be Lagrangian function.

Generalising Momentum

We defined $P_\alpha=\frac{\partial T}{\partial \dot{q}_\alpha}$

to be the generalised momentum associated with generalised co-ordinate q_α. We usually refer P_α as the momentum conjugate to q_α or the conjugate to momentum.

In case the system is conservative, we have

$$T=L+V\Rightarrow(\partial T/\partial \dot{q}_\alpha)=(\partial L/\partial \dot{q}_\alpha)+(\partial V/\partial \dot{q}_\alpha)=(\partial L/\partial \dot{q}_\alpha)$$

because V, the P.E. system does not depend upon q_α

$$\therefore q_\alpha=(\partial L/\partial \dot{q}_\alpha)$$

Lagrange's Equation with Impulsive Forces

$$\Rightarrow\left(\frac{\partial T}{\partial \dot{q}_\alpha}\right)-\frac{\partial T}{\partial q_\alpha}=\Sigma I_v \bullet \frac{\partial r_v'}{\partial q_\alpha}=p_\alpha \text{ (say)}$$

where subscripts 1 and 2 denote respective quantities before and after the application of the impulsive forces.

These equations are known as Lagrange's equations under impulsive forces.

SOLVED EXAMPLES

Example 1:

A hollow cylindrical garden roller is fitted with a counterpoise which can turn on the axis of the cylinder; the system is placed on the rough horizontal plane and oscillates under gravity, if $\frac{2\pi}{p}$ be the time of a small oscillation, show that p is given by the equation $p^2[2M + M')k^2 - M'h^2] = (2M + M')gh$ where M and M′ are the masses of the roller and counterpoise, k is the radius of gyration of M′ about the axis of the cylinder and h is the distance of its centre of mass from the axis.

Solution:

Suppose O′ is the centre of the cylindrical roller and G is the centre of gravity of the counterpoise, whose mass is M′. Obviously the line O′ K is a line fixed in the roller and the line O′ G is fixed in the counterpoise which were initially vertical. Suppose they make angles θ and ϕ with the vertical at the time t. Initially K was at O.

There is no slipping between the cylinder and the horizontal plane, hence OB = arc KB = aθ. Now assuming O as origin and the horizontal and vertical through O as co-ordinates axes, we have co-ordinates of G:-

$x_G = a\theta + h\sin\phi,\ y_G = a - h\cos\phi,$

$\therefore$ (velocity)^2 of $G = \dot{x}_G^2 + \dot{y}_G^2 = (a\dot{\theta} + h\cos\phi\dot{\phi})^2 + (h\sin\phi\dot{\phi})^2$

$= a^2\dot{\theta}^2 + h^2\dot{\phi}^2 + 2ah\dot{\theta}\dot{\phi}$ [neglecting higher powers of ϕ]

But radius of gyration of the counterpoise about O′ is k, where O′G = h. Hence the square of radius of gyration about G is $(k^2 - h^2)$.

Also (velocity)^2 of O′ = $(a\dot{\theta})^2 = a^2\dot{\theta}^2$. [$\therefore$ co-ordinates of O′ are (aθ, a)]

Now let T, be the kinetic energy and W, the work function of the system then we get

$$T = \frac{1}{2}M(a^2\dot{\theta}^2 + a^2\dot{\theta}^2) + \frac{1}{2}M'[(k^2 - h^2)\dot{\phi}^2 + a^2\dot{\theta}^2 + h^2\dot{\phi}^2 + 2ah\dot{\theta}h]$$

$$= \frac{M + M'}{2}a^2\dot{\theta}^2 + \frac{1}{2}M'k^2\dot{\phi}^2 + M'ah\dot{\theta}\dot{\phi}$$

$M = M' gh\cos\phi + D.$

$\therefore$ Lagrange's θ-equation gives $\frac{d}{dt}[(2M + M')a^2\dot{\theta} + M' ah\dot{\phi}] = 0$

$\Rightarrow (2M + M')a\ddot{\theta} + M' h\ddot{\phi} = 0$ ...(1)

while Lagrange's ϕ-equation gives

$\frac{d}{a}[M'k^2\dot{\phi} + M'ah\dot{\theta}] = -M' gh\phi \Rightarrow k^2\ddot{\phi} + ah\ddot{\theta} = -gh\phi$

$\Rightarrow ah\ddot{\theta} + k^2\ddot{\phi} = -gh\phi,$ ...(2)

$O'O = a, \quad O'G = h$

Now eliminating $\ddot{\theta}$ between (1) and (2), we get

$[Mh^2 - (2M + M')k^2]\ddot{\phi} = (2M + M')gh\phi.$

or $[(2M + M')k^2 - M'h^2]\ \ddot{\phi} = -(2M + M')gh\phi$ ...(3)

But, if $\frac{2\pi}{p}$ is the time of small oscillation, then we assume $\phi = A\cos(pt + b)$

$\therefore \dot{\phi} = -pA\sin(pt + B)$ and $\ddot{\phi} = -p^2A\cos(pt + B) = -p^2\theta$

$\therefore$ Equation (3) gives

$-p^2[(2M + M')k^2 - M'h^2] = -(2M + M')gh \qquad [\because \phi \neq 0]$

Example 2:

Two equal rods AB and BC; each of length l, smoothly jointed at B, are suspended from A and oscillate in a vertical plane through A. Show that the periods of normal oscillations are

$$\frac{2\pi}{n} \text{ where } n^2 = \left(3 \pm \frac{6}{\sqrt{7}}\right)\frac{g}{l}.$$

Solution:

At any time t, let the rod AB and BC be inclined at angles θ and ϕ to the vertical.

$\therefore$ Co-ordinates of G_1 are $\left(\frac{1}{2}\sin, \frac{1}{2}\cos\theta\right)$

$$\therefore v_{G_1}^2 = \left(\frac{1}{2}\cos\theta\dot{\theta}\right)^2 + \left(-\frac{1}{2}\sin\theta\dot{\theta}\right)^2 = \frac{l^2}{4}\dot{\theta}^2$$

$$x_{G_s} = l\sin\theta + \frac{1}{2}\sin\phi, y_{G_s} = l\cos\theta + \frac{1}{2}\phi$$

$$\therefore x_{G_s}^2 = l^2\dot{\theta}^2 + \frac{1}{2}l^2\dot{\phi}^2 + l^2\dot{\theta}\dot{\phi}\cos(\theta - \phi) = l^2\theta^2 + \frac{1}{4}l^2\phi^2 + l^2\dot{\theta}\dot{\phi}$$

[∵ θ and ϕ are small]

now, let T, be the kinetic energy and W, the work function, of the system, then

we have $W = mgl\cos\theta + mg\left(l\cos\theta + \frac{1}{2}\cos\phi\right)$

$$= \frac{1}{2}mgl(3\cos\theta + \cos\phi)$$

and $T = \frac{1}{2}m\left[\frac{1}{12}l^2\dot{\theta}^2 + \frac{1}{2}l^2\dot{\theta}^2\right] + \frac{1}{2}m\left\{\frac{1}{12}l^2\dot{\phi}^2 + l^2(\dot{\theta}^2 + \frac{1}{4}\dot{\phi}^2 + \dot{\theta}\dot{\phi})\right\}$

$$= \frac{1}{2}ml^2\left(\frac{4}{3}\dot{\theta}^2 + \frac{1}{3}\dot{\phi}^2 + \dot{\theta}\dot{\phi}\right)$$

∴ Lagrange's θ-equations is

$$\frac{d}{dt}\left(\frac{\partial T}{\partial\dot{\theta}}\right) - \frac{\partial T}{\partial\theta} = \frac{\partial W}{\partial\theta} \Rightarrow \frac{d}{dt}\left(\frac{4}{d}l\dot{\theta} + \frac{1}{2}l\dot{\phi}\right) = -\frac{3}{2}g\theta$$

$$\Rightarrow 8\ddot{\theta} + 3\ddot{\phi} = -\frac{9g}{l}\theta$$

While Lagrange's ϕ equation gives

$$\frac{d}{dt}\left(\frac{1}{3}l\dot{\phi} + \frac{1}{2}l\dot{\theta}\right) = -\frac{1}{2}g\phi \Rightarrow 3\ddot{\theta} + 2\ddot{\phi} = -\frac{3g}{l}\phi$$

Equations (1) and (2) can be written as

$(8D^2 + 9c)\,\theta + 3D^2\phi = 0$

and $3D^2\theta + (2D^2 + 3c)\,\phi = 0$ where $c = \left(\frac{g}{l}\right)$

Eliminating ϕ between (3) and (4), we get

$(8D^2 + 9c)\,(2D^2 + 3c)\ - 9D^4]\theta = 0$

i.e. $(7D^2 + 42cD^2 + 27c^2)\,\theta = 0$

Let us assume, $\theta = A\cos(nt + B) = -\ n^2\theta$ and $D^4\theta = n^4\theta$

Substituting these values in (5), we get

$(7D^4 + 42cD^2 + 27c^2)\,\theta = 0$ *i.e.* $7n^2 - 42cn^2 - 27c^2) = 0 \quad [\because \theta \neq 0]$

$$i.e.\ n^2 = \frac{42c \pm \sqrt{[\{(42c)^2 - 4(7)(27c^2)\}]}}{14}$$

$$= \left[3 \pm \frac{6}{14}\sqrt{(28)}\right]c = \left(3 \pm \frac{6}{\sqrt{7}}\right)c = \left\{3 \pm \frac{6}{\sqrt{7}}\right\}\frac{g}{l}.$$

Example 3(a):

A uniform rod AB, of length 2a can turn freely about a point distance c from its centre, and is at rest at an angle α to the horizon when a particle is hung by a light string of length l from one end. If the particle be displaced slightly in the vertical plane of the rod show that it will oscillate in the same time as a simple pendulum of length

$$l\frac{a^2 + 3ac\cos^2\alpha + 3c^2\sin^2\alpha}{2^2 + 3ac}$$

Solution:

Let M be the mass and G the centre of gravity of the rod. Further let O be the point about which it nac turn where OG = c. Let BP be a string of length l and m the mass of the particle tied to the end P of the string BP.

Initially rod was making an angle α with the horizontal and the string was vertical.

Let B_0 be the initial position of B so that $\angle B_0OX = \alpha$, After a time t let the rod make an angle θ with OB_0 *i.e.* an angle)α + θ) with the horizontal and let the string be inclined at an angle θ with the vertical.

Initially the system was at rest, hence taking moments about O, (in this position), we get Mc = m(a – c).

Assuming O as origin and the horizontal and the vertical through O as co-ordinates axes, we obtain the following results

$x_p = (a - c)\cos(\alpha + \theta) + l\sin\phi.\ y_p = (a - c)\sin(\alpha + \theta) + l\cos\phi.$

$\therefore (\text{velocity})^2 \text{of } P = \{-(a - c)\sin(\alpha + \theta)\dot{\theta} + l\sin\theta\dot{\theta}\}^2$

$\qquad + \{(a - c)\cos(\alpha + \theta)\dot{\theta} - l\sin\phi\dot{\phi}\}^2$

$= (a - c)^2\dot{\theta}^2 + l^2\dot{\phi}^2 - 2l(a - c)\dot{\theta}\dot{\phi}\sin(\alpha + \theta + \phi)$

$= (a - c)^2\dot{\theta}^2 + l^2\dot{\phi}^2 + 2l(a - c)\dot{\theta}\dot{\phi}\sin\alpha$

(neglecting small quantities of the higher order)

Also co-ordinates of G (the C.G. of the rod) are

$\{-c\cos(\alpha + \theta), -c\sin(\alpha + \theta)\}$

$\therefore$ (velocity)^2 of G = $\{c\sin(\alpha+\theta)\dot{\theta}\}^2 + \{-c\cos(a+\theta)\dot{\theta}\}^2 = c^2\dot{\theta}^2$.

Now let T, be the kinetic energy and W, the work function of the system, then we easily obtain

$$W = mg\,\{(a-c)\sin(\alpha+\theta) + l\cos\phi] - Mgc\sin(\alpha+\theta) + D$$

$$= mgl\cos\phi + D$$

[the first and the last terms cancel, because m (a – c) = Mc]

$$\text{and } T = \frac{1}{2}M\left[\frac{a^2}{3}\dot{\theta}^2 + c^2\dot{\theta}^2\right] + \frac{1}{2}m[(a-c)^2\dot{\theta}^2$$

$$+l^2\dot{\phi}^2 - 2l(a-c)\dot{\theta}\dot{\phi}\sin\alpha]$$

$\therefore$ Lagrange's θ-equation gives

$$\frac{d}{dt}\left[\left\{M\left(\frac{a^2}{3}+c^2\right)+m(a-c)^2\right\}\dot{\theta} - ml(a-c)\ddot{\phi}\sin\alpha\right] = 0$$

$$\Rightarrow \left\{M\left(\frac{a^2}{3}+c^2\right)+m\,(a-c)^2\right\}\dot{\theta} - ml(a-c)\ddot{\phi}\sin\alpha = 0.$$

$$\Rightarrow \left\{M\left(\frac{a^2}{3}+c^2\right)+mc\,(a-c)\right\}\ddot{\theta} - mcl\ddot{\phi}\sin\alpha = 0 \qquad [\because m(a-c) = Mc]$$

$\Rightarrow (a^2+3ac)\ddot{\theta} - 3lc\sin\alpha\ddot{\phi} = 0.$ While Lagrange's ϕ equation gives

$$\frac{d}{dt}\left\{ml^2\dot{\phi} - ml(a-c)\dot{\theta}\sin\alpha\right\} = -mgl\phi$$

$$\Rightarrow l\dot{\phi} - (a-c)\ddot{\theta}\sin\alpha = -g\phi$$

Eliminating θ we have

$$l\ddot{\phi} - (a-c)\sin\alpha\frac{3lc\sin\alpha}{a^2+3ac}\ddot{\phi} = -g\phi$$

$$i.e. \Rightarrow \frac{a^2+3ac-3c(a-c)\sin^2\alpha}{a^2+3ac}l\ddot{\phi} = -g\phi$$

$$\Rightarrow \ddot{\phi} = -\frac{g}{l(a^2+3ac\cos^2\alpha+3c^2\sin^2\alpha)}\phi = -\mu\phi \text{ (say)}$$

$\therefore$ Length of the simple equivalent pendulum = $\dfrac{g}{\mu}$

$$= \frac{a^2+3ac\cos^2\alpha+3c^2\sin^2\alpha}{a^2+3ac}l.$$

Example 3(b):

A uniform rod AB of length 8a is suspended from a fixed point O by means of light inextensible string, of length 13a, attached to B. If the system is slightly displaced in a vertical plane, show that $(\theta + 3\phi)$ *and* $(12\theta - 13\phi)$ *are principal co-ordinates where* θ and ϕ *are the angles which the rod and string respectively make with the vertical. Also show that periods of small oscillations are*

$$2\pi\sqrt{\left(\frac{a}{g}\right)} \text{ and } 2\pi\sqrt{\left(\frac{52}{3g}\right)}.$$

Solution:

We have

$$x_G = 13a \sin\phi + 4a \sin\theta$$

and $y_G = 13a\cos\phi + 4a\cos\phi$

$$\therefore \dot{x}_G^2 + \dot{y}_G^2 = 169a^2\dot{\phi}^2 + 16a^2\dot{\theta}^2$$

$$+104a^2\dot{\theta}\dot{\phi}\cos(\theta - \phi)$$

Thus $T = \frac{1}{2}m[k^2\dot{\theta}^2 + (\dot{x}_G^2 + \dot{y}_G^2)] = \frac{1}{2}ma^2\left[\frac{64}{3}\dot{\theta}^2 + 169\dot{\phi}^2 + 104\dot{\theta}\dot{\phi}\right]$

and the work function

W = mg (13acos ϕ + 4a cos θ)

$\therefore$ Lagrange's θ-equation gives

$$\frac{d}{dt}\left(\frac{\partial T}{\partial\dot{\theta}}\right) - \frac{\partial T}{\partial\theta} \Rightarrow 61\ddot{\theta} + 39\ddot{\phi} = -\frac{3g}{a}\theta \qquad ...(1)$$

While Lagrange's θ-equation gives $4\ddot{\theta} + 13\ddot{\phi} = -\frac{g}{a}\phi$ ...(2)

Equations (1) and (2) $\Rightarrow D^2(\theta + 3\phi) = -\frac{3g}{52a}(\theta + 3\phi)$

and $D^2(12\theta - 13\phi) = -\frac{g}{a}(12\theta + 13\phi)$

Now putting $q + 3\phi = X$ and $12\theta - 13\theta = Y$ in these equations, we get D^2X

$$= -\frac{3g}{52a}X \text{ and } D^2Y = -\frac{g}{a}\cdot Y;$$

which obviously represent two independent simple harmonic motions. Hence X and Y are principal co-ordinates, that is

$(\theta + 3\phi)$ and $(12\theta - 13\phi)$ are principal co-ordinates.

Also, periods of small oscillations are given by

$$2\pi\sqrt{\left(\frac{52a}{3g}\right)} \text{ and } 2\pi\sqrt{\left(\frac{a}{g}\right)}$$

Example 3(c):

A particle of mass m moves in a conservative force field. Find a the Lagrangian function, (b) the equations of motion in cylindrical co-ordinates (ρ, ϕ, z).

Solution:

We have $OP = OP_0 + P_0P = OA + Ap_0 + P_0P = \vec{\rho}$ (say)

$\therefore \vec{\rho} = \rho \sin\phi j + \rho\cos\phi i + zk$ where i, j, k

are the unit vector along OX , OY and OZ respectively.

Hence the unit vector along the direction of ρ increasing is

given by $\vec{\rho_1} = \frac{\partial \vec{\rho}}{\partial \rho} / \left|\frac{\partial \vec{\rho}}{\partial \rho}\right| = \sin\phi j + \cos\phi j$

Similarly $\vec{\phi_1} = \frac{\partial \vec{\rho}}{\partial \phi} / \left|\frac{\partial \vec{\rho}}{\partial \phi}\right|$

$$= \frac{\rho\cos\phi j - \rho\sin\phi i}{\rho} = -\sin\phi i + \cos\phi j$$

Now $v = \frac{d\vec{\rho}}{dt} = \frac{d}{dt}(\rho\sin\phi j + \rho\cos\phi i + zk)$

$$= \rho\cos\phi\dot{\phi}j + \dot{\rho}\sin\phi j - \rho\sin\phi\dot{\phi}i + \dot{\rho}\cos\phi i + k\dot{z}$$

$$= \dot{\rho}\cos\phi i + \dot{\rho}\sin\phi j + \rho\dot{\phi}(\cos\phi j - \sin\phi i) + z\dot{k}$$

$$= \rho\vec{\rho_1} + \rho\dot{\phi}\vec{\phi} + \dot{z}k$$

$$\therefore T = \frac{1}{2}m\{\dot{\rho}^2 + \rho^2\dot{\phi}^2 + \dot{z}^2\} \text{ and } V = V(\rho, \phi, z)$$

(a) Hence the Lagrangian function is

$$L = T - V = \frac{1}{2}m\left[\dot{\rho}^2 + \rho^2\dot{\phi}^2 + \dot{z}^2\right] - V(\rho, \phi, z)$$

(b) Lagrange's equations are

$$\frac{d}{dt}\left(\frac{\partial L}{\partial \dot{\rho}}\right)-\frac{\partial L}{\partial \rho}=0 \text{ i.e. } \frac{d}{dt}(m\dot{\rho})-\left(m\rho\dot{\phi}^2\frac{\partial V}{\partial \rho}\right)=0$$

$$\text{i.e. } m(\ddot{\rho}-\rho\dot{\phi}^2)=-\frac{\partial V}{\partial \rho} \qquad ...(1)$$

$$\frac{d}{dt}\left(\frac{\partial L}{\partial \dot{\phi}}\right)-\frac{\partial L}{\partial \phi}=0 \text{ i.e. } \frac{d}{dt}(m\rho^2\dot{\phi})+\frac{\partial V}{\partial \phi}=0,$$

$$\text{or } \frac{d}{dt}(\rho^2\dot{\phi})=-\frac{\partial V}{\partial \phi} \qquad ...(2)$$

$$\text{and } \frac{d}{dt}\left(\frac{\partial L}{\partial \dot{z}}\right)-\frac{\partial L}{\partial z}=0 \ \frac{d}{dt}(m\dot{z})+\frac{\partial V}{\partial z}=0 \text{ or } m\ddot{z}=-\frac{\partial V}{\partial z} \qquad ...(3)$$

Example 4:

A particle Q moves on a smooth horizontal circular wire of radius, a which is free to rotate about a vertical axis through a point 0, distance c from the centre C. If the $\angle QCA = \theta$, show that

$$a\ddot{\theta}+\dot{\omega}(a-c\cos\theta)=c\omega^2\sin\theta.$$

where ω is the angular velocity of the wire.

Solution:

Let $0Q = r$, and $\angle AOQ = \alpha$

$$\Rightarrow r^2 = a^2+c^2-2ac\cos\theta \qquad ...(1)$$

$$r\cos(\alpha-\theta)=a-c\cos\theta. \qquad ...(2)$$

The particle Q moves on circle of radius a, *so its velocity along the tangent QT will be aθ but Q revolves about O with angular velocity ω, which causes a velocity aw at the right angles to 0Q.*

$\Rightarrow v\frac{1}{2}=$ (velocity)^2 of the particle at Q

$$=a^2\dot{\theta}^2+r^2\omega^2+2ar\dot{\theta}\cos(\alpha-\theta)\omega$$

$$\angle AOQ=\alpha, \angle CQO=\alpha-\theta$$

Now $T=\frac{1}{2}mV_q^{\ 2}=\frac{1}{2}[a^2\dot{\theta}^2+r^2\omega^2$

$+2a\,r\,\omega\theta\cos(\alpha-\theta)]$

$$\left[\begin{array}{l}\angle NQT=\alpha-\theta, OQ=r\\ HQ=a-c\cos\theta\end{array}\right.$$

Note: *If r is the position vector of the particle at any time t, then $\partial r / \partial r$ is the vector tangent to the curve q = constant *i.e.* a vector in the direction of r (increasing r). A unit vector in this direction is thus given by

$$r_1 = \frac{\partial r}{\partial r} / \left| \frac{\partial r}{\partial r} \right|$$

Similarly, $\partial r / \partial \theta$ is the vector tangent to the curve r = constant, A unit vector in this direction is given by $\vec{\theta}_1 = \frac{\partial r}{\partial \theta} / \left| \frac{\partial r}{\partial \theta} \right|$.

$$= \frac{1}{2} m[a^2 \dot{\theta}^2 + (a^2 + c^2$$

$$= r \cos(\alpha - \theta)]$$

$-2ac \cos\theta)\omega^2 + 2a\omega\dot{\theta}(a - c \cos\theta)$ and work function = 0

(∵ weight does no work)

∴ Lagrange's θ equation

$$\Rightarrow \frac{d}{dt}\left(\frac{\partial T}{\partial \dot{\theta}}\right) - \frac{\partial T}{\partial \theta} = 0.$$

$$\Rightarrow \frac{d}{dt}[a^2\dot{\theta} + a\omega(a - c \cos\theta)] - ac\,\omega^2 \sin\theta - ac\,\omega\dot{\theta} \sin\theta = 0$$

$$\Rightarrow a^2\ddot{\theta} + a\dot{\omega}(a - c\cos\theta)] + a\omega c\dot{\theta}\sin\theta - ac\,\omega^2 \sin\theta - ac\,\omega\dot{\theta}\sin\theta = 0$$

$$\Rightarrow a\ddot{\theta} + \dot{\omega}(a - c \cos\theta) = c\,\omega^2 \sin\theta$$

Example 5(a):

A uniform rod, of mass 3m and length 2l, has its middle point fixed and a mass m attached at one extremity. The rod when in horizontal position is set rotating about a vertical axis through its centre with on angular velocity equal to $\sqrt{\left(\frac{2ng}{l}\right)}$ show that the heavy end of the rod will fall till the inclination of the rod to the vertical is $\cos^{-1}[\sqrt{(n^2+1)} - n$ and will the rise again.

Solution:

The mass is attached at L. On the rod ML. Take a point p such that

$OP = \xi$, the element $PQ = d\xi$. Further at any time t, let the plane through it and the vertical have turned through an angle ϕ from its initial position and let the rod be inclined at an angle θ to the vertical. Taking O, the mid point of the rod, as the origin and OX, OY (a line perpendicular to the plane of the paper) and OZ as axes of reference then co-ordinates of the point P on the rod are:

$$x = \xi \sin\theta \cos\phi, y = \zeta \sin\theta \sin\phi, z = \xi \cos\theta$$

$$\therefore \dot{x} = \xi \cos\theta \cos\dot{\theta}, - \sin\phi\dot{\phi}\theta.$$

$$\dot{y} = \xi \cos\theta \sin\phi\dot{\theta}, + \xi \sin\theta\dot{\theta}. \text{ Thus,}$$

$$v_\rho^2 = (\text{velocity})^2 \text{ of } P = \dot{x}^2 + \dot{y}^2 + \dot{z}^2 = \xi^2(\dot{\theta}^2 + \dot{\phi}^2 \sin^2\theta).$$

$$\therefore v_L^2 = l^2(\dot{\theta}^2 + \dot{\phi}^2 \sin^2\theta) = (\text{velocity})^2 \text{ of mass m},$$

Now mass of the element $PQ = \dfrac{3m}{2l} d\xi = dm, \text{say}.$

$\therefore$ Its kinetic energy

$$= \frac{1}{2} dm.v_\rho^2 = \frac{1}{2} \cdot \frac{3m}{2l} d\xi(\dot{\theta}^2 + \dot{\phi}^2 \sin^2\theta)\xi^2$$

$$= \frac{3m}{4l}(\dot{\theta}^2 + \dot{\phi}^2 \sin^2\theta)\xi^2 d\xi$$

and K.E. of the rod $= \dfrac{3m}{4l}(\dot{\theta}^2 + \dot{\phi}^2 \sin^2\theta)\displaystyle\int_{-l}^{l} \xi^2 d\xi$

$$= \frac{1}{2} m(\dot{\theta}^2 + \dot{\phi}^2 \sin^2\theta) l^2.$$

Again, (velocity)² of the particle $m = l^2(\dot{\theta}^2 + \dot{\phi}^2 \sin^2\theta)$.

$\therefore$ Kinetic energy of the particle of mass $m = \dfrac{1}{2} ml^2(\dot{\theta}^2 + \dot{\phi}^2 \sin^2\theta)$.

$\therefore$ Total K.E. = T = K.E. of the rod + K.E. of the particle

$$= \frac{1}{2} ml^2(\dot{\theta}^2 + \dot{\phi}^2 \sin^2\theta) + \frac{1}{2} ml^2(\dot{\theta}^2 + \dot{\phi}^2 \sin^2\theta)$$

i.e. $T = ml^2(\dot{\theta}^2 + \dot{\phi}^2 \sin^2\theta)$

Also the work function is given by W = mgl cos θ + C.

Lagrange's φ-equation is $\dfrac{d}{dt}\left(\dfrac{\partial T}{\partial \dot{\phi}}\right) - \dfrac{\partial T}{\partial \phi} = \dfrac{\partial W}{\partial \phi}$

which gives $\dfrac{d}{dt}(2ml^2 \dot{\phi} \sin^2\theta) = 0.$

Integrating it, we get $\dot{\phi} \sin^2\theta = K$ (constant).

Initially, $\theta = \frac{\pi}{2}$ and $\dot{\phi} = \sqrt{\left(\frac{2ng}{l}\right)}$. $\therefore K = \sqrt{\left(\frac{2ng}{l}\right)}$.

$$\therefore \dot{\phi}\sin^2\theta = \sqrt{\left(\frac{2ng}{l}\right)}. \quad ...(1)$$

and Lagrange's θ-equation is $\frac{d}{dt}\left(\frac{\partial T}{\partial \dot{\theta}}\right) - \frac{\partial T}{\partial \theta} = \frac{\partial W}{\partial \theta}$

i.e. $\frac{d}{dt}(2ml^2\dot{\theta}) - 2ml^2\dot{\phi}^2 \sin\theta\cos\theta = mgl\sin\theta$

$$\text{or } 2l\ddot{\theta} - 2l\dot{\phi}^2 \sin\theta\cos\theta = -g\sin\theta. \quad ...(2)$$

Substituting value of $\dot{\phi}$ from (1) in (2), we have

$$2l\ddot{\theta} - 4ng\cot\theta\,\text{cosec}^2\theta = -g\sin\theta. \quad ...(3)$$

Integration provides us $2l\dot{\theta}^2 + 4ng\cot^2\theta = 2g\cos\theta + k$.

Initially $\theta = \frac{\pi}{2}, \dot{\theta} = 0, \therefore k = 0$.

$$\therefore 2l\dot{\theta}^2 + 4ng\cot^2\theta = 2g\cos\theta. \quad ...(4)$$

The rod will fall till $\dot{\theta} = 0$.

i.e. $4ng\cot^2\theta = 2g\cos\theta$ or $2n\cos^2\theta - \cos\theta\sin^2\theta = 0$.

$\therefore$ either $\cos\theta = 0 \Rightarrow \theta = \frac{\pi}{2}$ which gives initial position.

as $2n\cos\theta - \sin^2\theta = 0 \Rightarrow \cos^2\theta + 2n\cos\theta - 1 = 0$.

Solving it, $\cos\theta = \frac{-2n \pm \sqrt{(4n^2+4)}}{2} = \left\{-n + \sqrt{(n^2+1)}\right\}$

[the other value being inadmissible because θ can not be obtuse]

or $\theta = \cos^{-1}[-n + \sqrt{(n^2+1)}]$.

This proves the required result. If we substitute this value of θ in equation (2′), then we find that θ comes out to be positive. Hence at that time the rod begins to rise.

Example 5(b):

A uniform rod, of mass 5m and length 2a, turns freely about one end which is fixed, to its other extremity is attached one end of a light string, of length 2a, which comes at its other end a particle of mass m, show that the periods of the small oscillations in a vertical plane are the same as those of simple pendulums of length $\frac{2a}{3}$ *and* $\frac{20a}{7}$

Solution:

Let the string BC and the rod AB make angles ϕ and θ with the vertical at any time t. The particle of mass m is tied to the end C of the string.

Now $x_c = 2a\sin\theta + 2a\sin\phi,$

$$\dot{x}_c = 2a(\cos\theta\dot{\theta} + \cos\phi\dot{\phi})$$

$$y_c = 2a\cos\theta + 2a\cos\phi.$$

$$\dot{y}_c = -2a(\sin\theta\dot{\theta} + \sin\phi\dot{\phi})$$

$\therefore$ (velocity)^2 of $m = \dot{x}_c^{\,2} + \dot{y}_c^{\,2}$

$$= 4a^2(\dot{\theta} + \dot{\phi}^2 + 2\dot{\theta}\dot{\phi}).$$

Again co ordinates of G are (a sin θ, a cos θ)

$\therefore$ (velocity)^2 of G $= a^2\dot{\theta}^2$

Now let T be the kinetic energy, and W the work function of the system, then we have

Total K.E. = K.E. of rod + K.E. of particle of mass m.

$$T = \frac{1}{2}5m\left(\frac{a^2}{3}\dot{\theta}^2 + a^2\dot{\theta}^2\right) + \frac{1}{2}m.4a^2(\dot{\theta}^2 + \dot{\phi}^2 + 2\dot{\theta}\dot{\phi})$$

$$= ma^2\left(\frac{16}{3}\dot{\theta}^2 + 2\dot{\phi}^2 + 4\dot{\theta}\dot{\phi}\right)$$

and W = 5m ga cos θ + mg.2a (cosθ + cosϕ)

$$= 7mga\cos\theta + 2mga\cos\phi = 7mag\left(1 - \frac{\theta^2}{2}\right) + 2mag\left(1 - \frac{\phi^2}{2}\right)$$

$\therefore$ Lagrange's θ equation is given by

$$\frac{d}{dt}\left(\frac{\partial T}{\partial \dot{\theta}}\right) - \frac{\partial T}{\partial \theta} = \frac{\partial W}{\partial \theta}$$

$$\Rightarrow \frac{d}{dt}\left(\frac{32}{3}\dot{\theta} + 4\dot{\phi}\right) = -\frac{7g}{a}\theta \Rightarrow 32\ddot{\theta} + 12\ddot{\phi} = -21\frac{g}{a}\theta \quad \text{...(1)}$$

Lagrange's ϕ equation is given by

$$\frac{d}{dt}(4\dot{\phi} + 4\dot{\theta}) = -\frac{2g}{a}\phi \Rightarrow \textit{i.e. } 2\ddot{\theta} + 2\ddot{\phi} = -\frac{g}{a}\phi \quad \text{...(2)}$$

$\therefore$ (1) and (2) $\Rightarrow (32D^2 + 21c)\theta\, 12D^2\phi = 0$ ...(3)

and $2D^2\theta + (2D^2 + c)\phi = 0$ where $\frac{f}{a} = c$

Now eliminating 'ϕ' between (3) and (4), we get

$[(32D^2 + 21c)(2D^2 + c) - 24D^2]\theta = 0$

or $[40D^4 + 74cD^2 + 21c^2]\theta = 0$ ...(5)

Now let $\theta = A\cos(pt + B) \Rightarrow D\theta = -pA\sin(pt + B)$.

$D^2\theta = -p^2A\cos(pt + B) = -p^2\theta$ and $D^4\theta = p^4\theta$.

substituting these in (5) we get

$$(40p^4 - 74cp^2 + 21c^2)\theta = 0 \text{ } i.e. \text{ } 40p^4 - 74cp^2 + 21c^2 = 0 \text{ as } \theta \neq 0$$

or $(2p^2 - 2c)(20p^2 - 7c) = 0$

i.e, $\left(2p^2 - \frac{3g}{a}\right)\left(20p^2 - 7\frac{g}{a}\right) = 0.$

$\Rightarrow p_1^2 = \frac{3g}{2a}$ and $\Rightarrow p_2^2 = \frac{7g}{20a}$.

*Hence length of equivalent pendulums are

$\frac{g}{p_1^2}$ and $\frac{g}{p_2^2}$ *i.e.* $\frac{2a}{3}$ and $\frac{20}{7}a$.

Example 6:

A uniform rod, of length 2a, can turn freely about one end, which is fixed. Initially it is inclined at an angle α, to the downward drawn vertical and it is set rotating about a vertical axis through its fixed end with angular velocity w. Show that, during the motion, the rod is always inclined to the vertical at an angle which is > or< α, according as $\omega^2 >$ or $< \frac{3}{4a\cos\alpha}$ *and that in each case its motion is inclined between the inclination α and*

$$\cos^{-1}[-n + \sqrt{(2n\cos\alpha + n^2)}] \text{ when } n = \frac{a\omega^2 \sin^2\alpha}{3g}$$

If it be slightly disturbed when revolving steadily at a constant angle α show that the time of a small oscillation is

$$2\pi\sqrt{\left[\frac{4a\cos\alpha}{3g(1 + 3\cos^2\alpha}\right]}$$

Solution:

The rod OA is turning about the end O. Take a point P on the rod such that $OP = \xi$, and the element $PQ = \xi$,

$\therefore$ mass of element $PQ = \frac{m}{2a} d\xi,$

where m is the mass of the rod Further at any time t, let the rod be inclined at an angle θ to the vertical and let the plane through the rod and the vertical have turned though angle ϕ from its initial position OX, then co-ordinates of the point P are

$$x_p = \xi \sin\theta \cos\phi,\ y_p = \xi \sin\theta \sin\phi,\ z = \xi \cos\theta$$

$$\therefore v_p^{\ 2} = (\text{velocity})^2 \text{of } p = \dot{x}_p^{\ 2} + \dot{y}_p^{\ 2} + \dot{z}_p^{\ 2} = \xi^2(\dot{\theta}^2 + \dot{\phi}^2 \sin^2\theta)$$

and kinetic energy of the element $PQ = \frac{1}{2}\frac{m}{2a} v^2{}_p$

$$= \frac{1}{2}\frac{m}{2a} d\xi(\dot{\theta}^2 + \dot{\phi}^2 \sin^2\theta)\xi^2$$

Now, let T, be the K.E. of the rod OA, the b we have

$$T = \frac{1}{2}\frac{m}{2a}(\dot{\theta}^2 + \dot{\phi} \sin^2\theta)\int_0^{2a} \xi^2 d\xi = \frac{2ma^2}{3}(\dot{\theta} + \dot{\phi}^2 \sin^2\theta)$$

or $T = \frac{2ma^2}{3}(\dot{\theta}^2 + \dot{\phi}^2 \sin^2\theta)$

Also the work function, $W = m\,ga \cos\theta + C.$

Lagrange's ϕ-equation gives

$$\frac{d}{dt}\left(\frac{4ma^2}{3}\dot{\phi}\sin^2\theta\right) = 0 \textit{ i.e. } \frac{d}{dt}[\dot{\phi}\sin^2\theta] = 0 \qquad \text{...(1)}$$

$$\Rightarrow \dot{\phi}\sin^2\theta = k \text{ (constant).} \qquad \text{...(2)}$$

Initially $\theta = \alpha, \dot{\phi} = \theta, \quad \therefore k = \omega \sin^2\alpha.$

Thus (2) gives $\dot{\phi}\sin^2\theta = \omega \sin^2\alpha,$

and Lagrange's θ-equation is $\frac{d}{dt}\left(\frac{\partial T}{\partial \theta}\right) - \frac{\partial T}{\partial \theta} = 0.$

* when $\theta = A(pt + B)$, the period of motion is given $T = \frac{2\pi}{p}$. If l is the length of the simple equivalent pendulum, we have

$$T = 2\pi\sqrt{(l/g)} \Rightarrow l = \frac{g}{p^2}$$

$$\Rightarrow \frac{d}{dt}\left(\frac{4ma^2}{3}\dot{\theta}\right) - \frac{2ma^2}{3}\dot{\phi}^2.2\sin\theta\cos\theta = -\, m\, ga\sin\theta.$$

$$\Rightarrow \ddot{\theta} - \dot{\phi}^2\sin\theta\cos\theta = -\frac{g}{4a}\sin\theta. \qquad ...(4)$$

Eliminating $\dot{\phi}$ between (4) and (3), we have

$$\ddot{\theta} - \frac{\omega^2\sin^4\alpha}{\sin^3\theta}\cos\dot{\theta} = -\frac{3g}{4a}\sin\theta. \qquad ...(5)$$

$$\Rightarrow \dot{\theta} + \frac{\omega^2\sin^4\alpha}{\sin^3\theta} = \frac{3g}{4a}\cos\theta + A. \qquad ...(6)$$

Initially $\theta = \alpha, \theta = 0, \therefore A = \omega^2\sin^2\alpha - \frac{3g}{2a}\cos\alpha$

Substituting this value of A in (6) , we get

$$\dot{\theta}^2 + \frac{\omega^2\sin^4\alpha}{\sin^2\theta} = \frac{3g}{2a}\cos\theta + \omega^2\sin^2\alpha - \frac{3g}{2a}\cos\alpha,$$

$$\text{or } \dot{\theta}^2 = \omega^2\sin^2\alpha\left(1 - \frac{\sin^2\alpha}{\sin^2\theta}\right) + \frac{3g}{2a}(\cos\theta - \cos\alpha)$$

$$= \frac{3ng}{a}\left(1 - \frac{\sin^2\alpha}{\sin^2\theta}\right) + \frac{3g}{2a}(\cos\theta - \cos\alpha)\left[\because n = \frac{a\omega^2\sin^2\alpha}{3g}\right]$$

$$= \frac{3g}{2a}.\frac{\cos\alpha - \cos\theta}{\sin^2\theta}[2n(\cos\alpha + \cos\theta) - \sin^2\theta]$$

$$i.e.\ \dot{\theta}^2 = \frac{3g}{2a}.\frac{\cos\alpha - \cos\theta}{\sin^2\theta}[(\cos^2\theta + 2n\cos\theta + 2n\cos\alpha - 1)] \qquad ...(7)$$

From (7) , we see that $\dot{\theta} = 0$, when

$(\cos\alpha - \cos\theta)[(\cos^2\theta + 2n\cos\theta + 2n\cos\alpha - 1)] = 0$

i.e. if either r $\cos\alpha - \cos\theta$ *i.e.* $\theta = \alpha$ (the initial position)

or $\cos^2 + 2n\cos\theta + 2n\cos\alpha - 1 = 0$

$$i.e.\ \cos\theta = \frac{-2n \pm \sqrt{[4n^2 + 4(1 - 2\cos\alpha)]}}{2}$$

$$\text{or } \cos\theta = -n + \sqrt{(1 - 3n\cos\alpha + n^2)} \qquad ...(8)$$

(the other value being inadmissible because that gives value of cos θ numerically greater that unity.

Hence the motion is included between $\theta = \alpha$ and $\theta = \theta_1$ where

$$\cos\theta_1 = \{\sqrt{(1 - 2n\cos\alpha + n^2)} - n\}$$

The rod will move above or below its initial position, if $\theta_1 >$ or $< \alpha$

or if $\cos\theta_1 <$ or $> \cos\alpha$

i.e. if $1 - 2n\cos\alpha + n^2 <$ or $> (n + \cos\alpha)^2$

i.e. if $\dfrac{3ng}{a\omega^2} <$ or $> 4n\cos\alpha$

i.e. if $\omega^2 >$ or $< \dfrac{3g}{4a\cos\alpha}$.

Small Oscillations about the Steady Motion : The motion will be steady if the rod goes round, inclined at the same angle α with the vertical or mathematically if $\theta = \alpha$(throughout the motion), then $\ddot{\theta} = 0$. Making these substitutions in (5), we get,

$$-\frac{\omega^2\sin^4\alpha}{\sin^3\theta}\cos\theta = -\frac{3g}{4a}\sin\theta \text{ i.e. } \omega^2 = \frac{3g}{4a\cos\alpha}$$

when ω^2 has this value and there are small oscillations about the position $\theta = \alpha$, then putting $\theta = \alpha + \psi$ in equation (5) we get

$$\ddot{\psi} = \frac{3g}{4a\cos\alpha}\frac{\sin^4\alpha}{\sin^3(\alpha+\psi)}\cos(\alpha+\psi) - \frac{3g}{4a}\sin(\alpha+\psi)$$

$$= \frac{3g}{4g}\left[\frac{\sin^4\alpha(\cos\alpha\cos\psi - \sin\alpha\sin\psi}{\cos\alpha(\sin\alpha\cos\psi + \cos\alpha\sin\psi)^3} - (\sin\alpha\cos\psi + \cos\alpha\sin\psi\right]$$

$$= \frac{3g}{4g}\left[\frac{\sin^4\alpha(\cos\alpha - \psi\sin\alpha\sin\psi}{\cos\alpha(\sin\alpha\cos\psi\cos\alpha)^3} - (\sin\alpha + \psi\cos\alpha\right], \text{ approximately}$$

$$= -\frac{3g\sin\alpha}{4a}[(1 - \psi\tan\alpha)(1 + \psi\cot\alpha)^{-3} - (1 + \psi\cot\alpha], \text{ approx.}$$

$$= -\frac{3g\sin\alpha}{4}(4\cot\alpha + \tan\alpha)\psi,$$

$$\text{app.} = \frac{-3g(1 + 3\cos^2\alpha)}{4a\cos\alpha}\psi = -\mu\psi \text{ say}$$

$\therefore$ time of small oscillation $= \dfrac{2\pi}{\sqrt{\pi}} = 2\pi\sqrt{\left(\dfrac{4a\cos\alpha}{3g(1+3\cos^2\alpha)}\right)}.$

Example 7:

A uniform bar of length 2 a is hung from a fixed point by a string of length b fastened to one end of the bar. Show that when the system makes small normal oscillations in a vertical plane, the length 1 of the equivalent simple pendulum is a root of the quadratic,

$$l^2 - \left(\frac{4}{3}a + b\right)l + \frac{ab}{3} = 0.$$

Solution:

At any time t, let the string OA and the rod AB make angles θ and ϕ with the vertical.

$x_G = b \sin\theta + a \sin\phi.$

$y_G = b \cos\theta + a \sin\phi.$

$\therefore \dot{x}_G^2 + \dot{y}_G^2 =$ (velocity)^2 of G

$= b^2\dot{\theta}^2 + a^2\dot{\phi}^2 + 2ab\dot{\theta}\dot{\phi}\cos(\theta - \phi)$

$= b^2\dot{\theta}^2 + a^2\dot{\phi}^2 + 2ab\dot{\theta}\dot{\phi}$ [$\because \theta$ and ϕ are small]

Now let T be the kinetic energy and W the work function of the system, then we easily obtain

$W = mg\,[b\cos\theta + a\cos\phi]$

and $T = \dfrac{1}{2}m\left[\dfrac{a^2}{3}\dot{\phi}^2 + b^2\dot{\theta}^2 + a^2\dot{\phi}^2 + 2ab\dot{\theta}\dot{\phi}\right]$

$= \dfrac{1}{2}m\left[\dfrac{4a^2}{3}\dot{\phi}^2 + b^2\dot{\theta}^2 + 2ab\dot{\theta}\dot{\phi}\right]$

$\therefore$ Lagrange's θ-equation is $\dfrac{d}{dt}\left(\dfrac{\partial T}{\partial\dot{\theta}}\right) - \dfrac{\partial T}{\partial\theta} = \dfrac{\partial W}{\partial\theta}$

$\Rightarrow \dfrac{d}{dt}\{m(b^2\dot{\theta} + ab\dot{\phi})\} = -\,mgb\theta.$ $\{\because \sin\theta = 0\}$

$\Rightarrow b\ddot{\theta} + a\dot{\phi} = -\,g\theta$ ...(1)

Lagrange's ϕ - equation is given by

$\dfrac{d}{dt}\left\{\left(m\dfrac{4}{3}a^2\dot{\phi} + ab\dot{\theta}\right)\right\} = -\,mag\phi$

$\Rightarrow 4a\ddot{\phi} + 3b\ddot{\theta} = -3g\phi.$ ...(2)

Equations (1) and (2) again can be written as

$(bD^2 + g)\,\theta + aD^2\,\phi = 0$ ...(3)

and $3bD^2\ \theta + (4\ aD^2\ 3g)\,\phi = 0.$ ...(4)

Eliminating θ between these equations, we obtain

$$[(bD^2 + g)(4aD^2 + 3g) - 3abD^4]\theta = 0$$

i.e. $[(abD^4 + (4a + 3b)gD^2 + 3g^2]\theta = 0.$ ...(5)

Now let $\theta = A\cos\left[\sqrt{\left(\frac{g}{l}\right)}t + B\right]$

where l is the length of the simple equivalent pendulum.

Then $D\theta = -\sqrt{\left(\frac{g}{l}\right)}A\sin\left[\sqrt{\left(\frac{g}{l}\right)}t + B\right]$

$$D^2\theta = -\frac{g}{l}A\cos\left[\sqrt{\left(\frac{g}{l}\right)}t + B\right] = \frac{g}{l}\theta \text{ and } D^2\theta = \frac{g^2}{l^2}\theta,$$

$$\therefore (5) \Rightarrow \left[ab\frac{g^2}{l^2} - (4a + 3b)\frac{g^2}{l} + 3g^2\right]\theta = 0$$

$\Rightarrow 3l^2 - (4a + 3b)\,l + ab = 0$ $\{\because \theta \neq 0\}$

$$\Rightarrow l^2 - \left(\frac{4}{3}a + b\right)l + \frac{ab}{3} = 0.$$

Example 8(a):

A uniform straight rod of length 2a, is freely movable about its centre and a particle of mass one-third that of the rod is attached by a light inextensible string, of length a, to one end of the rod; show that one period of principle oscillation is $\left(\sqrt{5+1}\right)\pi\sqrt{\left\{\frac{a}{g}\right\}}.$

Solution:

At time t, let θ and ϕ be the inclinations of the rod and the string to the vertical. Co - ordinates of C are

$x_C = a\sin\theta + a\sin\phi$ and

$y_C = a\cos\theta + a\cos\phi.$

$\therefore \dot{x}_C = a\cos\theta\dot{\theta} + a\cos\phi\dot{\phi}$ and $\dot{y}_C = -a\sin\theta\theta - a\sin\phi\dot{\phi}$

$$\Rightarrow \dot{x}_C^2 + \dot{y}_C^2 = a^2\dot{\theta}^2 + a^2\dot{\phi}^2 + 2a^2\cos(\phi - \phi)\dot{\theta}\dot{\phi}$$

$$= a^2\,\dot{\theta}^2 + a^2\,\dot{\phi}^2 + 2a^2\dot{\theta}\dot{\phi}$$

[neglecting higher powers of small quantities]

$\therefore$ (velocity)^2 of the particle C = $v_C^2 = a^2\dot{\theta}^2 + a^2\,\dot{\phi}^2 + 2a^2\dot{\theta}\dot{\phi}$.

And velocity of the C.G. of the rod *i.e.* of O, is zero.

Now let T, be the kinetic energy and W, the work function of the system, then we easily get $W = \frac{mg}{3}(a\cos\theta + a\cos\phi) + C$

and $T = \frac{1}{2}m\frac{a^2}{3}\dot{\theta}^2 + \frac{1}{2}\left(\frac{m}{3}\right)[a^2\dot{\theta}^2 + a^2\dot{\phi}^2 + 2a^2\dot{\theta}\dot{\phi}]$

$$= \frac{ma^2}{6}[2\dot{\theta}^2 + \dot{\phi}^2 + 2\dot{\theta}\dot{\phi}]$$

$\therefore$ Lagrange's θ-equation is given by

$$\frac{d}{dt}\left(\frac{2ma^2}{3}\dot{\theta} + \frac{ma^2}{3}\dot{\phi}\right) = -\frac{mga}{3}\theta \Rightarrow 2\ddot{\theta} + \ddot{\phi} = -\frac{g}{a}\theta \quad ...(1)$$

While Lagrange's ϕ-equation gives $\frac{d}{dt}\left[\frac{ma^2}{3}\dot{\phi} + \frac{ma^2}{3}\dot{\theta}\right] = -\frac{mga}{3}\theta$

i.e. $\ddot{\theta} + \ddot{\phi} = -\frac{g}{a}\phi.$

Equations (1) and (2) again give

$(2D^2 + c)\,\theta + D^2\,\phi = 0$ ...(3) and $D^2\,q + (D^2 + c)\,\phi = 0$ ...(4)

where $c = \frac{g}{a}$.

Eliminating ϕ in between (3) and (4), we get

$\{d^2 + c)\,(2D^2 + c) - D^4\}\,\theta = 0$ *i.e.* $[D^4 + 3cD^2 + c^2]\,\theta = 0.$ ...(5)

To solve (5), let $\theta = A\cos(pt + B)$ $\therefore$ $D\,\theta = pA\sin(pt + B)$,

$D^2\theta = -p^2A\cos(pt + B) = -p^2\theta$ and $D^4\theta = p^4\theta$.

With these substitutions, (5) gives

$(p^4 - 3cp^2 + c^2)\,\theta = 0 \Rightarrow p^4 - 3cp^2 + c^2 = 0$ $(\because \theta \neq 0)$

$$\therefore p^2 = \frac{3c \pm \sqrt{(9c^2 - 4c^2)}}{2} = \left(\frac{3 \pm \sqrt{5}}{2}\right)c = \frac{(3 \pm \sqrt{5}}{2}\cdot\frac{g}{a}.$$

$\therefore$ one period of principal oscillations corresponding to p_1, is given by

$$\frac{2\pi}{p_1} = 2\pi\sqrt{\left(\frac{2}{3-\sqrt{5}\cdot}\frac{a}{g}\right)} = 2\pi\sqrt{\left(\frac{a}{g}\right)}\sqrt{\left\{\frac{2(3+\sqrt{5}}{9-5}\right\}}$$

$$= 2\pi\sqrt{\left(\frac{a}{g}\right)}\sqrt{\left(\frac{6+2\sqrt{5}}{4}\right)} = 2\pi\sqrt{\left\{\frac{a}{g}\right\}}\sqrt{\left\{\frac{(\sqrt{5}+1)^2}{4}\right\}}$$

$$= (\sqrt{5}+1)\pi\sqrt{\left(\frac{a}{g}\right)}.$$

Example 8(b):

A mass m hangs from a fixed point by a light string of length l and a mass m′ hangs from m by a second string of length l′. For oscillations in a vertical plane, show that the periods of the principal oscillations are the values of $\frac{2\pi}{n}$ where n is given by the equation

$$n^4 - gn^2\frac{m+m'}{m}\left(\frac{1}{l}+\frac{1}{l'}\right) + g^2\frac{m=m'}{ml'l} = 0.$$

Solution:

At any time t, let the strings be inclined at angles θ and ϕ to the vertical. Co-ordinates of m are (l sin θ, l cos θ).

∴ (velocity)² of $m = l^2\dot{\theta}^2$ while co-ordinates of m′ are

$x_B = l\sin\theta + l'\sin\phi, \Rightarrow$

$\dot{x}_B = l\cos\theta\dot{\theta} + l'\cos\phi\dot{\phi}$

$y_B = -\,l\cos\theta + l'\cos\phi \Rightarrow$

$\dot{y}_B = -(l\sin\theta\dot{\theta} + l'\sin\phi\dot{\phi})$

∴ (velocity)² of $m' = \dot{x}_B^2 + \dot{y}_B^2 = l^2\dot{\theta}^2 + l^2\dot{\phi}^2 + 2ll\theta\dot{\phi}$ [∵ θ find ϕ are small]

Now let T, be the kinetic energy and , W the work function, of the system, then we have

W = mgl cos θ + m′g (l cos θ + l ′ cos ϕ)

= gl (m + m′) cos θ + m′ gl′ cos ϕ

and $T = \frac{1}{2}ml^2\dot{\theta}^2 + \frac{1}{2}m'[l^2\dot{\theta}^2 + l'\theta^2 + 2ll\dot{\theta}\dot{\phi}]$

$\frac{1}{2}[(m+m')l^2\dot{\theta}^2 + m'l'^2\dot{\theta}^2 + 2m'\,ll'\dot{\theta}\dot{\phi}$

Lagrange's θ-equation is given by $\frac{d}{dt}\left(\frac{\partial T}{\partial \dot{\theta}}\right)-\frac{\partial T}{\partial \theta}=\frac{\partial W}{\partial \theta}$

$$=\frac{d}{dt}[(m+m')\, l^2\dot{\theta}+m'll'\dot{\phi}]=-gl(m+m')\,\theta.$$

$$\Rightarrow \{m+m'\}\, l\ddot{\theta}+m'l'\ddot{\phi}=-g\,(m+m')\theta \quad ...(1)$$

While Lagrange's ϕ - equation gives

$$\frac{d}{dt}[m'l'^2\dot{\phi}+m'll'\dot{\theta}]=-m'gl'\phi$$

$$\Rightarrow l'\ddot{\phi}+l\ddot{\theta}=-g\phi \quad ...(2)$$

Equation (1) and (2) again give

$$(m+m')(lD^2+g)\theta+m'l'D^2\phi=0 \quad ...(3)$$

$$lD^2\theta+(l'D^2+g)\phi=0 \quad ...(4)$$

Eliminating ϕ, we get $[(m+m')(l'D^2+g)-m'll'D^4]\theta=0$

$$\textit{i.e.}\ [mll'D^4+(m+m')(l+l')gD^2+(m+m')g^2]\theta=0 \quad ...(5)$$

Now let θ = A cos (nt + B); ∴ $D\theta=-nA\sin(nt+B)$

$$D^2\,\theta=-n^2A\ \text{cps}\ (nt+B)=-m^2\theta \text{ and } D^4\theta=n^4\theta \quad ...(6)$$

∴ (5) and (6) give mall′ $n^4-(m+m')\,(l+l')\,gn^2+(m+m')\,g^2=0$

$$\text{or } n^4-\frac{m+m'}{m}\left(\frac{1}{l}+\frac{1}{l'}\right)gn^2+\frac{(m+m')g^2}{mall'}=0. \quad ...(7)$$

Example 8(c):

A mass M hangs from a fixed point at the end of a very long string whose length l is a, to M is suspended a mass m by a string whose length l is small compound with a; prove that the time of a small oscillation of m is

$$2\pi\sqrt{\left(\frac{M}{M+m}\cdot\frac{1}{g}\right)}$$

Solution:

Here, we have m = M, m′ = m, ! = a, l′ = l.

$$\therefore n^4-\frac{M+m}{M}\left(\frac{1}{a}+1\right)\frac{g}{l}n^2+\frac{(M+m)g^2}{Ml^2}=0$$

$$\textit{i.e.}\ \therefore n^4-\frac{M+m}{M}\left(\frac{1}{a}+1\right)\frac{g}{l}n^2+\frac{(M+m)g^2}{Ml^2}.\frac{1}{a}=0 \quad ...(8)$$

But a is large compared to $1 \therefore \frac{1}{a} \to 0$

Hence the equation (8), gives

$$n^4 - \frac{M+m}{M}.\frac{g}{l}.n^2 = 0 \text{ i.e. } n^2 = \frac{M+m}{M}.\frac{g}{l}$$

$\therefore$ Time of a small oscillation $= \frac{2\pi}{n} = 2\pi\sqrt{\left\{\frac{M}{M+m}.\frac{l}{g}\right\}}$.

Example 9:

At the lowest point of a smooth circular tube, of mass M and radius a, is placed a particle of mass M', the tube hangs in a vertical plane from its highest point, which is fixed, and can turn freely in its own plane about this point. If the system be slightly displaced, show that the periods of the two independent oscillations of the system are

$$2\pi\sqrt{\left(\frac{2a}{g}\right)} \text{ and } 2\pi\sqrt{\left(\frac{Mag^{-1}}{M+M'}\right)}.$$

And that for one principal mode of oscillations, the particle remains at res relative to the tube and for the other, the centre of gravity of the particle and the tube and for the other, the centre of gravity of the particle and the tube remain at rest.

Solution:

Let C be the centre of the tube and A the position of the particle M′ at time t when OC and CA make angle 3 θ and ϕ with the vertical

$\therefore x_A = a \sin\theta + a \sin\phi,$

$y_A = a\cos\theta + a\cos\phi.$

and (velocity)^2 of A $= \dot{x}_A^2 + \dot{y}_A^2$

$$= (a\cos\theta\theta + a\cos\phi\dot{\phi})^2 + (-a\sin\theta\dot{\theta} - a\sin\phi\dot{\phi})^2$$

$$= a^2\dot{\theta}^2 + a^2\,\dot{\phi}h2 + 2a^2\dot{\theta}\dot{\phi}\cos(\theta-\phi) = a^2\dot{\theta}^2 + 2a^2\dot{\theta}\dot{\phi}$$

[neglecting small quantities of the higher order].

Also $C \equiv (a\sin\theta, a\cos\theta)$

(velocity)^2 of $C = (a\cos\theta\dot{\theta})^2 + (-a\sin\theta\dot{\theta})^2 = a^2\dot{\theta}^2$ ...(1)

Now let T, be the kinetic energy and W the work function of the system then we readily obtain $W = Mga\cos\theta + M'g\ = (M+M')\,ga\cos\theta + M'ga\cos\phi + K$

T = K.E. of circular tube + K.E. of particle

$$=\frac{1}{2}M(a^2\dot{\theta}^2 + a^2\dot{\theta}^2) + \frac{1}{2}M(a^2\dot{\theta}^2 + a^2\dot{\phi}^2 + 2a^2\dot{\theta}\dot{\phi})$$

$$=\frac{2M+M'}{2}a^2\dot{\theta}^2 + \frac{1}{2}M'a^2\dot{\phi}^2 + M'a^2\dot{\theta}\dot{\phi}. \quad ...(2)$$

$\therefore$ Lagrange's θ-equation gives

$$\frac{d}{dt}[(2M+M')\,a^2\dot{\theta} + M'a^2\dot{\phi}] = -(M+M')ga\theta.$$

$$\Rightarrow (2M+M')\ddot{\theta} + M'\ddot{\phi} + = -(M+M')\frac{g}{a}\theta. \quad ...(3)$$

Also Lagrange's ϕ - equation gives

$$\frac{d}{dt}[M'\,a^2\dot{\phi} + M'\,a^2\dot{\theta}] = -M'ga\,\ddot{\phi} \Rightarrow \phi + \ddot{\theta} = -\frac{g}{a}\phi \quad ...(4)$$

Equations (3) and (4) can be re-written as

$$[(2M + M')\,D^2 + (M + M')\,c]\,q + M'\,D^2\,\phi = 0 \quad ...(5)$$

and $D^2\theta + (D^2 + c)\,\phi = 0$ where $c = \frac{g}{a}$.

Eliminating ϕ between these two equations, we get

$$[\{(2M + M')\,D^2 + (M + M')c\}\,(D^2 + c) - M'\,D^4]\,\theta = 0$$

i.e. $[2MD^4 + c\,(3M + 2M')\,D^2 + c^2\,(M + M')]\,\theta = 0.$ ...(7)

To solve (7),

let $\theta = A\cos(pt + B);\ D\theta = -pA\sin(pt + B)$

$D^2\,\theta = -p^2A\cos(pt + B) = -p^2\theta$ and $D^4\theta = \theta^4\theta$ ...(8)

$\therefore$ (7) and (8) give $[2Mp^4 - c\,(3M + 2M')\,p^2 + c^2\,(M + M')\,\theta = 0$

i.e. $2Mp^4 - c\,(3M + 2M')\,p + c^2\,(M + M') = 0.$ $[\because \theta \neq 0]$

which again gives $(2p^2 - c)\,[Mp^2 - c\,(M + M')] = 0$

$\therefore p_1^2 = \frac{c}{2}$ and $p_2^2 = \frac{c\,(M+M')}{M}\frac{g}{a}.$ $\left(\because c = \frac{g}{a}\right)$

Hence periods of oscillations are given by

$$\frac{2\pi}{p_1} \text{ and } \frac{2\pi}{p_2} \text{ i.e. by } 2\pi\sqrt{(2a/g)} \text{ and } 2\pi\sqrt{\left(\frac{M}{(M+M'}\frac{a}{g}\right)}.$$

Multiplying (6) by λ and adding to (5), we have

$$D^2\{(2M + M' + l)\,\theta + (M' + \lambda)\,\phi\} = -\ \{(M + M')\,\theta + \lambda\phi\} \quad ...(9)$$

Now chose λ such that

$$\frac{2M+M'+\lambda}{M'+\lambda}=\frac{M+M'}{\lambda}\Rightarrow \lambda = M' \text{ and } \lambda = -(M+M').$$

taking $\lambda = M'$, equation (9) reduces to

$$D^2\{(M+M')\,\theta + M'\,\phi\} = -\frac{1}{2}c\{M+M')\theta + M'\phi\}$$

and when $\lambda = -(M + M')$, equation (9) reduces to

$$D^2\,\{\theta - \phi) = -\,\frac{M+M'}{M}c(\theta - \phi)$$

∴ Principal co-ordinates are $q - f$ and $\{(M + M')\,\theta + M\text{–}\phi\}$. For the first mode, $\theta - \phi = 0$, *i.e.* $\theta = \phi$. This shows that the particle is at rest relative to the tube. For the second mode we have $(M + M')\,\theta + M\,'\phi = 0$. Further, the x-coordinates of C.G. of the particle and the tube

$$=\frac{M a\sin\theta + M'(a\sin\theta + a\sin\phi)}{M+M'}$$

$$=\frac{a}{M+M'}\{M\theta + M'(\theta+\phi)\}\ \text{(since } \theta \text{ and } \phi \text{ are small,}$$

$\therefore \sin\theta = \theta$ and $\sin\phi = \phi$)

$$=\frac{a}{M+M'}\{(M+M')\theta + M'\phi\} = 0\ \text{[using above results]}$$

⇒ The common C.G. of the particle and the tube remains at rest.

Example 10:

A perfectly rough sphere lying inside a hollow cylinder, which rests on a perfectly rough plane, is slightly displaced from its position of equilibrium. Show that the time of a small oscillation is

$$2\pi\sqrt{\left(\frac{a-b}{g}\cdot\frac{4M}{10M+7m}\right)}$$

where a is the radius of the cylinder, b that of the sphere, and M, m are the masses of the cylinder and the sphere.

Solution:

Let CA be a line fixed in the cylinder and C'N, a line fixed in the sphere, which were initially vertical. Let these lines, after a time t, make angles ψ and φ with the vertical. Further let θ be the angle which the line joining centre makes with the vertical.

Initially O was the point of contact with the horizontal plane, which is regarded as origin. There is no slipping; between the cylinder, horizontal plan e and between the cylinder, sphere; therefore, we get OB = are AB = aψ and are AP = arc NP

i.e. $a(\theta + \psi) = b(\theta + \phi)$, *i.e.* $b\phi = (a - b)\,\dot{\theta} + a\dot{\psi}$.

Referred to the horizontal and vertical through O as co-ordinate axes, we have

$$xc' = a\psi + a(a-b)\sin\theta,\ yc' = a - (a-b)\cos\theta. \quad \text{(co-ordinates of } c'$$

$\therefore$ (velocity)^2 of c′

$$\dot{x}c'^2 + \dot{y}c^2 = \{a\dot{\psi} + (a-b)\cos\theta\dot{\theta}\}^2 + \text{man}\{a-b)\sin\theta\dot{\theta}\}^2$$

$$= a^2\dot{\psi}^2 + (a-b)^2\dot{\theta}^2 + 2a(a-b)\dot{\psi}\dot{\theta}\cos\theta$$

$$= a^2\dot{\psi}^2 + (a-b)^2\dot{\theta}^2 + 2a(a-b)\dot{\psi}\dot{\theta}. \qquad ...(1)$$

Also co-ordinates of C are (a ψ, a).

$$\therefore \text{(velocity)}^2\ (a\dot{\psi})^2 = a^2\dot{\psi}^2. \qquad ...(2)$$

Now let T be the kinetic energy and $\dot{W}$, the work function. Then we have

$$W = (a-b)mg\cos\theta + C \qquad ...(3)$$

$$\text{and } T = \frac{1}{2}M\left(v^2C + a^2\dot{\psi}^2\right) + \frac{1}{2}m\left(v^2c' + \frac{2b^2}{5}\dot{\phi}^2\right)$$

$$\Rightarrow T = \frac{1}{2}M(a^2\dot{\psi}^2 + a^2\dot{\psi}^2)$$

$$+\frac{1}{2}m\left[\frac{2b^2}{5}\dot{\phi} + a^2\dot{\psi}^2 + (a-b)^2\dot{\theta}^2 + 2a(a-b)\dot{\psi}\dot{\theta}\right]$$

$$= Ma^2\dot{\psi}^2 + \frac{1}{2}m[\frac{2}{5}\{(a - b\theta + a\dot{\psi}\} + a^2\dot{\psi}^2 + (a-b)^2\dot{\theta}^2 + 2a(a-b)\dot{\psi}\dot{\theta}]$$

$$= \frac{10M + 7m}{10}a^2\dot{\psi}^2 + \frac{7m}{10}[(a-b)^2\theta^2 + 2a(a-b)\dot{\psi}\dot{\theta}] \qquad ...(4)$$

$\therefore$ Lagrange's θ-equation gives

$$\frac{d}{dt}\left[\frac{7m}{5}\{(a-b)^2\dot{\theta} + a(a-b)\dot{\psi}\}\right] = -(a-b)mg\sin\theta$$

$$\Rightarrow 7(a-b)\ddot{\theta} + 5a\ddot{\psi} = -5g\theta \text{ (as } \theta \text{ is small).} \qquad ...(5)$$

Also Lagrange's ψ-equation $\dfrac{d}{dt}\left[\dfrac{10M + 7m}{5}a^2\dot{\psi} + \dfrac{7m}{5}a(a-b)\theta\right] = 0$

$$\Rightarrow (10M + 7m)a\ddot{\psi} + 7m(a - b)\ddot{\theta} = 0$$

$$\Rightarrow (7m(a - b)\ddot{\theta} + (10M + 7m)a\ddot{\psi} = 0 \qquad ...(6)$$

Eliminating $\ddot{\psi}$ between(5) and (6) ,we get

$$[7(10M + 7m) - 49m](a - b)\ddot{\theta} = - 5(10M + 7m)g\theta$$

or $70M(a - b)\ddot{\theta} = -5(10M + 7m)g\theta$

or $\ddot{\theta} = -\dfrac{(10M + 7m)}{14M} \cdot \dfrac{g}{a - b)}\theta = -\mu\theta$ say,

$\therefore$ time of a small oscillation $= \dfrac{2\pi}{\sqrt{\mu}} = 2\pi\sqrt{\left(\dfrac{a - b}{g} \cdot \dfrac{14M}{10M + 7m}\right)}$.

Example 11:

A perfectly rough sphere, of mass m and radius b, rest at the lowest point of a fixed spherical cavity of radius a. To the highest point of the movable sphere is attached a particle of mass m and the system is disturbed. Show that the oscillations are the same as those of a simple pendulum of length

$$a - b)\frac{4m' + \frac{7}{5}m}{m + m'\left(2 - \frac{a}{b}\right)}$$

Solution:

O is the centre of the fixed cavity, C the centre of the sphere and at B, a particle of m′ is attached. Initially NB was vertical, and N coincided with A.

After the time t, let BN or CN, a line fixed in the sphere make an angle ϕ with the vertical. But there is no slipping between the cylinder and the sphere.

$\therefore$ arc AP = arc NP,

i.e. $a\theta = b(\theta + \phi)$

i.e. $b\phi = (a - b)\theta \Rightarrow b\dot{\phi} = (a - b)\dot{\theta}$

Now co-ordinates of iB (i.e. of m ′) are

$x_B = (a - b)\sin\theta + b\sin\phi, y_B = (a - b)\cos\theta - b\cos\phi,$

and co-ordinates of the centre C are

$$\therefore (\text{velocity})^2 \text{ of } B = \dot{x}_B^{\ 2} + \dot{y}_B^{\ 2}$$

$$= \{(a - b)\cos\theta\dot{\theta} + b\cos\theta\dot{\theta}\}^2 + \{-(a - b)\sin\theta\dot{\theta} + b\sin\theta\dot{\theta}\}^2$$

$$= (a-b)^2\dot{\theta}^2 + b^2\phi^2 + 2b(a-b)\dot{\theta}\dot{\phi}$$

[neglecting small quantities of second and higher orders]

and (velocity)^2 of C = $\{(a-b)\cos\theta\dot{\theta}\}^2 + \{-(a-b)\sin\theta\dot{\theta}\}^2$

$$= (a-b)^2\dot{\theta}^2$$

Now let T, be the kinetic energy and W the work function of the system, then we readily obtain

$$W = mg(a-b)\cos\theta + m'g\{(a-b)\cos\theta - b\cos\phi\} + D$$

$$= (a-b)(m+m')g\cos\theta - m'gb\cos\phi + D$$

$$= (a-b)(m+m')g\cos\left(\frac{b}{a-b}\phi\right) - m'gb\cos\phi + D \qquad \{\because (a-b)\theta = b.\phi\}$$

and $T = \frac{1}{2}m\left[\frac{2b^2\dot{\phi}^2}{5} + (a-b)^2\dot{\theta}^2\right] + \frac{1}{2}m'\{(a-b)^2\dot{\theta}^2 + b^2\dot{\phi}^2 + 2b(a-b)\dot{\theta}\dot{\phi}$

$$= \frac{1}{2}m\left[\frac{2b^2\dot{\phi}^2}{5} + b^2\dot{\phi}^2\right] + \frac{1}{2}m'\{b^2\dot{\phi}^2 + b^2\dot{\phi}^2 + 2b^2\dot{\phi}^2] \qquad [\because (a-b)\dot{\theta} = b\dot{\phi}]$$

$$= \frac{b^2}{10}(7m + 20m')\dot{\phi}^2.$$

Further Lagrange's ϕ-equation gives $\frac{d}{dt}\left[\frac{b^2}{5}(7m+20m')\dot{\phi}\right]$

$$= -(a-b)(m+m')g\frac{b}{a-b}\sin\left(\frac{b}{a-b}\phi\right) + m'gb\sin\phi$$

$$= -(m+m')bg\sin\frac{b\phi}{a-b} + m'gb\sin\phi$$

$$\Rightarrow \frac{b^2}{5}(7m+20m')\phi = -(m+m')bg.\frac{b}{a-b}\phi + m'gb\phi \text{ app.}$$

$$\Rightarrow \left(4m' + \frac{7m}{5}\right)\ddot{\phi} = -\frac{g}{a-b}\left[m + m'\left(2 - \frac{a}{b}\right)\right]\phi, \text{ app.}$$

$$\Rightarrow \ddot{\phi} = -\frac{g}{a-b}\cdot\frac{m + m'\left(2-\frac{a}{b}\right)}{4m' + \frac{7}{5}m}\phi = -\mu\phi \quad [\text{form } \ddot{x} = -\mu x]$$

$\therefore$ Length of the simple equivalent pendulum $= \frac{g}{\mu}$

$$= (a - b)\frac{4m' + \frac{7}{5}m}{m + m'\left(2 - \frac{a}{b}\right)}.$$

Example 12:

A perfectly tough sphere rests at the lowest point of a fixed spherical cavity of double its own radius. To the highest point of the movable sphere is attached a particle of mass 7/20 times that of the sphere and the system is disturbed. Show that the oscillations are the same as those of a simple pendulum of length 14/5 times the radius of the sphere.

Solution:

Put proceed similarly to get the required result.

Example 13:

A plank, of mass M, radius of gyration k and length 2b, can swing like a sea-saw across a perfectly rough fixed cylinder of radius a. At its ends hang two particles each of mass m, by strings of length l. Show that, as the system swings, the lengths of its simple equivalent pendulum are

$$l \text{ and } \frac{Mk^2 + 2mb^2}{(m + 2m)a.}$$

Solution:

Let the strings be AP and BQ such that their free extremities are having particles of mass m. Initially G was at D and AP, BQ were vertical. After a time t, when the plank ψ with the horizontal. Let the strings be inclined at θ and ϕ to the vertical.

But there is no slipping between the plank and the cylinder, so we have

$$\text{Now } xp = a\sin\psi + (b - a\psi)\cos\psi + l\sin\theta = b + l\theta - \frac{1}{2}b\psi^2 \qquad ...(1)$$

$$yp = a\cos\psi - (b - a\psi)\sin\psi - l\cos$$

$$= a(1 - \frac{1}{2}\psi^2) - (b - a\psi)\psi - l(1 - \frac{1}{2}\theta^2) \qquad ...(2)$$

$$\therefore \dot{x}_p = l\dot{\theta} - b\psi\dot{\psi} \text{ and } \therefore \dot{y}_p = -b\dot{\psi} + a\psi\dot{\psi} + l\theta\dot{\theta}$$

$$\therefore (\text{Velocity})^2 \text{ of } = \dot{x}_p{}^2 + \dot{y}_p{}^2 = l^2\dot{\theta}^2 + b^2\dot{\psi}^2 \, p$$

$$\text{Thus kinetic energy of m at } P = \frac{1}{2}m(l^2\dot{\phi}^2 + b^2\dot{\psi}^2)$$

Similarly kinetic energy of m at $Q = \frac{1}{2}m(l^2\dot{\phi}^2 + b^2\dot{\psi}^2)$

To obtain the corresponding quantities for the point Q.

We shall write–ψ for ψ and ϕ for θ.

Also y-coordinate of Q $= a - l + b\psi + \frac{1}{2}a\psi^2 + \frac{1}{2}l\phi^2 = yq$ say

[put–ψ for ψ and ϕ for θ in (2)]

the co-ordinates of G are {a sin ψ –aψ cosψ, acosψ + aψ sinψ} (neglecting higher powers of ψ as it is small)

$$i.e. \left\{\frac{a\psi^2}{2}, a + \frac{a\psi^2}{2}\right\}.$$

∴ (velocity)²of G = $a^2\psi^2\dot{\psi}^2 [\because \dot{x} = a\dot{\psi}\psi, \dot{y} = a\psi\dot{\psi}]$

(neglecting smaller, quantities). Thus kinetic energy of the plank

$$= \frac{1}{2}M(k^2\dot{\psi}^2 + a^2\psi^2\dot{\psi}^2) = \frac{1}{2}Mk^2\dot{\psi}^2, \text{ approximately}$$

Now let T be the kinetic energy and W, the work function of the system, then we obtain

$$W = -mg\left(a - l - b\psi + \frac{a\psi^2}{2} + l\frac{\theta^2}{2}\right)$$

$$-mg\left(a - l + b\psi + \frac{a\psi^2}{3} + l\frac{\theta^2}{2}\right) - mg\left(a + \frac{a\psi^2 2}{}\right)$$

$$= K - \frac{1}{2}a(M + 2m)g\psi^2 - \frac{1}{2}m\,gl\left(\theta^2 - \phi^2\right)$$

and $T = \frac{1}{2}m(l^2\dot{\theta}^2 + b^2\dot{\psi}^2) + \frac{1}{2}m(l^2\dot{\phi}^2 + b^2\dot{\psi}^2) + \frac{1}{2}MK^2\dot{\psi}^2$

Lagrange's θ-equation is $\frac{d}{dt}\left(\frac{\partial T}{\partial \dot{\theta}}\right) - \frac{\partial T}{\partial \theta} = \frac{\partial W}{\partial \theta}$

$$\Rightarrow \frac{d}{dt}(ml^2\dot{\theta}^2) = -mgl\theta \Rightarrow \ddot{\theta} = -\frac{g}{l}\theta,$$

∴ length of the simple equivalent pendulum = l

Also Lagrange's ψ – aqⁿ gives

$$\frac{d}{dt}\{(Mk^2 + 2mb^2)\dot{\psi}\} = -a(M + 2m)g\psi \Rightarrow \ddot{\psi} = -ag\frac{M + 2m}{Mk^2 + 2mb^2}\psi$$

Example 14(a):

A solid uniform sphere has a light rod rigidly attached to it which passes through its centre. This rod is so jointed to a fixed vertical axis that the angle q between the rod and the axis may alter but the rod must turn with the axis. If the vertical axis be forced to revolve constantly with uniform angular velocity, show that the equation of motion is of the form

$$\dot{\theta}^2 = n^2(\cos\theta\ 0\cos\beta)(\cos\alpha - \cos\theta)$$

Show also that the total energy imparted to the sphere as q increase from θ_1 to θ_2 *varies as* $\cos^2\theta_1 - \cos^2\theta_2$.

Solution:

First of all, we reduce the revolving axis OZ to rest by introducing an additional force $m\omega^2$ NO′, along NO′ *i.e.* a force $m\omega^2 l \sin\theta$ along NO′.

After the interval of time t. Let the rod make an angle q with the vertical ONZ (taken as Z-axis. (Further let ϕ be the angle through which the sphere turn in the horizontal plane, then the co-ordinates of C.G. of the sphere, are given by

$$x_{0'} = l\sin\theta\cos\phi.\ y_0{}' = l\sin\theta\sin\phi,\ z_{0'} = l\cos\theta$$

$$\therefore \dot{x}'_0 = l\dot{\theta}\cos\theta\cos\phi - l\dot{\phi}\sin\theta\sin\phi,$$

$$\dot{y}_0 = l\dot{\theta}\cos\theta\cos\phi + l\dot{\phi}\sin\theta\cos\phi,\ \dot{z}'_0 = -\,l\dot{\theta}\sin\phi$$

$$\Rightarrow \dot{x}'^2_0 + \dot{y}'^2_0 + \dot{z}'^2_0 = l^2\dot{\theta}^2 + l^2\sin^2\theta\dot{\phi}^2$$

$$\therefore T = \frac{1}{2}m\left[\frac{2a^2}{5}\dot{\theta}^2 + \dot{x}_{0'}{}^2 + \dot{y}_{0'}{}^2 + \dot{z}_{0'}{}^2\right]$$

$$= \frac{1}{2}m\left[\frac{2a^2}{5}\dot{\theta}^2 + l^2\dot{\theta}^2 + l^2\dot{\phi}\sin^2\theta\right]$$

$$= \frac{1}{2}m\left[\left(\frac{2a^2}{5} l^2\right)\dot{\theta}^2 + l^2\omega^2\sin^2\theta\right] \quad (\because \phi \neq \omega \text{ given})$$

and $W^* = mgl\cos\theta + m\omega^2 l\sin\theta = mgl\cos\theta + m\omega^2 . 2\sin\theta\cos\theta$

$= -\,mgl\sin\theta + 2\,m\omega^2 l^2\sin\theta\cos\theta$

$$\Rightarrow \left(\frac{2a^2}{5} + l^2\right)\ddot{\theta} = -\,mgl\sin\theta + 3m\omega^2 l^2\sin\theta\cos\theta$$

Multiplying both sides by $2\dot{\theta}$ and integrating, we get

$$\left(\frac{2}{5}a^2 + l^2\right)\dot{\theta}^2 = 2gl\cos\theta - 3l^2\omega^2\cos^2\theta + D$$

$$\Rightarrow \dot{\theta}^2 = n^2(\cos\theta - \cos\beta)(\cos\alpha - \cos\theta)$$

where n is constant and α, β are the values of θ for which θ vanish.

Second Part:

K.E. + Potential Energy = T + K, when P.E. is K and is given by

$$K = T - W + C = \frac{1}{2}m\left[\left(\frac{2a^2}{5} + l^2\right)\dot{\theta}^2 + l^2\omega^2\sin^2\theta\right]$$

$$- mgl\cos\theta - ml^2\omega^2\sin^2\theta + C$$

$$= \frac{1}{2}m[\{2gl\cos\theta - 3l^2\omega^2\cos^2\theta + D\} + l^2\omega^2\sin^2\theta]$$

$$- mgl\cos\theta - ml^2\omega^2\sin^2\theta + C$$

$$= -\frac{1}{2}m\omega^2l^2\cos^2\theta - \frac{1}{2}m\omega^2 l + A \text{ where A is constant.}$$

$$= -\frac{1}{2}m\omega^2l^2\cos^2\theta + B \text{ where B is constant}$$

∴ Total energy imparted

$$= \left[-\frac{1}{2}m\omega^2l^2\cos^2\theta + B\right]_{\theta_1}^{\theta_2} = \frac{1}{2}m\omega^2l^2(\cos^2\theta_1 - \cos^2\theta_2)$$

which varies as $(\cos^2\theta_1 - \cos^2\theta_2)$.

Example 14(b):

Four uniform rods; each of length 2a, are hinged at their ends so as to form a rhombus ABCD. The angles B and D are connected by an elastic string and the lowest end A rests on a horizontal plane while the end C slides on a smooth vertical wire passing through A; in the position of equilibrium the string is stretched to twice its natural length and the angle BAD is 2α. Show that the time of a small oscillation about this position is

$$2\pi\left\{\frac{2a(1+3\sin^2\alpha)}{3g\cos 2\alpha}\cos\alpha\right\}^{1/2}$$

Solution:

In the position of equilibrium, rods are making angles α with the vertical. When the system is slightly displaced from the position of equilibrium, let the rods make angle (α + θ) with the vertical θ being a small displacement. Now assuming the fixed and A as origin and the horizontal and vertical lines through it as co-ordinate axes, the co-ordinates of G_2 are {a sin (α + θ), 3a cos (α + θ)}

$$\therefore (\text{velocity})^2 \text{ of } G_2 = \{a\cos(\alpha+\theta)\dot{\theta}\}^2 + (-3a\sin(\alpha+\theta)\dot{\theta}\}^2$$

$= a^2[(1 + 8\sin^2(\alpha + \theta)]\dot{\theta}^2$

Co-ordinates of G_1 are $\{a\sin(\alpha + \theta), a\cos(\alpha + \theta)\}$

$\therefore$ (velocity)^2 of $G_1 = a^2\dot{\theta}^2$

$\therefore$ Kinetic energy of the four taken together is

$$T = 2.\frac{1}{2}m.\left[\frac{a^2}{3}\dot{\theta}^2 + a^2\dot{\theta}^2\right]$$

$$+ 2.\frac{1}{2}m\left[\frac{a^2}{3}\dot{\theta}^2 + a^2\{1 + 8\sin^2(\alpha + \theta)\}\dot{\theta}\right]$$

$$= \frac{8ma^2}{3}\left[1 + \sin^2(\alpha + \theta)\right]\dot{\theta}^2 \ (\because v_{G_1} = v_{G_4} \text{ and } v_{G_2} = v_{G_3})$$

The work function W is given by $W = 2\{-mga\cos(\alpha + \theta)\}$

$$+2\{-mg3a\cos(\alpha + \theta)\} - 2\int_0^{2a\sin(\theta + \alpha)} \lambda\left(\frac{y - c}{c}\right)dy$$

$$= -8mga\cos(\alpha + \theta) - \frac{\lambda}{c}\{2a\sin(\alpha + \theta) - \}^2$$

Lagrange's θ equation gives

$$\frac{d}{dt}\left|\frac{16ma^2}{3}\{1 + 3\sin^2(\alpha + \theta)\}\dot{\theta} - 16ma^2\sin(\alpha + \theta)\cos(\alpha + \theta)\dot{\theta}^2\right.$$

$$= 8mga\sin(\alpha + \theta) - \frac{4\lambda}{c}a\cos(\alpha + \theta)\{2a\sin(\alpha + \theta) - c\}$$

$$\Rightarrow \frac{16ma^2}{3}(1 + 3\sin^2(\alpha + \theta)\}\ddot{\theta}$$

$$= 8mga\sin(\alpha + \theta) - \frac{4\lambda a}{c}\cos(\alpha + \theta)\{2a\sin(\alpha + \theta) - c\} \quad ...(1)$$

Initially when $\theta = 0, \dot{\theta} = 0, \ddot{\theta} = 0, c = a\sin\alpha$, hence (1) gives

$$\lambda = \frac{2mgc}{a\cos\alpha.}$$

Putting this value of λ in equation (1), we get

$$\frac{16ma^2}{3}\{1 + 3\sin^2(\alpha + \theta)\}\ddot{\theta}$$

$$= mga8\sin(\alpha + \theta) - \frac{8mg\cos(\alpha + \theta)}{\cos\alpha}\{2a\sin(\alpha + \theta) - c\}$$

*The force $m\omega^2 l\sin\theta$ also contributes to W. The distance of the point of application at O′ of this force from the vertical OZ is equal to l sin θ, hence the contribution $m\omega^2 l^2 \sin^2\theta$ to W is as given in (2).

** If W is the work function of the system , then P.E. = C –W.

$$i.e.\ \frac{16ma^2}{3}\{1+3\sin^2\alpha\}\theta = 8mga(\sin\alpha+\theta\cos\alpha)$$

$$-\frac{8mg}{\cos\alpha}(\cos\alpha-8\sin\alpha)\{2a(\sin\alpha+\theta\cos\alpha)-a\sin\alpha\}.$$

$$=-\frac{8mg\cos 2\alpha}{\cos\alpha}\theta \text{ app.} \Rightarrow \ddot{\theta} = -\frac{3g\cos 2\alpha}{2a\cos\alpha(1+3\sin^2\alpha)}\theta \text{ app.}$$

∴ Time of a small oscillation about the position of equilibrium is given by

$$2\pi\sqrt{\left\{\frac{2a\cos\alpha(1+3\sin^2\alpha}{3g\cos^2\alpha}\right\}}$$

Example 14(c):

A smooth circular wire, of mass 8 m and radius a, swings in a vertical plane, being suspended by an inextensible string of length a attached to one point of it, a particle of mass m can slide on the wire. Prove that the periods of small oscillations are

$$2\pi\sqrt{\left(\frac{8a}{3g}\right)}, 2\pi\sqrt{\left(\frac{a}{3g}\right)}, 2\pi\sqrt{\left(\frac{8a}{9g}\right)},$$

Solution: .

At any time t, let the string OA, and the radius AC be inclined at angles θ and ϕ with he vertical and further let the radius to the particle (m) be inclined at an angle ψ with the vertical.

Now co-ordinates of C are

$$(a\sin\theta + a\sin\phi, a\cos\theta + a\cos\phi).$$

$$\therefore (\text{velocity})^2 \text{ of } C = a^2\dot{\theta}^2 + a^2\dot{\phi}\cos(\theta-\phi)$$

$$= a^2\dot{\theta}^2 + a^2\dot{\phi}^2 + 2a^2\dot{\theta}\dot{\phi} \text{ approximately } \left[\begin{array}{l}\because \dot{x}_C = a\cos\theta\dot{\theta} + a\cos\phi\dot{\phi} \\ \dot{y}_C = -a(\sin\theta\dot{\theta}+\sin\phi\dot{\phi}\end{array}\right]$$

Also co-ordinates of the particle m (*i.e.* of the pt. P) are

$x_P = a\,(\sin\theta + \sin\phi + \sin\psi)$,

$y_P = a\,(\cos\phi \div \cos\phi + \cos\psi)$.

$\because$ (celocity)^2 of $m = a^2(\dot{\theta}^2 + \dot{\phi}^2 + \dot{\psi}^2 + 2\dot{\theta}\dot{\phi} + 2\dot{\theta}\dot{\phi} + 2\dot{\phi}\dot{\psi} + 2\dot{\psi}\dot{\theta})$ app,

Let T, be the kinetic energy and W, the work function of the system, then we readily get

$$W = 8mg\,(a\cos\theta + a\cos\phi) + mg\,(a\cos\theta + a\cos\phi + a\cos\psi)$$

$$= mga\,[9\cos\theta + 9\cos\phi + \cos\psi)$$

$$\text{and } T = \frac{1}{2}.8m[a^2\dot{\phi}^2 + (a^2\dot{\theta}^2 + a^2\dot{\phi}^2 + a^2\dot{\phi}^2 + 2a^2\dot{\theta}\dot{\phi})] \qquad ...(1)$$

$$+\frac{1}{2}ma^2[\dot{\theta}^2 + \dot{\phi}^2 + \dot{\psi}^2 + 2\dot{\theta}\dot{\phi} + 2\dot{\phi}\dot{\psi} + 2\dot{\psi}\dot{\theta}]$$

$$i.e.\ T = \frac{1}{2}m[9\dot{\theta}^2 + 17\dot{\phi}^2 + \dot{\psi}^2 + 18\dot{\theta}\dot{\phi} + 2\dot{\phi}\dot{\psi} + 2\dot{\psi}\dot{\theta}] \qquad ...(2)$$

$\therefore$ Lagrange's q and y equations give

$$9\ddot{\theta} + 9\ddot{\phi} + \ddot{\psi} = -9\frac{g}{a}\theta \qquad ...(3)$$

$$9\ddot{\theta} + 17\ddot{\phi} + \ddot{\psi} = -9\frac{g}{a}\phi \qquad ...(4)$$

$$\text{and } \ddot{\theta} + \ddot{\phi} + \ddot{\psi} = \frac{g}{a}\psi,$$

which can be rewritten as

$$(9D^2 + 9c)\,\theta + 9D^2\,\phi + D^2\,\psi = 0 \qquad ...(6)$$

$$9D^2\theta + (17D^2 + 9c)\,\phi + D^2\psi = 0 \qquad ...(7)$$

$$\text{and } D^2\theta + D^2\,\phi + (D^2 + c)\,\psi = 0 \qquad ...(8)$$

Eliminating ϕ and ψ in ...(6), (7) and (8), we get

$$\begin{vmatrix} 9D^2 + 9c & 9D^2 & D^2 \\ 9D^2 & 17D^2 + 9c & D^2 \\ D^2 & D^2 & D^2 + c \end{vmatrix} \theta = 0$$

$$i.e.\ (8D^2 + 9c)\,[9c(2D^2 + c) + D^2(8D^2 + 9c)]\theta = 0$$

$$i.e.\ (8D^2 + 9c)\,[8D^4 + 27cD^2) + 9c^2]\theta = 0$$

$$i.e.\ (8D^2 + 9c)\,(8D^4 + 3c)\,D^2 + 3c]\theta = 0 \qquad ...(9)$$

Now let $\theta = A\cos(pt + B)$, then

$D\theta = -pA\sin(pt + B)$, $D^2\theta$

$$= -p^2A\cos(pt + B) = -p^2\theta \qquad ...(10)$$

$\therefore$ (9) gives $(8p^2 - 9c)(8p^2 - 3c)(p^2 - 3c) = 0$ $[\because \theta \neq \theta]$

$$\Rightarrow \left(8p^2 - \frac{9g}{a}\right)\left(8p^2 - \frac{3g}{a}\right)\left(p^2 - \frac{3g}{a}\right) = 0.$$

i.e. $p_1^2 = \frac{9g}{8a}, p_2^2 = \frac{3g}{8g}, p_3^2 = \frac{3g}{a}$

Thus periods of small oscillations are $\frac{2\pi}{p_1}, \frac{2\pi}{p_2}, \frac{2\pi}{p_3}$

i.e. $2\pi\sqrt{\left(\frac{8a}{9g}\right)}, 2\pi\sqrt{\left(\frac{8a}{3g}\right)}, 2\pi\sqrt{\left(\frac{a}{9g}\right)}$

Example 15:

A plank, 2a feet long, is placed symmetrically across a light cylinder, of radius a which rests and is free to roll on a perfectly rough horizontal plane. A heavy particle whose mass is an times that of plank is embedded in the cylinder at its lowest point. If the system is slightly displaced, show that its periods of oscillation are the values of

$$\frac{2\pi}{p}\sqrt{\left(\frac{a}{g}\right)}$$

given by the equation $4p^4 - (n + 12)p^2 + 3(n - 1) = 0.$

Solution:

C is the centre of the cylinder whose radius is a.

Let the mass of the plank be m, therefore mass of the particle embedded at the lowest point of the cylinder is nm.

Suppose θ and ϕ are the angles through which the cylinder and the plank have turned in time t, from their initial positions.

Initially G was coinciding with K which was at the highest point of the cylinder.

There is no slipping between the plank and the cylinder, hence Further there is no sliding between the cylinder and the horizontal plane.

$PG = \text{arc } KP = a(\phi = \theta)$

hence $OC_1 = \text{arc } C_1S = a\theta$, where O was the initial point of contact of the cylinder and the horizontal plane.

Taking O as origin and the horizontal and vertical through O as co-ordinates axes, we obtain $x_G = aq + a \sin f - a(f - q) \cos f$

$= a\theta + a\phi - a(\phi - \theta) = 2a\theta$

and $y_G = a + a\cos\phi + a(\phi - \theta)\sin\phi = 2a - a\theta\phi + a\,\dfrac{\phi^2}{2}$

Further co-ordinates of the particle nm (*i.e.* of the point S)

arc $\equiv \{a\theta - a\sin\theta, a - \cos\theta\}$

$\therefore$ (velocity)^2 of (nm) $= (a\dot{\theta} - a\cos\theta\dot{\theta})^2 + (a\sin\theta\dot{\theta})^2,$

$$= (a^2 + a^2 - 2a^2\cos\theta)\dot{\theta}^2.$$

Now let T, be the kinetic energy and W, the work function of the system, then we readily obtain

$$W = -n\,mg\,(a - a\cos\theta) - mg\left(2a - a\theta\phi + a\frac{\phi^2}{2}\right)$$

$$= -\ nmga\frac{\theta^2}{2} - mga\left(\frac{\phi^2}{2} - \theta\phi\right) - 2mga$$

$$\text{and } T = \frac{1}{2}m\left[\frac{a^2}{3}\dot{\phi}^2 + 4a^2\dot{\theta}^2\right] + \frac{1}{2}nm[2a^2 - 2a^2\cos\theta]\dot{\theta}^2$$

$$= \frac{1}{2}m\left[\frac{a^2}{3}\dot{\phi}^2 + 4a^2\dot{\theta}^2\right] + \frac{1}{2}nm.2a^2\dot{\theta}^2\frac{\theta^2}{2}$$

$$= \frac{1}{2}m\left[\frac{a^2}{3}\dot{\phi}^2 + 4a^2\dot{\theta}^2\right], \text{ approximately, (neglecting the second part)}$$

$\therefore$ Lagrange's q - equation is given by $\dfrac{d}{dt}\left(\dfrac{\partial T}{\partial \dot{\theta}}\right) - \dfrac{\partial T}{\partial \theta} = \dfrac{\partial W}{\partial \theta}$

$$\Rightarrow \frac{d}{dt}(4ma^2\dot{\theta}) = -\,nmga\theta + mga\phi \Rightarrow 4\ddot{\theta} = -\frac{g}{a}(n\theta - \phi) \qquad ...(1)$$

Further Lagrange's f-equation gives

$$\frac{d}{dt}\left(\frac{ma^2}{3}\dot{\phi}\right) = -\,mg(\phi - \theta) \Rightarrow \ddot{\phi} = -\frac{3g}{a}(\phi - \theta) \qquad ...(2)$$

Equations (1) and (2) can be re-written at $(4D^2 + nc)\,\theta - c\phi = 0$ ...(3)

and $3cq - (D^2 + 3c)\,\phi = 0$ where $c = (g/a)$ ...(4)

Eliminating f in between (3) and (4), we get

$[(4D^2 + nc)\,(D^2 + 3c) - 3c^2\,]\,\theta = 0$

$\Rightarrow [4D^2 + c(n + 12)\,D^2 + 3\,(n - 1)\,c^2]\,\theta = 0$ ...(5)

Now if the periods of oscillation are the values of

$$\frac{2\pi}{p}\sqrt{\left(\frac{a}{g}\right)} \text{ i.e. values of } \frac{2\pi}{\sqrt{\left(\frac{g}{a}p^2\right)}},$$

then the solution of the above equation must be of the form

$$\theta = A\cos\left[\sqrt{\left(\frac{g}{a}\right)}pt + B\right]$$

$$\therefore D\theta = -\sqrt{\left(\frac{g}{a}\right)}pA\sin\left[\sqrt{\left(\frac{g}{a}\right)}pt + B\right]$$

$$D^2\theta = -\left(\frac{g}{a}\right)pA\cos\left[\sqrt{\left(\frac{g}{a}\right)}pt + B\right]$$

$$= -\left(\frac{g}{a}\right)p^2\theta = -cp^2\theta \text{ and } D^4\theta = c^2p^4\theta.$$

Substituting these values in (5) we get

$$[4c^2p^4 - c^2(n + 12)p^2 + 3(n - 1)c^2]\,\theta = 0$$

$$\Rightarrow 4p^4 - (n + 12)p^2 + 3(n - 1) = 0 \qquad (\because \theta \neq 0)$$

Example 16:

To a point of a solid homogeneous sphere, of mass M, is freely hinged one end of a homogeneous rod, of mass nM; and the other end is freely hinged to a fixed point. If the system makes small oscillations under gravity about the position of equilibrium, the centre of a sphere and the rod being always in a vertical plane passing through the fix point, show that the periods of the principal oscillations are the values of $\frac{2\pi}{p}$ *given by the equation*

$$2ab(6 + 7n)p^4 - p^2g\{10a(3 + n) + 21b(2 + n)\} + 15g^2(2 + n) = 0,$$

where a is the length of the rod and b is the radius if the sphere.

Solution:

At any time t, let the rod and the sphere have turned through angel θ and ϕ to vertical, where OA = a, AC = b.

Now co - ordinates of C are

$$x_C = a\sin\theta + b\sin\phi,\; y_C = a\cos\theta + b\cos\phi$$

$$\therefore (\text{velocity})^2 \text{ of } C = \dot{x}_C^2 + \dot{y}_c^2 = a^2\dot{\theta}^2 + b^2\dot{\phi}^2 + 2ab\dot{\theta}\dot{\phi}\cos(\theta - \phi)$$

$= a^2\dot{\theta}^2 + b^2\dot{\phi}^2 + 2ab\dot{\theta}\dot{\phi}$ [$\because$ θ and ϕ are small]

Also the co-ordinates of G are $\left(\frac{a}{2}\sin\theta, \frac{a}{2}\cos\theta\right)$

$\therefore$ (velocity)^2 of G = $\frac{a^2}{4}\dot{\theta}^2$

Now let T be, the kinetic energy and W, the work function of the system, then we readily obtain

$$W = n\ Mg\frac{a}{2}\cos\theta + Mg\,(a\cos\theta + b\cos\phi)$$

$$= \frac{1}{2}a(n+2)\,Mg\cos\theta + bMg\cos\phi$$

and $T = \frac{1}{2}M\left[\frac{2b^2}{5}\dot{\phi}^2 + (a^2\dot{\theta}^2 + b^2\dot{\theta}^2 + 2ab\dot{\theta}\dot{\phi}\right] + \frac{1}{2}n.M\left[\frac{a^2}{12}\dot{\theta}^2 + \frac{a^2}{4}\dot{\theta}^2\right]$

$$= \frac{1}{6}a^2(n+3)\,M\dot{\theta}^2 + \frac{7b^2}{10}M\dot{\phi}^2 + abM\dot{\theta}\dot{\phi}$$

$\therefore$ Lagrange's q-equation is given by $\frac{d}{dt}\left(\frac{\partial T}{\partial\dot{\theta}}\right) - \frac{\partial T}{\partial\theta} = \frac{\partial W}{\partial\theta}$

$$\Rightarrow \frac{d}{dt}\left\{\frac{1}{3}a^2(n+3)\,M\dot{\theta} + abM\dot{\phi}.\right\} = -\frac{1}{2}a(n+2)Mg\,\theta \text{ app.}$$

$$\Rightarrow 2a(n+3)\dot{\theta} + 6b\ddot{\phi} = -3(n+2)g\theta \qquad ...(1)$$

While Lagrange's ϕ-equation gives

$$\frac{d}{dt}\left\{\frac{7b^2}{5}M\dot{\phi} + abM\dot{\theta}\right\} = -3(n+2)g\theta$$

$$\Rightarrow 7b\ddot{\phi} + 5a\ddot{\theta} = -5g\phi \qquad ...(2)$$

These can be re-written as

$$\{2a\,(n+3)D^2 + 3\,(n+2)\,g\}\,\theta + 6b\,D^2\,\phi = 0 \qquad ...(3)$$

and $5aD^2\theta + (7bD^2 + 5g)\,\phi = 0$ ...(4)

Eliminating ϕ between (3) and (4), we get

$$[2ab\,(7n+6)\,D^4 + \{10a\,(n+3) + 21b\,(n+2)\}\,gD^2 + 15\,(n+2)\,g^2]\,\theta = 0 \qquad ...(5)$$

To solve (5), let us put

$\theta = A\cos(pt + B), \therefore D\theta = -pA\sin(pt + B).$

$D^2\theta = -p^2 A\cos(p\dot{t} + \dot{B}) = -p^2\theta$ and $D^2\theta = p^4\theta$.

Substituting these values in (5), we have

$$2ab(7n+6)p^4 - \{10(n+3)a + 2lb(n+2)\}p^2 g + 15(n+2)g^2 = 0.$$

Example 17(a):

Three equal uniform rods AB, BC, CD, each of mass m and length 2a, are at rest in a straight line smoothly jointed at B and C. A blow I is given to the middle rod at a distance c from the centre O in a direction perpendicular to it, show that the initial velocity of O is $\frac{2I}{3T'}$ *and that the initial angular velocities of the rods are*

$$\frac{(5a+9c)I}{10ma^2}, \frac{6cI}{5ma^2} \text{ and } \frac{(5a-9c)I}{10ma^2}$$

Solution:

Just after the blow let $\dot{y}$ be the linear velocity of O, and let θ, φ, ψ be the inclination of BC, CD and AB to their initial positions. Now let T be K.E. of the system, then we have

$$T = \frac{1}{2}\left(\frac{a^2}{3}\dot{\theta}^2 + \dot{y}^2\right) + \frac{1}{2}m\left\{\frac{a^2}{3}\dot{\psi}^2 + (\dot{y} - a\dot{\theta} + a\dot{\psi})^2\right\}$$

$$+\frac{1}{2}m\left\{\frac{a^2}{3}\dot{\phi}^2 + (\dot{y} - a\dot{\theta} + a\dot{\phi})^2\right\}$$

$$= \frac{1}{2}m\left[3\dot{y}^2 + \frac{7a^2}{3}\dot{\theta}^2 + \frac{4a^2}{3}\dot{\psi}^2 + \frac{4}{3}a^2\dot{\phi}^2 + 2a\dot{y}\dot{\psi} + 2a\dot{y}\dot{\phi} + 2a\dot{\theta}\dot{\phi} - 2a\dot{\theta}\dot{\psi}\right]$$

Also just before the action of the impulse, we have T = 0

∴ Lagrange's equation for blows gives

$$\left(\frac{\partial T}{\partial \dot{y}}\right)_2 - \left(\frac{\partial T}{\partial \dot{y}}\right)_1 = I \Rightarrow 3m\dot{y} + ma\dot{\phi} + ma\dot{\psi} = I$$

$$\Rightarrow 3\dot{y} + a\dot{\theta} + a\dot{\psi} = \frac{1}{m}.$$

Similarly Lagrange's θ, φ, ψ equations give

$$\frac{7}{3}a\dot{\theta} + a\dot{\phi} - a\dot{\psi} = \frac{Ic}{ma'}\dot{y} + \frac{4}{3}a\dot{\phi} + a\dot{\theta} = 0, \dot{y} - a\dot{\theta} + \frac{4}{3}a\dot{\psi} = 0$$

Last two equations $\Rightarrow 3\dot{y} + 2a\dot{\phi} + 2a\dot{\psi} = 0 3\dot{y} = \frac{2I}{m'}$ or $\dot{y} = \frac{2I}{3m}$

Also $2a\dot{\theta} = \frac{4a}{3}(\dot{\psi} - \dot{\phi})$, i.e. $\frac{3}{2}a\dot{\theta} + a\dot{\phi} - \dot{\psi} = 0$

$$\therefore \frac{5a}{6}\dot{\theta} = \frac{Ic}{ma} \text{ i.e. } \dot{\theta} = \frac{6Ic}{5ma^2}$$

Substituting now the values of $\dot{y}$ and $\dot{\theta}$ in the equation

$\dot{y} + \frac{4}{3}a\dot{\phi} + a\dot{\theta} = 0$ we get

$$\frac{2I}{3m} + \frac{4}{3}a\dot{\phi} + \frac{6Ic}{5ma} = 0 \text{ or } \dot{\phi} = \frac{(5a - 9c}{10ma^2}.$$

Also substituting values of $\dot{y}$ and $\dot{\theta}$ in

equation $\dot{y} - a\dot{\theta} + \frac{4}{3}a\dot{\psi} = 0,$

we get $\frac{I}{3m} - \frac{6Ic}{5ma} + \frac{4}{3}a\dot{\psi} = 0$ or $\dot{\psi} = \frac{(5a - 9c)I}{10ma^2}.$

Example 17(b):

Six equal uniform rods form a regular hexagon, loosely jointed at the angular points, and rest on a smooth table; a blow is given perpendicular to one of them at its middle point, find the resulting motion and show that the opposite rod begins to move with one tenth of the velocity of the rod that is struck.

Solution:

The impulse is given at the middle point of AB in the direction as shown in the figure so the ensuing motion of BC, AF and CD, EF will be symmertrical.

Therefore they have after impulse same angular velocity say ω.

Also the rods AB and DE will begin to move at right angles to themselves. Now let $\dot{x}$ and $\dot{y}$ be their velocities just after the action of impulse.

$\therefore$ Velocity of C in the direction of $\dot{x}$

= velocity of B + velocity of C relative to B

$= \dot{x} + 2a\omega \sin 30^\circ = \dot{x} + a\omega$ ...(1)

and velocity of C in the same direction (with respect to the rod CD)

= velocity of D + velocity of C relative to D,

$= \dot{y} - 2a\omega \sin 30^\circ = \dot{y} - a\omega$ ...(2)

$\therefore$ (1) and (2) $\Rightarrow \dot{x} + a\omega = \dot{y} - a\omega \Rightarrow a\omega = \frac{1}{2}(\dot{y} - \dot{x})$

Now actual velocity of the C.G of the rod BC is $a\omega\cos 30^\circ$ parallel to AB and $\dot{x} + a\omega \sin 30^\circ$ at right angles to AB

$\therefore$ (velocity)^2 of $G_1 = (a\omega\cos 30^\circ)^2 + (\dot{x} + a\omega\sin 30^\circ)^2$

Thus K.E.of BC

$$= \frac{1}{2}m[\frac{1}{3}a^2\omega^2 + \{(a\omega\cos 30)^2 + (\dot{x} + a\omega\sin 30)^2\}]$$

$$= \frac{1}{2}m(\frac{4}{3}a^2\omega^2 + \dot{x}^2 + a\omega\dot{x}) = \frac{1}{12}m[5\dot{x}^2 + 2\dot{y}^2 - \dot{x}\dot{y}] = T_1 \text{say}$$

$$[\because a\omega = \frac{1}{2}(\dot{y} - \dot{x})]$$

Also Kinetic energy of CD

$$= \frac{1}{2}m[\frac{1}{3}a^2\omega^2 + \{(a\omega\cos 30)^2 + (\dot{y} - a\omega\sin 30)^2\}]$$

$$= \frac{1}{2}m(\frac{4}{3}a^2\omega^2 + \dot{y}^2 - a\omega\dot{y}] = \frac{1}{12}m[2\dot{x}^2 + 5\dot{y}^2 - \dot{x}\dot{y}] = T_2 \text{say}$$

$\therefore$ Total kinetic energy T is given by $T = 2T_1 + 2T_2 + T_3 + T_4$

where T_3 and T_4 are K.E.'s of AB and DE respectively.

$$\therefore T = \frac{1}{2}m\dot{x}^2 + \frac{1}{2}m\dot{y}^2 + 2.(m/12)[5\dot{x}^2 + 2\dot{y}^2 - \dot{x}\dot{y}]$$

$$+ 2.(m/12)[5\dot{y}^2 + 2\dot{x}^2 - \dot{x}\dot{y}]$$

or $T = \frac{1}{3}m(5\dot{x}^2 + 5\dot{y}^2 - \dot{x}\dot{y})$ $\therefore$ Lagrange's x-equation is

$$\left(\frac{\partial T}{\partial \dot{x}}\right)_2 - \left(\frac{\partial T}{\partial \dot{x}}\right)_1 = I \Rightarrow \frac{1}{3}m(10\dot{x} - \dot{y}) = I$$

Lagrange's y-equation gives $\frac{1}{3}m(10\dot{y} - \dot{x}) = 0$

Solving (1) and (2), we get $\dot{x} = \frac{10}{33}\frac{1}{m}$ and $\dot{y} = \frac{1}{33m}$

Also $\dot{y} = (x/10)$ and $a\omega = \frac{1}{2}(\dot{y} - \dot{x}) \Rightarrow \omega = -\frac{31}{22m}a$

Example 17(c):

Use Lagrange's equations to find the differential equation for a compound pendulum which oscillates in a vertical plane about a fixed horizontal axis.

Solution:

Let the plane of oscillation be represented by xy-plane, where B is its intersection with the axis of rotation and G is the centre of gravity. Let the mass of the pendulum be M and let its moment of inertia about the axis of rotation be

MK^2. Then potential energy relative to the horizontal plane through N is V = – Mgh cos.

$$T=\frac{1}{2}Mk^2\dot{\theta}^2 \therefore L=T-V=\frac{1}{2}Mk^2\dot{\theta}^2+Mgh\cos\theta \qquad ...(1)$$

$$\Rightarrow \frac{\partial L}{\partial\dot{\theta}}=Mk^2\theta \text{ and } \frac{\partial L}{\partial\theta}=-Mgh\sin\theta \qquad ...(2)$$

Now Lagrange's θ equation gives

$$\frac{d}{dt}\left(\frac{\partial L}{\partial\theta}\right)-\frac{\partial L}{\partial\theta}=0\Rightarrow\frac{d}{dt}(Mk^2\theta)+Mgh\sin\theta=0$$

i.e. $Mk^2\ddot{\theta}+Mgh\sin\theta=0\Rightarrow\ddot{\theta}=-\frac{gh}{k^2}\sin\theta$

When θ is small, we have $D^2\theta=-\frac{gh}{k^2}\theta \quad (\because \sin\theta=\theta)$

or $\left(D^2+\frac{gh}{k^2}\right)\theta=0.$

This is the differential equation of the penduium.

Example 18:

A ring slides on a smooth circular hoop of equal mass and of radius a which can turn in a vertical plane about a fix point O in its circumference. If θ and φ be the inclinations to the vertical of the radius through O and of the radius through the ring, prove that principal co-ordinates are (2θ + φ) *and* φ – θ), *and the periods of small oscillations are*

$$2\pi\sqrt{\left(\frac{a}{2g}\right)} \text{ and } 2\pi\sqrt{\left(\frac{2a}{g}\right)}.$$

Solution:

Let the mass of the hoop and the ring be m. In a displaced position, let O′ be the centre of the hoop and P the position of the ring which slides on the hoop.

∴ Co-ordinates of the point P are given by

$x_p = a(\sin\theta+\sin\phi)$

$y_p = a(\cos\theta+\cos\phi)$

$$\Rightarrow \dot{x}_P^2+\dot{y}_P^2=a^2(\dot{\theta}^2+\dot{\phi}^2+2\dot{\theta}\dot{\phi})$$

when θ and φ are small.

also co-ordinates of $O'=a^2\dot{\theta}^2$

Thus T = Energy of circular hoop + Energy of ring

$$= \frac{1}{2}m(a^2\dot{\theta}^2 + a^2\dot{\theta}^2) + \frac{1}{2}ma^2(\dot{\theta}^2 + \dot{\phi}^2 + 2\dot{\theta}\dot{\phi})$$

$$= \frac{1}{2}ma^2(3\dot{\theta}^2 + \dot{\phi}^2 + 2\dot{\theta}\dot{\phi})$$

Also the work function

W = mga cos θ + mg (a cos θ + a cos ϕ) = mga (2 cos θ + cos ϕ).

∴ Lagrange's θ-equation gives $3\ddot{\theta} + \ddot{\phi} = -\frac{2g}{a}\theta$ ...(1)

Lagrange's f-equation gives $\ddot{\theta} + \ddot{\phi} = -\frac{g}{a}\phi$ ...(2)

Now adding (1) and (2) and multiplying (2) by 2 and subtracting (1) from that, we get $D^2(2\theta + \phi) = -\frac{g}{2a}(2\theta + \phi)$

and $D^2(\phi - \theta) = -\frac{2g}{a}(\phi - \theta)$

Taking 2θ + ϕ = X and ϕ – θ = Y, we get

$$D^2X = -\frac{g}{2a}X \text{ and } D^2Y = -\frac{2g}{a}Y;$$

which obviously represent two independent simple harmonic motions. Thus X and Y are principal co-ordinates, that is (2θ + ϕ) and (ϕ – θ) are principal co-ordinates.

Also, periods of small oscillations are given by

$$2\pi\sqrt{\left(\frac{2a}{g}\right)} \text{ and } 2\pi\sqrt{\left(\frac{a}{2g}\right)}.$$

Example 19:

A square ABCD formed by four rods each of length 2l and mass hinged at their ends, rests on a horizontal frictionless table. An impulsive of magnitude I is applied to the vertex A in the direction AD Find the equations of motion, and prove that the K.E. developed immediately after the application of the impulsive forces is

$$T = (I^2/2m)$$

Solution:

When the square is struck, its shape will in general be a rhombus. Suppose that at any time t, the angels made by sides AD (or BC) and BA (or CD) with the x-axis are θ_1 and θ_2 respectively, while the co-ordinates of the centre M are (x, y).

*Suppose that the force F_V acting on a system are such that $\lim_{\tau\to 0}\int_0^{\tau} F_V dt = I_v$ where t represents a time interval, Then we call F_v impulsive forces and I_v are called Impulse.

** If we call P_α to be the generalised impulse, (3) can be written as

Hence x, y, θ_1, θ_2, are the generalised co-ordinates.

From the adjoining diagram, we see that the position vectors of the centre E,F, G.H of the rods are given respectively by

$r_E = (x - l\cos\theta_1)i + (\dot{y} - l\sin\theta_1)j,$

$r_F = (x + l\cos\theta_2)i + (y - l\sin\theta_2)j,$

$r_G = (x + l\cos\theta_2)i + (y - l\sin\theta_1)j,$

and $r_H = (x - l\cos\theta_2)i + (y + l\sin\theta_1)j.$

$\because v_E = \dot{r}_E = (\dot{x} - l\sin\theta_1\dot{\theta}_1)i + (\dot{y} - l\cos\theta_1\dot{\theta}_1)j,$

$v_F = \dot{r}_F = (\dot{x} - l\sin\theta_2\dot{\theta}_2)i + (\dot{y} - l\cos\theta_2\dot{\theta}_2)j,$

$v_G = \dot{r}_G = (\dot{x} - l\sin\theta_1\dot{\theta}_1)i + (\dot{y} + l\cos\theta_1\dot{\theta}_1)j,$

$v_H = \dot{r}_H = (\dot{x} + l\sin\theta_2\dot{\theta}_2)i + (\dot{y} + l\cos\theta_2\dot{\theta}_2)j.$

Now, K.E. of the rod $AB = \frac{1}{2}m\dot{r}_E^2 + \frac{1}{3}ml^2\dot{\theta}_1^2.\frac{1}{2} = T_{AB}$ say

K.E. of the rod $CB = \frac{1}{2}m\dot{r}_F^2 + \frac{1}{3}ml^2\dot{\theta}_1^2.\frac{1}{2} = T_{BC}$ say

K.E. of the rod $CD = \frac{1}{2}m\dot{r}_G^2 + \frac{1}{3}ml^2\dot{\theta}_2^2.\frac{1}{2} = T_{CD}$ say

K.E. of the rod $DA = \frac{1}{2}m\dot{r}_H^2 + \frac{1}{3}ml^2\dot{\theta}_1^2.\frac{1}{2} = T_{DA}$ say

$\therefore$ K.E. of system $\left(\frac{\partial T}{\partial}\right)_2 = (4m\dot{x})_2 = 4m\dot{x}, \left(\frac{\partial T}{\partial \dot{y}}\right)_2 = (4m\dot{y})_2 = 4m,$

$\frac{1}{2}m(4\dot{x}^2 + 4\dot{y}^2 + 2l^2\dot{\theta}_1^2 + 2l^2\dot{\theta}_2^2) + \frac{ml^2}{3}(\dot{\theta}_1^2 + \dot{\theta}_2^2).$

$$= 2m(\dot{x}^2 + \dot{y}^2) + \frac{4}{3}ml^2(\dot{\theta}_1^2 + \dot{\theta}_2^2). \quad ...(1)$$

Let us now assume that initially the rhombus is a square at rest with its sides parallel to the co-ordinate axes and its centre located at the origin.

Then, we get $x = 0 = y,\ \theta_1 = \frac{\pi}{2}, \theta_2 = 0, \dot{x} = 0, \dot{y} = 0, \dot{\theta}_1 = 0, \dot{\theta}_2 = 0.$

Now if we use the notation, $(\ )_1$ and $(\ )_2$ to denote quantities before and after the impulse is applied, we have

$$\left(\frac{\partial T}{\partial \dot{x}}\right)_1 = (4m\dot{x})_1 = 0 \left(\frac{\partial T}{\partial \dot{y}}\right)_1 = (4m\dot{y})_1 = 0$$

$$\left(\frac{\partial T}{\partial \dot{\theta}}\right)_1 = \left(\frac{8}{3}ml^2\dot{\theta}_1\right) = 0, \left(\frac{\partial T}{\partial \dot{\theta}_2}\right) = \left(\frac{8}{3}ml^2\dot{\theta}_2\right) = 0$$

and $\left(\frac{\partial T}{\partial \dot{x}}\right)_2 = (4m\dot{x})_2 = 4m\dot{x}, \left(\frac{\partial T}{\partial \dot{y}}\right)_2 = (4m\dot{y})_2 = 4m,$

$$\left(\frac{\partial T}{\partial \dot{\theta}_1}\right)_2 = \left(\frac{8}{3}ml^2\dot{\theta}_1\right)_2 = \frac{8}{3}ml^2\dot{\theta}_1 \left(\frac{\partial T}{\partial \dot{\theta}_2}\right)_2 = \frac{8}{3}ml^2\dot{\theta}_2.$$

Hence Lagrange's equations are

$$\left(\frac{\partial T}{\partial \dot{x}}\right)_2 - \left(\frac{\partial T}{\partial \dot{x}}\right)_1 = p_x \Rightarrow 4m\dot{x} = P_x \qquad ...(2)$$

$$\left(\frac{\partial T}{\partial \dot{y}}\right)_2 - \left(\frac{\partial T}{\partial \dot{y}}\right)_1 = p_y \Rightarrow 4m\dot{y} = P_y \qquad ...(3)$$

$$\left(\frac{\partial T}{\partial \dot{\theta}_1}\right)_2 - \left(\frac{\partial T}{\partial \dot{\theta}_1}\right)_1 = p_{\theta_1} \Rightarrow \frac{8}{3}ml^2\dot{\theta}_1 = P_{\theta_1}. \qquad ...(4)$$

$$\left(\frac{\partial T}{\partial \dot{\theta}_1}\right)_2 - \left(\frac{\partial T}{\partial \dot{\theta}_2}\right)_1 = p_{\theta_1} \Rightarrow \frac{8}{3}ml^2\dot{\theta}_2 = P_{\theta 2}. \qquad ...(5)$$

Now we shall find the value of P_x, P_y, p_{θ_1} and p_{θ_2}

we have $P_\alpha = \sum I_v \bullet \frac{\partial r_v}{\partial q_\alpha}$ where I_v are the impulsive forces.

$$\therefore P_x = I_A \bullet \frac{\partial r_A}{\partial x} + I_B \bullet \frac{\partial r_B}{\partial x} + I_c \bullet \frac{\partial r_c}{\partial x} + I_D \bullet \frac{\partial r_D}{\partial x} \qquad ...(6)$$

$$P_y = I_A \bullet \frac{\partial r_A}{\partial y} + I_B \bullet \frac{\partial r_B}{\partial y} + I_c \bullet \frac{\partial r_c}{\partial y} + I_D \bullet \frac{\partial r_D}{\partial y} \qquad ...(7)$$

$$P_{\theta_1} = I_A \bullet \frac{\partial r_A}{\partial \theta_1} + I_B \bullet \frac{\partial r_B}{\partial \theta_1} + I_c \bullet \frac{\partial r_c}{\partial \theta_1} + I_D \bullet \frac{\partial r_D}{\partial \theta_1} \qquad ...(8)$$

$$P_{\theta 2} = I_A \bullet \frac{\partial r_A}{\partial \theta_2} + I_B \bullet \frac{\partial r_B}{\partial \theta_2} + I_c \bullet \frac{\partial r_c}{\partial \theta_2} + I_D \bullet \frac{\partial r_D}{\partial \theta_1} \qquad ...(9)$$

where

$$r_A = (x - l\cos\theta_1 - l\cos\theta_2)\,i + (y - l\sin\theta_1 + l\sin\theta_2)j \qquad ...(10)$$

$$r_B = (x - l\cos\theta_1 + l\cos\theta_2)\,i + (y - l\sin\theta_1 - l\sin\theta_2)j \qquad ...(11)$$

$$r_C = (x + l\cos\theta_1 + l\cos\theta_2)\,i + (y + l\sin\theta_1 - l\sin\theta_2)j \qquad ...(12)$$

$$r_D = (x + l\cos\theta_1 - l\cos\theta_2)\,i + (y + l\sin\theta_1 + l\sin\theta_2)j \qquad ...(13)$$

But initially the impulsive force at A is in the direction of the positive y-axis, we have

$$I_A = Ij. \qquad ...(14)$$

$\therefore$ Equation (6), (7), (8) and (9) $\Rightarrow$

$$P_x = 0, P_y = I, P_{\theta_1} = -Il\cos\theta_1 \text{and} P\theta_2 = Il\cos\theta_2. \qquad ...(15)$$

Thus equation (2) and (3),(4) and (5) give

$$4m\dot{x} = 0, 4m\dot{y} = I, \frac{8}{3}ml^2\dot{\theta}_1 = -Il\cos\theta_1, \frac{8}{3}ml^2\dot{\theta}_2 = Il\cos\theta_2. \qquad ...(16)$$

Second Part

We have $\dot{x} = 0, \dot{y} = \frac{1}{4m'}\dot{\theta}_1 = -\frac{3I}{8ml}\cos\theta_1$ and $\dot{\theta}_2 = \frac{3I}{8ml}\cos\dot{\theta}_2.$

$$\therefore T = 2m(x^2 + y^2) + \frac{4}{3}ml^2(\dot{\theta}^2_1 + \dot{\theta}_2^2)$$

$$= 2m\left(0 + \frac{I^2}{16m^2}\right) + \frac{4}{3}ml^2\left[\frac{9I^2}{64m^2l^2}\cos^2\theta_1 + \frac{9I^2}{64m^2l^2}\cos^2\theta_2\right] \qquad ...(17)$$

But immediately after the application of the impulsive forces,

$\theta_1 = \frac{1}{2}\pi$ and $\theta_2 = 0$ approximately, so (17) gives $T = \frac{I^2}{2m}\cdot$

Example 20:

Three equal uniform rods, AB, BC, CD are freely joined at B and C and the ends A and B are fastened to smooth, fixed pivots whose distance apart is equal to the length of either rod. The frame being at rest in the form of the square, a blow J is given perpendicular to AB at its middle point and in the plane of square. Show that the energy set up is $\frac{3J^2}{40m}$ *where m is the mass of each rod.*

Find also the blows at the joints B and C.

Solution:

The blow J is given at G_1 the C.G. of AB, at right angles to AB

The rods AB and CD will turn through the same angle θ but the rod BC will remain parallel to AD.

Now let T be the K.E. of the system, then we have

$$T = 2.\frac{1}{2}m\left[\frac{a^2}{3}\dot{\theta}^2 + a^2\dot{\theta}^2\right] + \frac{1}{2}m(2a\dot{\theta})^2 = \frac{10ma^2}{3}\dot{\theta}^2.$$

∴ Lagrange's θ-equation gives

$$\left(\frac{\partial T}{\partial\dot{\theta}}\right) - \left(\frac{\partial T}{\partial\dot{\theta}}\right) = J \Rightarrow \frac{20ma^2}{3}\dot{\theta} = Ja$$

$$i.e.\ \dot{\theta} = \frac{3J}{20ma} \therefore T = \frac{10ma^2}{3}.\frac{9J^2}{400m^2a^2} = \frac{3J^2}{40m}.$$

Second Part:

Let P and Q be the impulses at B and C respectively. With directions, as shown in the figure, then considering the motion of AB and taking moments about A, we get

Change in the angular momentum about the axis through A

= moments of the impulses about this axis

$$i.e.\ m\frac{4a^2}{3}\dot{\theta} = Ja - P2a \Rightarrow P = \frac{1}{2}J - \frac{2ma}{3}$$

$$\therefore P = \frac{1}{2}J - \frac{2ma}{3}\frac{3J}{20ma} = \frac{2J}{5}$$

Again considering the motion of CD and taking moments about D, we obtain

$$m = \frac{4a^2}{3}\dot{\theta} = Q2a \Rightarrow Q = \frac{2ma}{3}\dot{\theta} = \frac{3J}{20ma} = \frac{1}{10}J.$$

Example 21:

A bear, of mass M, slides on a smooth fixed wire, whose inclination to the vertical is α, and has hinged to it a rod, of mass m and length 2l, which can move freely in the vertical plane through the wire.

If the system starts from rest with the rod hanged vertically, show that

$$\{4M + m(1 + 3\cos^2\theta\}l\dot{\theta}^2 = 6(M + m)\,g\sin\alpha(\sin\theta - \sin\alpha)$$

where θ is the angle between the rod and the lower part of the wire.

Solution:

Let OL be the fixed wire. At any time t. Let the bead of mass M be at A where OA = x, also let θ be the angle which the rod AB makes with the lower part of the fixed wire.

Take O as origin and the fixed wire OL as x axis and a line though O and prep to OL as y axis; the co-ordinates of G, the C.G. of the rod AB, are

$$\{x + l\cos\theta, l\sin\theta\}$$

i.e. $x_G = x + l\cos\theta)$ and $y_G = l\sin\theta$.

$$\ddot{x}_G = (\dot{x} - l\sin\theta\dot{\theta});\ \dot{y}_G = l\cos\theta\dot{\theta}$$

$$\therefore (\text{velocity})^2 \text{ of } G = v_G^2 = \dot{y}_G^2 + \dot{y}_G^2 = (\dot{x} - l\sin\theta\dot{\theta})^2 + (l\cos\theta\dot{\theta})^2.$$

Now let T be the kinetic energy and W the work function of the work function of the system Then we easily get

Total energy = T = K.E. of the bead + K.E, of the rod

$$= \frac{1}{2}M\dot{x}^2 + \frac{1}{2}m\left[\frac{l^2}{3}\dot{\theta}^2 + (\dot{x} - l\sin\theta\dot{\theta})^2 + (l\cos\theta\dot{\theta}^2\right]$$

$$= \frac{1}{2}(M+m)\dot{x}^2 - ml\dot{x}\dot{\theta}\sin\theta + \frac{2}{3}ml^2\dot{\theta}^2.$$

Also, the work function is given by

$$W = M\,g\,x\cos\alpha + mg\ \{x\cos\alpha + l\cos(\theta - \alpha)\}$$

$$= (M + m)\,gx\cos\alpha + mgl\cos\cos(\theta - \alpha)$$

$\therefore$ Lagrange's x-equation gives, $\frac{d}{dt}\left(\frac{\partial T}{\partial \dot{x}}\right) - \frac{\partial T}{\partial x} = \frac{\partial W}{\partial x}$

i.e $\frac{d}{dt}[M+m)\,\dot{x} - ml\dot{\theta}\sin\theta] = (M+m)g\cos\alpha$

or $(M+m)\ddot{x} - ml\ddot{\theta}\sin\theta - ml\dot{\theta}^2\cos\theta = (M+m)go\cos\alpha.$ ...(i)

Again Lagrange's θ – equation gives

$$\frac{d}{dt}\left(\frac{\partial T}{\partial \dot{\theta}}\right) + \frac{\partial T}{\partial \theta} = \frac{\partial W}{\partial \theta}$$

i.e. $\frac{d}{dt}[-ml\dot{x}\sin\theta + \frac{4}{3}ml^2\dot{\theta}] + ml\dot{x}\dot{\theta}\cos\theta = -mgl\sin(\theta - \alpha),$

or $-\ddot{x}ml\sin\theta - \dot{x}\dot{\theta}ml\cos\theta + \frac{4}{3}ml^2\ddot{\theta} + \dot{x}\dot{\theta}ml\cos\theta = -mgl\sin(\theta - \alpha).$

or $-\ddot{x}\sin\theta + \frac{4}{3}l\ddot{\theta} = -g\sin(\theta - \alpha)$ ...(ii)

Eliminating $\ddot{x}$ between (i) and (ii), we get

$$\ddot{\theta}[-ml\sin^2\theta + \frac{4}{3}(M+m)l] - ml\dot{\theta}^2\sin\theta\cos\theta$$

$$= (M + m)\, g\, [\cos a \sin\theta - \sin(\theta - a)]$$

or $l\ddot{\theta}[3M + m + 3m\cos^2\theta] - 3ml\dot{\theta}^2\sin\theta\cos\theta$

$$= 3\,(M + m)\, g \cos\theta \sin\alpha.$$

Whence on integrating, we get

$$l\dot{\theta}^2[4M + m + 3m\cos^2\theta] = 6\,(M + m)\, g \sin a \sin\theta + C \qquad \text{...(iii)}$$

When $\theta = a, \dot{\theta} = 0$,

$$\therefore C = -6(M + m)g\sin^2\alpha.$$

Putting the value of C in (iii), we get

$$l\dot{\theta}^2(4M + m + 3m\cos^2\theta) = 6\,(M + m)g\sin\alpha\,(\sin\theta - \sin\alpha).$$

Example 22(a):

A uniform rod, of length 2a, which has one end attached to a fixed point by a light inextensible string, of length $\frac{5}{12}a$, performing small oscillations in a vertical plane about its position of equilibrium. Find the position at any time, and show that the period of its principal oscillations are $2\pi\sqrt{\left(\frac{5a}{3g}\right)}$ and $\pi\sqrt{\left(\frac{a}{3g}\right)}$.

Solution:

Figure is self explanatory at any time t, let the string and the rod be inclined at θ and ϕ to the vertical OY.

Co-ordinates of G are given by

$$x_G = \frac{5}{12}a\sin\theta + a\sin\phi,$$

$$y_G = \frac{5}{12}a\cos\theta + a\cos\phi,$$

$$\dot{x}_G = \frac{5a}{12}a\cos\theta\dot{\theta} + a\cos\phi\dot{\phi},$$

$$\dot{y}_G = -\left(\frac{5a}{12}\sin\theta\dot{\theta} + a\sin\phi\dot{\phi}\right)$$

$\therefore \dot{x}_G^{\,2} + \dot{y}_G^{\,2} = (\text{velocity})^2 \text{of } G$

$= \frac{25a^2}{144}\dot{\theta}^2 + a^2\dot{\phi}^2 + \frac{5a^2}{6}\dot{\theta}\dot{\phi}\cos(\theta+\phi) = \frac{25a^2}{144}\dot{\theta}^2 + a^2\dot{\phi}^2 + \frac{5}{6}a^2\dot{\theta}\dot{\phi}$

[∵ θ and ϕ are small so cos (θ + ϕ) = 1]

Again let T, be the kinetic energy and W, the work function of the system, then we easily get

$$T = \frac{1}{2}m\left[\frac{a^2}{3}\dot{\phi}^2 + \left(\frac{25}{144}a^2\dot{\theta}^2 + a^2\dot{\phi}^2 + \frac{5}{6}a^2\dot{\theta}\dot{\phi}\right)\right]$$

$$= \frac{ma^2}{288}[25\dot{\theta}^2 + 192\dot{\phi}^2 + 120\dot{\theta}\dot{\phi}]$$

and $W = mg\left[\frac{5}{12}a\cos\theta + a\cos\phi\right]$

∴ Lagrange's θ-equation gives

$$\frac{d}{dt}\left(\frac{\partial T}{\partial \dot{\theta}}\right) - \frac{\partial T}{\partial \theta} = \frac{\partial W}{\partial \theta} \Rightarrow \frac{d}{dt}\left\{\frac{ma^2}{144}(25\dot{\theta} + 60\dot{\phi})\right\} = -\frac{5mga}{12}\theta$$

{sin θ = θ is small}

$\Rightarrow 5\dot{\theta} + 12\ddot{\phi} = -\frac{12g}{a}\theta.$ ...(1)

and Lagrange's ϕ- equation gives

$\frac{d}{dt}\left\{\frac{ma^2}{144}(192\dot{\phi} + 60\dot{\theta})\right\} = -mga\phi \Rightarrow 5\ddot{\theta} + 16\ddot{\phi} = -12\frac{g}{a}$ ...(2)

Equations (1) and (2) ⇒ $(5D^2 + 12c)\,\theta + 12D^2\theta = 0$ ...(3)

and $5D^2\theta + (6D^2 + 12c)\,\phi = 0$ where $(g/a) = c$. ...(4)

Now eliminating f between these two equations, we get

$[(5D^2 + 12c)(16D^2 + 12c) - 60D^4]\,\theta = 0.$ ...(5)

Let $\theta = A\cos(pt + B)$ ∴ $D\theta = -pA\sin(pt + B)$,

$D^2\theta = -p^2A\cos(pt + B) = -p^2\theta$ and $D^4\theta = p^4\theta$.

Substituting these values in (5), we get

$(5p^4 - 63c^2 - 36c^2)\,\theta = 0 \Rightarrow (5p^4 - 63cp^2 + 36c^2) = 0$ (∵ θ ≠ 0)

$\Rightarrow (5p^2 - 3c)(p^2 - 12c) = 0 \Rightarrow \left(5p^2 - \frac{3g}{a}\right)\left(p^2 - \frac{12g}{a}\right) = 0$

$\therefore p_1^2 = \frac{3g}{5a}$ and $p_2^2 = \frac{12g}{a}$

The periods of oscillations are $\frac{2\pi}{p_1}$ and $\frac{2\pi}{p_2}$

i.e. $2\pi\sqrt{\left(\frac{5a}{3g}\right)}$ and $2\pi\sqrt{\left(\frac{a}{12g}\right)}$ *i.e.* $2\pi\sqrt{\left(\frac{5a}{3g}\right)}$ and $\pi\sqrt{\left(\frac{a}{3g}\right)}$

EXERCISES

1. A rhombus of smoothly jointed equal uniform rods of length 2a lies on a smooth horizontal table. It is struck by a blow J perpendicular to one of its sides at a point distant c from the middle point , show that the angular velocity of the side instantly becomes $\frac{3jc}{8ma^{2,}}$ where m is the mass of each side.

2. A hollow circular roller of radius a mass M has attached to the centre of its axis a particle of mass m by a light a rod of length b. The particle is free to swing about the axis a plane perpendicular to the axis. The roller is at rest on a horizontal plane rough enough to prevent slipping, with rod held at an angle a with the downward vertical. If the rod is then released prove that the centre of the roller will oscillate through a distance

$$\frac{2mba^2 \sin\alpha}{M(a^2 + k^2) + ma^2.}$$

where k is the radius of gyration of the roller about the axis.

3. A homogeneous rod OA, of mass m_1 and length 2a, is freely at O to a fixed point , as its other end is freely attached another homogeneous rod AB, of mass m_2 and length 2b; the system moves under gravity, find equations to determine motion.

4. A thin circular ring of a radius a and M lies on a smooth horizontal plane and two light elastic strings are attached to it at ends of a diameter. The other ends of the string being fastened fixed points in the diameter produced. Show that for small oscillations in the plane of the ring the periods are the values of $2\pi/p$ given by

$$\frac{Mlp^2}{2T} - = 0 \text{ or } \frac{b}{1-b} \text{ or } \frac{1}{a}$$

where b is the natural length l the equilibrium length .and T the equilibrium tension of each string.

5. A heavy uniform rod of mass m and length 2a rotating in a vertical plane falls and strikes a smooth horizontal plane. If u and ω βε its linear and

angular velocities and θ the inclination to the vertical just before impact, prove that the impulse J is given by

$$(1+3\sin^2\theta)j = m(u + a\omega\sin\theta).$$

6. A square ABCD formed of four freely jointed uniform rods, is rotating about its centre of gravity on smooth horizontal table with angular velocity ω when the joint A is suddenly fixed. Show that the initial angular velocities of the rods are equal and that 3 / 5 of the kinetic energy immediately lost.
7. A uniform straight rod of length 2a, is freely movable about, its on the and a particle of mass one fourth that of the rod is attached by a light inextensible string of length 3a, to one end of the rod, find the period of principle oscillations.
8. Obtain Lagrange's equations of motion for a system under finite forces and hence deduce the same under impulsive forces.

2

Variational Principle with Hamilton's Methods

INTRODUCITON

Variational (mimimal) principles has been the greatest fascination of old generation physicists and in this pursuit several and varied variational principles have been put forth in the various branches of physics, such as, in optics, the Fermat's principles of *'lest time'*, in mechanics, the Gauss's principles of *'least constraint'*, the Hertz's principle of *'lest curvature'* and most important of all the Hamilton's *variational principle'* which will be our main centre of attraction here in mechanics.

The great value of these variational principles; lies in their extreme economy of expression.

Let T, V be the kinetic and potential energies of a. Conservative holonomic dynamical system defined by n generalised co-ordinates q_1, q_2, ...,q_n at time t, writing L = T – V, we know that Lagrange's equations for the motion of the system are:

$$(\partial L/\partial q_i) - (d/dt)(\partial L/\partial \dot{q}_i) \qquad (i = 1, 2, ...,n)$$

Now applying previous article we can say that these are the n Euler-Lagrange's equation arising from the variational problem.

$$\int_{t_1}^{t_1} L(q_1, q_2, ..., q_n; \dot{q}_1, \dot{q}_2, ..., \dot{q}_n; t)\ dt = \text{stationary } (\textit{i.e.} \text{ stat})$$

where t_1, t_2 are fixed. Thus we have established that

During the motion of a 'conservative holonomic dynamical system over a fixed time interval, the time integral over that interval of the difference between the kinetic and potential energies is stationary. This is Hamilton's principle.

DERIVATION OF HAMILTON EQUATIONS FROM THE VARIATIONAL PRINCIPLE

Hamilton's principle implies

$$\delta I = \delta \int_{t_1}^{t_2} L dt = 0 \text{ or } \delta\, \delta \int_{t_1}^{t_2} [\sum_i p_i \dot{q}_1 - H(q,p,t)]\, dt = 0$$

$$[\because L = \sum_i p_i \dot{q}_i - H(q,p,t)]$$

This implies $\delta \sum_i \int_{q_1}^{q_2} p_1 d\, q_1 - \delta \int_{t_1}^{t_2} H\, dt = 0.$ ...(1)

Equation (1) is called the modified Hamilton's principle.

Now $\delta I = \dfrac{\partial I}{\partial \alpha} d\alpha = d\alpha \dfrac{\partial}{\partial \alpha} \int_{t_1}^{t_2} (\sum_i p_i \dot{q}_i - H(q,p,t) dt = 0$

where $\delta = d\alpha(\partial / \partial_\alpha)$

$$\Rightarrow d\alpha \int_{t_1}^{t_2} \frac{\partial}{\partial \alpha} [\Sigma p_i, \dot{q}_i - H(q,p,t) dt = 0$$

[$\because$ the times t_1, t_2 are not varied not so they are not functions of α, thus the differentiation can be interchanged].

$$\Rightarrow d\alpha \int_{t_1}^{t_2} \sum_i \left[\frac{\partial p_i}{\partial \alpha} \dot{q}_i + \frac{\partial \dot{q}_i}{\partial \alpha_i} p_i - \frac{\partial H}{\partial q_i} \frac{\partial q_i}{\partial \alpha} - \frac{\partial H}{\partial p_i} \frac{\partial p_i}{\partial \alpha} \right] dt = 0 \qquad ...(2)$$

Further, we have $\displaystyle\int_{t_1}^{t_2} \frac{\partial \dot{q}_i}{\partial \alpha} p_i dt = \int_{t_1}^{t_2} p_i \frac{d}{dt}\left(\frac{\partial q_i}{\partial \alpha} \right) dt = p_i \left[\frac{\partial p_i}{\partial \alpha} \right]_{t_1}^{t_2} - \int_{t_1}^{t_2} \dot{p}_i \frac{\partial q_i}{\partial \alpha} dt.$

But all the varied paths have the same end points. Hence $\dfrac{\partial \dot{q}_i}{\partial \alpha}$ vanishes for t_1 and t_2. $\Rightarrow \displaystyle\int_{t_1}^{t_2} \frac{\partial q_i}{\partial \alpha} p_i dt = - \int_{t_1}^{t_2} \dot{p}_i \frac{\partial q_i}{\partial \alpha} dt = \int_{t_2}^{t_1} p_i \frac{\partial q_i}{\partial \alpha} dt.$

Making this substitution in equations (2), we have

$$d\alpha \int_{t_1}^{t_2} \Sigma \left[\frac{\partial q_i}{\partial \alpha} \dot{q}_i - \frac{\partial H}{\partial q_i} \frac{\partial q_i}{\partial \alpha} - \frac{\partial H}{\partial p_i} \frac{\partial p_i}{\partial \alpha} - \frac{\partial q_i}{\partial \alpha} \right] dt = 0 \qquad ...(2)$$

$$\Rightarrow \int_{t_1}^{t_2} \Sigma \dot{q}_i \frac{\partial \dot{q}_i}{\partial \alpha} q\alpha - \frac{\partial H}{\partial q_i} \frac{\partial q_i}{\partial \alpha} q\alpha - \frac{\partial H}{\partial p_i} \frac{\partial p_i}{\partial \alpha} d\alpha - p_i \frac{\partial q_i}{\partial \alpha} d\alpha\, dt = 0 \qquad ...(3)$$

But $\delta p_i = d\alpha.(\partial p_i / \partial \alpha)$ and $\delta q_i = (\partial q_i / \partial \alpha) d\alpha$

$$\therefore (3) \Rightarrow \int_{t_1}^{t_2} \Sigma \left[\delta p_i \left(\dot{q}_i - \frac{\partial H}{\partial p_i} \right) + \delta q_i \left(\frac{\partial H}{\partial q_i} - \dot{p}_i \right) \right] dt = 0.$$

The variation δq_i and δp_i are independent to each other, hence equation (4) holds good only when the coefficients of δp_i and δq_i vanish separately.

The (4) $\Rightarrow \dot{q}_i = (\partial H / \partial p_i), \dot{p}_i = -(\partial H / \partial q_i)$. These are Hamilton's equations.

PRINCIPAL OF LEAST ACTION

Principle of least action states that if T is kinetic energy, at time t, of a conservative, holonomic dynamical system specified by the generalised co-ordinates, then the integral $I = \int_{t_1}^{t_2} 2T dt$

has necessary an extreme value, minimum or maximum, on actual path as compared with varied path as the systems passes from one configuration at time t_0 to another configuration at time t_1. We know that L = T – V, *i.e.* Lagrangian = K.E. – P.E. and T + V = E (const), since system is conservative.

But by Hamilton's principle, we know that

$$\int_{t_0}^{t_1} \delta L \, dt = 0 \Rightarrow \int_{t_0}^{t_1} \delta(T - V) dt = 0 \int_{t_0}^{t_1} \delta(2T - E) dt = 0$$

$$\Rightarrow \int_{t_0}^{t_2} \delta(2T) \, dt = 0 \Rightarrow \delta \int_{t_0}^{t_1} (2T) \, dt = 0. \quad ...(1)$$

[$\because$ $\delta E = 0$ as E, the total energy is const.]

Result (1), is known as *Principle of least action.*

Equation (1) can also be written as $\delta A = 0$, where $A = \int_{t_0}^{t_1} 2T \, dt$ and is defined by action as follows:

This implies that principal of least action states that the action in the actual path is minimum compared with the varied path, as the system passes from one configuration to another.

Distinction Between Hamilton's Principle and Principle of Least Action

In Hamilton's principle, the time of description $t_1 - t_0$ is prescribed (fixed) as the body moves from one configuration to another configuration, while in the principle of least action there is no such restriction on the time $t_1 - t_0$ but the total energy between the end points A and B is prescribed.

Deduction of Largrange's Equation from Hamilton's Principle

By Hamilton's principle, we have $\int_{t_0}^{t_1} \delta L dt = 0$

$$\therefore (1) \Rightarrow \int_{t_0}^{t_1} \sum_{i=1}^{n} \left(\frac{L}{\partial q_i} \delta q_i + \frac{L}{\partial \dot{q}_i} \delta \dot{q}_i \right) dt = 0$$

$$\Rightarrow \int_{t_0}^{t_1} \sum_{i=1}^{n} \frac{\partial L}{\partial q_i} \delta q_i \, dt + \int_{t_0}^{t_1} \sum_{i=1}^{n} \frac{\partial L}{\partial \dot{q}_i} \frac{d}{dt} (\delta q_i) \, dt = 0$$

Now integrating by parts, we have

$$\int_{t_0}^{t_1} \sum_{i=1}^{n} \frac{\partial L}{\partial q_i} \delta q_i \, dt + \left\{ \sum_{i=1}^{n} \frac{\partial L}{\partial \dot{q}_i} \delta q_i \right\}_{t_0}^{t_1} - \int_{t_0}^{t_1} \left\{ \sum_{i=1}^{n} \frac{d}{dt} \left(\frac{\partial L}{\partial \dot{q}_i} \right) \delta q_i \right\} dt = 0$$

$$\Rightarrow \int_{t_0}^{t_1} \left[\sum_{i=1}^{n} \right] \left\{ \frac{d}{dt} \left(\frac{\partial L}{\partial \dot{q}_i} \right) - \frac{\partial L}{\partial q_i} \delta q_i \right\} dt = 0$$

[$\because$ middle term vanishes as all $\delta q_i = 0$ at t_0 and t_1]

$$\Rightarrow \sum_{i=1}^{n} \left\{ \frac{d}{dt} \left(\frac{\partial L}{\partial \dot{q}_i} \right) - \frac{\partial L}{\partial q_i} \right\} \delta q_i = 0.$$

Here δq_i's are arbitrary and independent to each other, so equating to zero their coefficients, we readily obtain

$$\frac{d}{dt} \left(\frac{\partial L}{\partial \dot{q}_i} \right) - \frac{\partial L}{\partial q_i} = 0$$

DEDUCTION OF HAMILTON'S PRINCIPLE

(Proof based on D'Alembert's principle).

Assume that the conservative holonomic dynamical system moves from A to C, where A and C, are the initial and final configurations of the system at time t_1 and t_2 respectively. Let ABC be the actual path and AB'C. AB"C the two neighbouring paths out of infinite number of possibilities. *In order to deduce the principle, the following two conditions must be satisfied.*

1. δt must be equal to zero at A and B *i.e.* at t_1 the particle must be at A, at t_2 the particle must be at C.
2. δr must be equal to zero at A and C $\Rightarrow$ that the points A, and C are fixed in space.

   Now assume that the system is acted upon by a number of forces represented by F.

Let the ith particle of system be acted upon by a force $F_i = m_i \ddot{r}_i$ is acceleration vector.

Again by D'A Alembert's principle, we have

$$\sum_i (F_i - m\ddot{r}_i)\cdot\delta r_i = 0, \text{ i.e. } \sum_i F_i\cdot\delta\dot{r}_i - \Sigma m_i \ddot{r}_i\cdot\delta r_i = 0 \quad ...(1)$$

If there is a little variation along the actual and neighbouring paths, then

$$\delta r_i = r_i - r_i \quad \text{(say)}.$$

$$\Rightarrow (d/dt)(\delta r_i) = (d/dt)(r_i - r_i) = (dr_i'/dt) - (dr_i/dt)$$

$$= \delta\ (dr_i/dt) = \delta(d\dot{r}_i) \quad ...(2)$$

where dashes have been used for neighbouring paths.

But $\ddot{r}_i\cdot\delta_i = (d/dt)\ddot{r}_i\cdot\delta r_i) - \dot{r}_i\cdot(d/dt)(\delta r_i)$. ...(3)

Using (2), (3) $\Rightarrow \ddot{r}_i\cdot\delta r_i = (d/dt)(\dot{r}_i\cdot\delta r_i) - \dot{r}_i\delta(\dot{r}_i)$ ...(4)

$$\therefore (1) \Rightarrow \sum_i F_i\cdot\delta r_i - \sum_i m_i\,[(d/dt)(r_i\cdot\delta r_i) - \dot{r}_i\cdot\delta\dot{r}_i] = 0$$

$$\text{or } \sum_i F_i\cdot\delta r_i - \sum_i m_i\,[(d/dt)(r_i\cdot\delta r_i) - \frac{1}{2}\delta(\dot{r}_i^2 = 0$$

$$\text{or } \sum_i F_i\cdot\delta r_i + \sum_i \frac{1}{2} m_i\delta(\dot{r}_i^2 = \sum_i (d/dt)(m_i\ddot{r}_i\cdot\delta r_i)$$

$$\text{or } \sum_i F_i\cdot\delta r_i + \delta\left(\sum_i \frac{1}{2} m_i\dot{r}_i^2\right) = \sum_i (d/dt)(m_i r_i\cdot\delta r_i) \quad ...(5)$$

But $\sum_i \frac{1}{2} m_i\dot{r}_i^2$ = kinetic energy of the system = T

and $\sum_i F_i\cdot\delta r_i$ = work done by the forces F_i during displacement δr_i δW (say).

$\therefore$ equation (5) $\Rightarrow \delta W + \delta T = \sum_i (d/dt)(m_i\dot{r}_i\cdot\delta r_i)$ ...(6)

Integrating (6) between the limits t_1 and t_2, we get

$$\int_{t_1}^{t_2}(\delta W + \delta T)\,dt = \int_{t_1}^{t_2}\sum_i \frac{d}{dt}(m_i\dot{r}_i\cdot\delta r_i)\,dt = \Sigma\int_{t_1}^{t_2} d\,(m_i r_i\cdot\delta r_i)$$

$= \Sigma\left[m^2 r_i^2\cdot\delta r^2\right]_A^C = 0$ [since $\delta r^2 = 0$ at the end points A anc C.

But we know that, for a conservative system, $\delta W = -\delta V$ where V is potential energy

$$\Rightarrow \int_{t_1}^{t_2}(-\delta V + \delta T)dt = 0, \text{i.e. } \delta\int_{t_1}^{t_2}(T - V)dt = 0$$

$$i.e. \quad \int_{t_1}^{t_2} (T - V)\, dt = 0 \quad \text{or} \quad \delta \int_{t_1}^{t_2} L\, dt = 0$$

$$\text{or} \quad \delta \int_{t_1}^{t_2} L\, dt = \text{stat} = \text{extremum.}$$ This is Hamilton's principle.

PRINCIPLE OF CONSERVATION OF LINEAR MOMENTUM (UNDER FINITE FORCES)

If the forces acting on a system be such that they have no component along a certain fixed straight line, throughout the motion, then the motion is such that the linear momentum resolved along this line is constant.

Let (x, y, z,) be the coordinates of any particle m of a body at time t, referred to fixed axes where the fixed straight line has been taken as the axis of x. Let X be the resolved part parallel to fixed line of the external forces acting on the particle, then

$$\text{or} \quad \frac{d}{dt} \Sigma m \frac{dx}{dt} = \Sigma X.$$

But from the given conditions $\Sigma X. = 0$, hence $\frac{d}{dt} \Sigma m \frac{dx}{dt} = 0$

$$i.e. \quad \Sigma m \frac{dx}{dt} = \text{constant}$$

$$\Rightarrow M \frac{d\bar{x}}{dt} = \text{constant} \qquad \left(\because \bar{x} = \frac{\Sigma mx}{M} \right) \qquad \text{...(1)}$$

where $\bar{x}$ is the x - coordinate of the centre of gravity of body and M its total mass.

From (1), we observe that the total momentum of the body parallel to the axis of x is constant throughout the motion.

PRINCIPLE OF CONSERVATION OF ENERGY

If a system moves under the action of finite forces and if the geometrical equations of the system do-not contain time explicity the change in the kinetic energy of the system in passing from one configuration to any other is equal to the corresponding work done by the forces.

Let (x, y, z) be the coordinates of a particle of mass m at the time t, and let the resolved parts of the external forces parallel to the axes at that time be X, Y, Z. Then, by D' Alembert's principle, the forces $X - m\ddot{x}, Y - m\ddot{y}, Z - m\ddot{z}$ acting at the point (x, y, z) and the similar forces acting at other particles of the system, form a system of forces in equilibrium.

If dx, dy, dz denote arbitrary small displacements of the particle of mass m at (x, y, z) consistent with geometrical conditions at any time t, then by the principle of virtual work, we have

$$\Sigma[X - m\ddot{x})\,\delta x + (Y - m\ddot{y})\,\delta y + (Z - m\ddot{z})\delta z] = 0 \quad ...(1)$$

If the geometrical conditions are expressed by the equations which do not contain time explicity, then the conditions which hold at time t, will also hold through the time δ t. Therefore, we can take the arbitrary displacements $\delta x, \delta y, \delta z$ as the actual displacements $\frac{dx}{dt}\delta t, \frac{dy}{dt}\delta t, \frac{dz}{dt}\delta t,$

Replacing δx by $\frac{dx}{dt}\delta t, \delta y$ by $\frac{dz}{dt}\delta t, \delta z$by $\frac{dz\cdot}{dt}\delta t$ in (1), we have

$$\Sigma\left[(X - m\ddot{x})\frac{dx}{dt} + (Y - m\ddot{y}\frac{dy}{dt} + (Z - m\ddot{z})\frac{dz}{dt}\right]\delta t = 0$$

$$\text{or } \Sigma m\left[\ddot{x}\frac{dx}{dt} + \ddot{y}\frac{dy}{dt} + \ddot{z}\frac{dz}{dt}\right] = \Sigma\left[X\frac{dx}{dt} + Y\frac{dy}{dt} + Z\frac{dz}{dt}\right]$$

On integration w. r.t. "t" we get

$$\Sigma\frac{1}{2}m\left[\dot{x}^2 + \dot{y}^2 + \dot{z}^2\right]_{t_1}^{t_2} = \Sigma\int_{t=t_1}^{t=t_2}(xdx + ydy + Zdz)$$

Thus we see that change kinetic energy of the system from time t_1 to t_2 equal to the work done by the external forces on the body from the configuration of the body at time t_1 to the configuration at time t_2.

PRINCIPLE OF CONSERVATION OF ANGULAR MOMENTUM (MOMENT OF MOMENTUM UNDER FINITE FORCES)

If the forces acting on a system be such that they have no moment about a certain fixed straight line, throughout the motion then the angular momentum about this straight line is constant.

Let the co-ordinates of a particles of mass m at time t referred to fixed axes be (x, y, z) where the fixed straight line has been taken as the axis of x. Let X, Y, Z be the resolved parts parallel to the axis of the external forces acting on particle m. The sum of the moments of the external force about the given fixed line is Σ (yZ - zY). By D- Alembert's principle, we have

$$\Sigma m\left(y\frac{d^2z}{dt^2} - z\frac{d^2y}{dt^2}\right) = \Sigma(yZ - zY).$$

But from the given condition that throughout the motion, the sum of the moments of the external forces about the axis of x is zero.

i.e. $\Sigma(yZ - zY) = 0$

$$\therefore \Sigma m\left(y\frac{d^2z}{dt^2} - z\frac{d^2y}{dt^2}\right) = 0 \text{ throughout the motion}$$

i.e. $$\frac{d}{dt}\Sigma m\left(y\frac{dz}{dt} - z\frac{dy}{dt}\right) = 0$$

i.e. $$\Sigma m\left(y\frac{dz}{dt} - z\frac{dy}{dt}\right) = \text{constant.} \qquad ...(2)$$

From (2), we observe that total moment of moment of momentum of the body about x - axis is constant throughout the motion.

PRINCIPLE OF CONSERVATION OF LINEAR MOMENTUM (IMPULSIVE FORCES)

If the sum of the impulsive forces parallel to a certain fixed line vanishes, the total momentum in that direction remains the same just before and after the application of impulses.

Let us take the fixed line as the axis of x, then

$$\Sigma.m\frac{d^2x}{dt^2} = \Sigma X \;\; i.e. \;\; \frac{d}{dt}\left[\Sigma.m\frac{dx}{dt}\right] = \Sigma X. \qquad ...(1)$$

If τ be the time (small during which the impulsive forces act on the body, then on integration (1),

$$\text{we get } \left[\Sigma\frac{dx}{dt}\right]_0^\tau = \Sigma\int_0^\tau X dt = \Sigma x_1$$

i.e. $$\Sigma m(v-u) = \Sigma X_1 \text{, when } X_1 = \int_0^\tau X dt \qquad ...(2)$$

where u value are the velocities before and after the impulse and X_1 is the impulse of the forces parallel to the x–axis . But if $\Sigma X_1 = 0$, we have from (2) $\Sigma m(v-u) = 0$ *i.e.* $\Sigma mv = \Sigma mu$),

i.e. total momentum after impulse = the total momentum before impulse

PRINCIPLE OF CONSERVATION OF ANGULAR MOMENTUM [IMPULSIVE FORCES]

If the sum of the moments of the impulses about a certain line vanishes then the angular momentum about it remains the same before and after the application of the impulse. Let us take the fixed line as axis of x. Let t be the time (small) during which the impulses act on the body. By taking moments about the fixed axis, we have

$$\Sigma m(y\ddot{z} - z\ddot{y}) = \Sigma(yZ - zY). \qquad ...(1)$$

integrating it w.r.t. "t" we have

$$\int_0^{\tau} \Sigma m(y\ddot{z} - z\ddot{y})dt = \int_0^{\tau} \Sigma(yz - zY)dt$$

$$\Rightarrow [\Sigma m(y\dot{z} - z\dot{y})]_0^{\tau} = \Sigma(yZ_1 - zY_1)$$

$$\text{where } Z_1 = \int_0^{\tau} Zd \text{ and } Y_1 = \int_0^{\tau} Ydt. \qquad ...(2)$$

L.H.S. of (2) is angular momentum after the impulse minus angular momentum before impulse.

R.H.S. of (2) is the sum of the moments of the impulses about x – axis. In case the sum of the moments of impulses about x – axis is zero

i.e., $\Sigma (yZ_1 - zY_1) = 0$, we have

Angular momentum after impulse = angular momentum before impulse.

CONSERVATIVE FORCES

If the forces acting on the body are such that $\Sigma (xdx + Y dy + Zdz)$ *is the complete differential of some quantity say V, then the forces are said to be conservative.*

Thus we see, If the forces are conservative, then

$$\Sigma(Xdx + Ydy + Zdz) = \int_{t_1}^{t_2} \{\Sigma(Xdx + Ydy + Zdz)\}$$

$$= \int_{t_1}^{t_2} dV = Vt_2 - V_{t_1} \qquad ...(1)$$

which implies that the potential of forces is V. Thus we note that L.H.S. of (1) is independent of the path followed by the body in moving from one position to the other and depends only on the configuration of the body at time t_1 and t_2.

Thus if A is the configuration at time t_1 and B the configuration at time t_2, then have K.E. at time t_2 – K.E. at time $t_1 = V_t - V_t$ *i.e.* K.E. at B – K.E. at A = $V_B - V_A$

When a body moves under the action of a system of Conservative faces, the sum of its kinetic and potential energies is constant throughout the motion. (Conservation of energy)

In the last article, we have shown that if A and B are the configuration of the body at times t_1 and t_2 then

K.E. at time t_2 *i.e.* at the position B – K.E. at t_1 *i.e.* at the position A

$$= \int_{t_1}^{t_2} dV = VB - V_A \qquad ...(1)$$

The potential energy of the body in any position A is the work done by the forces in moving the body form position A to some standard position C (say)

$\therefore$ Potential energy at time t_1

$$= \Sigma \int_A^C (Xdx + Ydy + Ydz) = \int_A^C dV = V_C - V_A \qquad ...(2)$$

and potential energy at time

$$t_2 = \Sigma \int_A^C (Xdx + Ydy + Ydz)$$

$$= \int_A^C dV = V_C - V_B \qquad ...(3)$$

Combining (2) and (3), we have P.E. at time t_1 – P.E. at time t_1

$$= V_C - V_A - (V_C - V_B) = V_B - V_A \qquad ...(4)$$

Comparing (1) and (4), we have P.E. at time t_1 – P.E. at time t_2

= K.E. at time t_2 – K.E. at time t_1

i.e. K.E. at time t_1 + P.E. at time t_1

= K.E. at time t_2 + P.E. at time t_2.

i.e. sum of the potential energy and kinetic energy is constant at all times.

The kinetic energy of a rigid body, moving in any manner, is at any instant equal to the kinetic energy of the whole mass, supposed to be collected at its centre of inertia and moving with it, together with the kinetic energy of the whose mass relative to its centre of inertia.

Consider a particle of mass m of the body. Let (x, y, z) be the coordinates of this particle at any time t. Let $(\bar{x}, \bar{y}, \bar{z})$ be the co-ordinates of G, the centre of gravity of the body. If (x', y', z') are the co-ordinates of some particle relative to G then, we have $x = \bar{x} + x', y = \bar{y} + y', z = \bar{z} + z'$ and $*Smx' = 0$, $Smy' = 0$, $\Sigma mz' = 0$

so that $\Sigma m \dfrac{dx'}{dt} = 0, \Sigma m \dfrac{dy'}{dt} = 0; \Sigma m \dfrac{dz'}{dt} = 0$

The total K.E. of the body

$$= \frac{1}{2} \Sigma m \left[\left(\frac{dx}{dt}\right)^2 + \left(\frac{dy}{dt}\right)^2 + \left(\frac{dz}{dt}\right)^2 \right]$$

$$=\frac{1}{2}\Sigma m\left[\left(\frac{d\bar{x}}{dt}+\frac{dx'}{dt}\right)^2+\left(\frac{d\bar{y}}{dt}+\frac{dy'}{dt}\right)^2+\left(\frac{d\bar{z}}{dt}+\frac{dz'}{dt}\right)^2\right]$$

$$=\frac{1}{2}\Sigma m\left[\left(\frac{d\bar{x}}{dt}\right)^2+\left(\frac{d\bar{y}}{dt}\right)^2+\left(\frac{d\bar{z}}{dt}\right)^2\right]+\frac{1}{2}\Sigma m\left[\left(\frac{dx'}{dt}\right)^2\right.$$

$$\left.+\left(\frac{dy'}{dt}\right)^2+\left(\frac{dz'}{dt}\right)^2\right]+\Sigma m\left[\frac{d\bar{x}}{dt}\frac{dx'}{dt'}+\frac{d\bar{y}}{dt}\frac{dy'}{dt}+\frac{d\bar{z}}{dt}\frac{dz'}{dt}\right]$$

$$=\frac{1}{2}\Sigma m\left[\left(\frac{d\bar{x}}{dt}\right)^2+\left(\frac{d\bar{y}}{dt}\right)^2+\left(\frac{d\bar{z}}{dt}\right)^2\right]$$

$$+\frac{1}{2}\Sigma m\left[\left(\frac{dx'}{dt}\right)^2+\left(\frac{dy'}{dt}\right)^2+\left(\frac{dz'}{dt}\right)^2\right]$$

$$+\frac{d\bar{x}}{dt}\Sigma m\frac{dx'}{dt}+\Sigma m\frac{dy'}{dt}\Sigma m\frac{dy'}{dt}+\frac{d\bar{z}}{dt}\Sigma m\frac{dz'}{dt}$$

$$=\frac{1}{2}\left[\left(\frac{d\bar{x}}{dt}\right)^2+\left(\frac{d\bar{y}}{dt}\right)^2+\left(\frac{d\bar{z}}{dt}\right)\right]^2\Sigma m$$

$$+\frac{1}{2}\Sigma m\left[\left(\frac{dx'}{dt}\right)^2+\left(\frac{dy'}{dt}\right)^2+\left(\frac{dz'}{dt}\right)\right]^2$$

$$=\frac{1}{2}\text{ (velocity of centre of gravity G)}^2\ \Sigma m$$

$$+\frac{1}{2}\text{ (velocity of G)}^2\times\Sigma m+\frac{1}{2}\Sigma m$$

$$=\frac{1}{2}Mv^2+\frac{1}{2}\Sigma m\text{ (velocity of m relative to G)}^2$$

Taking G as the origin the co-ordinates of general point are (x', y', z'). Then co-ordintes of the C.G. of the body are given as x coordinate of $C.G=\frac{\Sigma mx'}{Zm}$ But x coordinate of G = 0 (since G is the origin) $\therefore\ \Sigma mx' = 0$ Similarly $\Sigma my' = 0$ $\Sigma mz' = 0$. where M is the total mass of the body and v is the velocity of the centre of gravity G. *To sum up: Total K.E. of the body = K.E. of mass M supposed to be collected at G + K. E. of the body relative to G.*

VIS-VIVA

Vis-Viva of a particle is the product of its mass and the square of its velocity.

If a particle of mass m is moving with velocity v, then its

Vis-Viva = $mv^2 = 2\cdot\left(\frac{1}{2}mv^2\right)$ = Twice its kinetic energy.

PRINCIPLE OF VIS-VIVA

If a system moves under the action of finite forces and if the geomertical conditions of the parts of the system be expressed by equations which do not contain time explicity, the change in the Vis - Viva of the system in passing from one position to any other is equal to twice the corresponding work done by the forces.

Note: Proof is the same as that of principle of conservation of energy

EXTENSION OF HAMILTON'S PRINCIPLE TO NON-CONSERVATIVE AND NON-HOLONOMIC SYSTEM

We can generalise Hamilton's principle to include non-conservative forces as well, so that an alternative form of the Lagrange's equation can be obtained. The extension of the Hamilton's principle is

$$\delta I = \delta \int_1^2 (T + W)\, dt = 0 \qquad ...(1)$$

where $$W = \Sigma F_i \cdot r_i \qquad ...(2)$$

We know that the variations δq_i or δr_i are identical with virtual displacements of the co-ordinates as there is no variation of time. Thus we can consider the varied path in configuration space as built up by a succession of virtual displacements from the actual path of motion C. Again each virtual displacements takes place at some definite time and at that time the forces acting on the body have definite values. δW represents the amount of work done by the forces on the system during the period of virtual displacement from the actual to the varied path. Thus the Hamilton's principle given by (1) can be said as *the integral of the variation of kinetic energy together with the amount of virtual work involved in the variation must be zero.*

We can evaluate the variations δr_i in terms of δq_i by making use of the equations of transformation between r and q, where each q depends on the path chosen through a parameter say α:

$r_i = r\ \{q_k\ (\alpha,t)t\}$

We can abbreviated the process by using the equivalence of δr_i with a virtual displacement. We know that

$\sum_i F_i \cdot \delta r_i = \sum_k Q_k \delta q_k$.

thus (1) may be put as $\delta \int_t^2 T\,dt + \int_1^2 \sum_k Q_k \delta q_k dt = 0.$ ...(3)

Now it can be shown easily that (3) reduces to ordinary from of Hamilton's principle, in case Q_k's are derivable form the generalised potential. Therefore the integral of virtual work, under these conditions becomes,

$$\int_t^2 \sum_k Q_k \delta q_k\, \delta q_k dt = -\int_1^2 \sum_k \delta q_k \left(\frac{\partial V}{\partial q_k} - \frac{d}{dt}\frac{\partial V}{\partial \dot{q}_k} \right) dt$$

Now reversing by integration-paths procedure; the above integral can be put as

$$-\int_t^2 \sum_k \left(\frac{\partial V}{\partial q_k} \delta q_k + \frac{\partial V}{\partial \dot{q}_k} \delta q_k \right) dt = -\delta \int_1^2 V\,dt$$

Thus (3) reduces to

$$\delta \int_1^2 T\,dt = \int_1^2 V\,dt = \delta \int_1^2 (T - V)\,dt\ \delta \int_1^2 L dt = 0,$$

which is Hamilton's principle.

For more general problem, the variation of the first integral in (3) can be written at once, as T like L for conservative systems is a function of q_k and $\dot{q}_k$

$$\therefore \delta \int_1^2 T\,dt = \int_1^2 \sum_b \left(\frac{\partial T}{\partial q_k} - \frac{d}{dt}\frac{\partial T}{\partial \dot{q}_k} \right) \delta q_k dt \qquad ...(4)$$

$$\Rightarrow \int_1^2 \sum_k \left(\frac{\partial T}{\partial q_k} - \frac{d}{dt}\frac{\partial T}{\partial \dot{q}_k} + Q_k \right) \delta q_k dt = 0 \qquad ...(5)$$

Further it is assumed that the constraints are holonomic, so the integral (5) can vanish if and only if the separate coefficients vanish, *i.e.*

$$\frac{d}{dt}\frac{\partial T}{\partial \dot{q}_k} - \frac{\partial T}{\partial q_k} = Q_k \qquad ...(6)$$

so that (1) represents the proper extension of Hamilton's principle, which yields that from of Lagrange's equation, in case the forces are not derived from a potential.

We can also extend the Hamilton's principle to cover certain categories of non-holonomic systems as well. While deriving Lagrange's equations from Hamilton's or D'Alembert's principle, application of holonomic constraints is made only in last step when the variation q_k are considered as independent of each other.

While in non-holonomic system the generalized co-ordinates are not independent to each other so they cannot be reduced further by means of equations of constraints of the form

$f[q_1, q_2, \ldots q_n; t] = 0$ Thus q_k's can not be as all independent.

We can further treat non-holonomic system, provided the equations of constraint can be putting the from

$$\sum_j q_{ij} da_j + a_{ij} dt = 0 \quad [j = 1,2, - \ldots m] \qquad \ldots(7)$$

which is a relation connecting the differentials of q's since in variation process used in Hamilton's principle, the time for each point on the path is taken constant, therefore the virtual displacements occurring in the variation must satisfy the equations of constraint of the form

$$\sum_j q_{ij} \delta q_j = 0, (j = 1,2,\ldots m) \qquad \ldots(8)$$

The equations (8) can very well be used to reduce the number of virtual displacement to independent ones. The method used for eliminating these extra virtual displacements is that of Lagrange's *undetermined multipliers.*

Also from (8), we get

$$\lambda_1 \sum_J a_{ij} \delta q_j = 0, (j = 1,2,\ldots m) \qquad \ldots(9)$$

where λ_1 are some undetermined constants, which in general, are functions of time.

Now first of all summing the equations (9) over l and then integrating the resulting equation from 1 to 2, we get

$$\int_1^2 \sum_{ji} \lambda_1 a_{ij} \delta q_j dt = 0 \qquad \ldots(10)$$

Again the equations corresponding to (5) in the case of conservative system are given as

$$\int_1^2 dt \sum_j \left(\frac{\partial L}{\partial q_j} - \frac{d}{dt} \frac{\partial L}{\partial \dot{q}_j} \right) \delta q_j = 0 \qquad \ldots(11)$$

Combining the equations (10) with (11), we get

$$\int_1^2 dt \sum_j \left(\frac{\partial L}{\partial q_j} - \frac{d}{dt} \frac{\partial L}{\partial \dot{q}_j} + \sum_l \lambda_l a_{ij} \right) \delta q_j = 0 \qquad \ldots(12)$$

It is to be noted here that δq_j's are still not independent. They are connected by the m relations given by (8.) This is, while the first n – m of these equations may be chosen independently, the last m are then fixed by the equations (8). Since the values of λ_l's are at our disposal, we choose them to be such that

$$\frac{\partial L}{\partial q_j} - \frac{d}{dt}\frac{\partial L}{\partial \dot{q}_j} + \sum_l \lambda_l a_{ij} = 0, (j = n - m + 1, ...n) \qquad ...(13)$$

which are in the nature of equations of motion for the last m of the q_j variable.

Using the value of λ_l obtained from (13) in (12), we get

$$\int_1^2 dt \sum_{j=1}^{n-m} \left(\frac{\partial L}{\partial q_j} - \frac{d}{dt}\frac{\partial L}{\partial \dot{q}_j} + \sum_l \lambda_l a_{ij} \right) \delta q_j = 0 \qquad ...(14)$$

Since δq_j's involved here are independent ones, therefore we have

$$\frac{\partial L}{\partial q_j} - \frac{d}{dt}\frac{\partial L}{\partial \dot{q}_j} + \sum_l \lambda_l a_{lj} = 0, (j = 1, 2, ...n - m) \qquad ...(15)$$

Now combining (13) with (15), we get

$$\frac{d}{dt}\frac{\partial L}{\partial \dot{q}_j x} - \frac{\partial L}{\partial q_j} = \sum_l \lambda_l a_{lj}, (j = 1, 2, ...n) \qquad ...(16)$$

which are known as Lagrange's equations for non holonomic systems. We now observe that (16) gives us a total of only n equations, where as the unknowns involved are (n + m) in number, namely then n co-ordinates q_l and the m, λ_l's. The additional equationals required are exactly the equations of constraint *i.e.*

$$\sum_j a_{ij}\dot{q}_j + a_{ij} + 0 \qquad ...(17)$$

The equations (16) together with (17) constitute (n + m) equations for n + m unknowns.

It is to be noted here that equation (7) is not the most general type of non-holonomic constraint. For example, it does not include equations of constraints in the form of inequalities. More over it includes holonomic constraints. An equation of the form

$$f(q_1, q_2, q_3, ...q_n ; t) = 0 \qquad ...(18)$$

it is of the same form as the equation 97) with the coefficients

$$a_{lj} = \frac{\partial f}{\partial q_j} \text{ and } a_{ji} = \frac{\partial f}{\partial t} \qquad ...(19)$$

It means that method of Lagrange's undetermined multipliers can be used also for holonomic constraints, when it is not easy to reduce all the q's to independent co-ordinates.

PHYSICAL SIGNIFICANCE OF THE HAMILTONIAN

If the Hamiltonian H is independent of t explicity, prove that it is (a) constant and (b) equal to the total energy of the system.

Proof:

(a) we have $\dfrac{dH}{dt} = \sum_{i=1}^{n} \dfrac{\partial H}{\partial qi}\dfrac{dqi}{dt} + \sum_{i=1}^{n} \dfrac{\partial H}{\partial pi}\dfrac{dpi}{dt}$

$$= \sum_{i=1}^{n} -(\dot{p}_i)\dot{q}_i + \sum_{i=1}^{n} \dot{q}_i \dot{p}_i \left(\because \dot{p}_i = -\frac{\partial H}{\partial qi} \text{ and } \dot{q}_i = \frac{\partial H}{\partial pi} \right) = 0$$

$\Rightarrow$ H = constant, say E.

(b) By Euler's theorem on homogeneous function, we have

$\Sigma \dot{q}_i \dfrac{\partial T}{\partial \dot{q}_i} = 2T$, where T is the K.E. of the system.

But L = T – V, $\therefore \dfrac{\partial L}{\partial \dot{q}_i} = \dfrac{\partial (T-V)}{\dot{q}_i} = \dfrac{\partial T}{\partial \dot{q}_i}$ (V does not depend on $\dot{q}_i$)

or $\Sigma \dot{q}_i \dfrac{\partial L}{\partial \dot{q}_i} = 2T \Rightarrow \Sigma \dot{q}_i\ p_i = 2T \left(\because p_i = \dfrac{\partial L}{\partial \dot{q}_i} \right)$

$\therefore H = \Sigma p_i \dot{q}_i - L = 2T - (T - V) = T + V = E.$

EQUATIONS OF MOTION WITH HAMILTON'S FORM

Here we shall obtain the differential equations of motion of a conservative holonomic dynamical system in a form which constitutes the basis of most of the advanced theory of dynamics. Let $(q_1, q_2, \ldots\ldots\ldots, q_n)$ be the generalised co-ordinates and let $L(q_1, q_2, \ldots q_n; \dot{q}_1, \dot{q}_2, \ldots \dot{q}_n; t)$, the kinetic potential of the system, so that the equations of motion in the Lagrangian form are

$$\frac{d}{dt}(\partial L/\partial \dot{q}_i) - (\partial L/\partial q_i) = 0; (i = 1, 2, \ldots., n) \qquad \ldots(1)$$

writing $p_i = (\partial L/\partial \dot{q}_i)$ we get $\dot{p}_i = (\partial L/\partial q_i)$ (i = 1, 2,...,n) ...(2)

hence from the former of these sets of equations we can regard either of the sets of quantities $(q_1, q_2 \ldots a_n)$ or $(p_1, p_2, \ldots, p_n)$ as functions of the other set.

Now, let δ denote the increment in any function of the variables $(q_1, q_2 \ldots, q_n; p_1, p_2, \ldots, p_n)$ or $(\dot{q}_1, q_2 \ldots, q_n; \dot{q}_1, \dot{q}_2, \ldots, \dot{q}_n)$; then we get

$$dL = \sum_{i=1}^{n}\left(\frac{\partial L}{\partial q_i}dq_i + \frac{\partial L}{\partial \dot{q}_i}d\dot{q}_i\right) + \frac{\partial L}{\partial t}\text{(when L contains t explicitly)}$$

$$dL = \sum_{i=1}^{n}\left((\dot{p}_1 dq_i + \dot{d}q_i)\frac{\partial L}{\partial t}\right)dt$$

$$= d\left(\sum_{i=1}^{n} p_i\dot{q}_i\right) + \sum_{i=1}^{n}(\dot{q}_1 dq_i - p_i dp_i) + \frac{\partial L}{\partial t}dt$$

$$\Rightarrow d\left[\sum_{i=1}^{n}(p_1\dot{q}_1) - L\right] = \sum_{i=1}^{n}(\dot{q}_i dp_i - p_i dq_i) - (\partial L/\partial t)dt. \qquad ...(3)$$

Thus if the quantity $\sum_{i=1}^{n}(p_i\dot{q}_i = L)$ when expressed in terms of $(q_1, q_2 ... q_n ; p_1, p_2, ..., p_n ; t)$ be denoted by H, we have

$$dH = \sum_{i=1}^{n}(q_1 dp_i - \dot{p}_i\, dq_i) - \frac{\partial L}{\partial t}dt$$

$$\Rightarrow \sum_{i=1}^{n}\frac{\partial H}{\partial q_i}dq_i + \sum_{i=1}^{n}\frac{\partial H}{\partial p_i}dp_i + \frac{\partial H}{\partial t}dt = \sum_{i=1}^{n}(\dot{q}_1\, d\,p_i - \dot{p}_i\, d\,q_i) - \frac{\partial L}{\partial t}dt$$

$$\Rightarrow \dot{p}t = -\frac{\partial H}{\partial pi}, \dot{p}i = -\frac{\partial H}{\partial qi}, \frac{\partial H}{\partial t} = -\frac{\partial L}{\partial t}$$

If H does not contain t explicity (*i.e.* does not contain t explicity) we have $\dot{p}i = -(\partial H/\partial qi)$ and $\dot{q}i = (\partial H/\partial pi)$. ...(4)

These equations are called as Hamilton's equations, or Hamilton's canonical equations and the function H is called Hamiltonian.

The total order of Hamilton's equations is the same as the total order of Lagrange's equations, namely 2n. But whereas Lagrange's equations present us with n equations each of the second order. Hamilton's equation are 2n equations, each of the first order. Hamilton's equations can also be written as

$$\frac{dpi}{-(\partial H/\partial qi)} = \frac{dqi}{(\partial H/\partial pi)} = dt.$$

VARIATIONAL METHODS

Techniques of Calculus of Variations

The calculus of variations arose out of the quest for the mathematical requirements in the solution of problems like the study of:

(i) The path followed by a body falling freely under gravity, (brachistochrone) first studied by Newton,

(ii) The equilibrium shape of a freely hanging homogeneous flexible cord between two horizontal points (catenary), first studied by Bernoulli.

Presently the technique of calculus of variations, serves as a mathematical preliminary (in the form of Euler-Lagrange's equation) in the study of a wide range of physical problems such a geodesics and minimal surface in Riemannian and differential geometrics and the various *variational (minimal)* principles in the different branches of physics, The technique is co-ordinates invariant and there in lies its great power. First we develop here this technique in a purely mathematical form.

Suppose A and B are fixed points (x_1, y_1), (x_2, y_2) in a cartesian plane.

Also suppose that (x, y, y′) is known functional form of the variables x, y, y′ {= dy/dx)}. Then if C is a curve joining A and B and having equation y = y (x), Then the integral

$$I = \int_{x_1}^{x_2} f[x, y, y']\, dx \qquad ...(1)$$

has a definite value whenever the function y (x) is prescribed. The value of I will change as we vary the from of the curve C through A, B. Consequently we may consider that in general there will be some curve

C through these fixed points such that the value of I *taken along it is stationary (in general a* maximum or minimum) compared with the value along neighbouring paths C. Calculus of variations, the branch of mathematics of concerned amongst other things with finding the form of y (x) from which this stationary property holds. Let the path $\bar{C}$ have the equation $y = \bar{y}(x)$ and let the equation of a neighbouring curve C by $y = \bar{y}(x) + \varepsilon\eta(x)$, and $\bar{P}X = \bar{y}(x) \Rightarrow P\bar{P} = \varepsilon\eta(x)$.

Hence the values of I taken along is thus a function of ε of the form

$$I(\varepsilon) = \int_{x_1}^{x_2} f[x, \bar{y} + \varepsilon\eta, \bar{y}' + \varepsilon\eta']\, dx = \int_{x_1}^{x_2} f[x, y, y']\, dx \qquad ...(2)$$

where $y = \bar{y} + \varepsilon\eta$ and $y' = \bar{y}' + \varepsilon\eta'$

But the curve $\bar{C}$, for which ε = 0, makes I stationary, so we must have I′ stationary, so we must have I′ (0) = 0. Now differentiating (2) w.r.t. ε, we get

$$I'(\varepsilon) = \int_{x_1}^{x_2} \left[\frac{\partial f}{\partial x}\frac{\partial x}{\partial \varepsilon} + \frac{\partial f}{\partial y}\frac{\partial y}{\partial \varepsilon} + \frac{\partial f}{\partial y'}\frac{\partial y'}{\partial \varepsilon}\right] dx \qquad [\because f = f[x, y, y']] \quad ...(3)$$

$$\Rightarrow I'(\varepsilon) = \int_{x_1}^{x_2} \left[\frac{\partial f}{\partial y}\eta(x) + \frac{\partial f}{\partial y'}\eta'(x)\right] dx \qquad [\because y = \bar{y} + \varepsilon\eta(x)]$$

$$= \int_{x_1}^{x_2} [\eta f_y . \eta(x) + f_{y'} . \eta'(x)]\, dx$$

$$I'(\varepsilon) = \int_{x_1}^{x_2} [\eta f_y (x, \bar{y} + \varepsilon\eta, \bar{y'} + \varepsilon\eta')] \quad + \eta' f_{y'}(x, \bar{y} + \varepsilon\eta, \bar{y} + \varepsilon\eta')] dx \qquad ...(4)$$

where $f_{y'}(x, \bar{y} + \varepsilon\eta, \bar{y} + \varepsilon\eta')]dx$

$$= (\partial / \partial y)\, f\,(x, y, y') \text{ when } y = \bar{y} + \varepsilon\eta,\ y' = \bar{y'} + \varepsilon\eta,$$

with a similar meaning for $f_{y'}\ (x,\ \bar{y} + \varepsilon\eta,\ \bar{y'} + \varepsilon\eta')$. Hence the stationary condition give $I'\ (0) = 0$.

$$\Rightarrow \int_{x_1}^{x_2} [\eta f_{y'} (x, \bar{y}, \bar{y'}) + \eta' f_{y'}\ (x, \bar{y}, \bar{y'})\, dx = 0 \qquad ...(5)$$

Now $\int_{x_1}^{x_2} \eta' f_{y'} (x, \bar{y}, \bar{y'})\, dx$

$$= \left[\eta f_{y'} (x, \bar{y}, \bar{y'}\right]_{x_1}^{x_2} - \int_{x_1}^{x_2} n \frac{d}{dx} f_{y'}\ (x, \bar{y}, \bar{y}')\, dx$$

$$= - \int_{x_1}^{x_2} n \frac{d}{dx} f_{y'}\ (x, \bar{y}, \bar{y}')\, dx\ (\because \eta(x_1) = \eta(x_2) = 0) \qquad ...(6)$$

Using (6), equation (5) implies

$$\int_{x_1}^{x_2} \eta(x) \left[f_y (x, \bar{y}, \bar{y'} - \frac{d}{dx} f_{y'} (x, \bar{y}, \bar{y'}) \right] dx = 0 \qquad ...(7)$$

But η (x) is arbitrary, subject to its being differentiable and vanishing at A, B; (7) implies that

$$f_y (x, \bar{y}, \bar{y'}) - (d / dx)\, f_{y'} (x, \bar{y}, \bar{y'}) = 0 \qquad ...(8)$$

or $$(\partial f / \partial y) - (d / dx)\, (\partial f / \partial y') = 0. \qquad ...(9)$$

This is called *Euler's Lagrange's equation.* Equation (9) can be proved from (7) by using the following lemma.

Lemma: If x_1 and x_2 ($>x_1$) *are fixed constants and ϕ (x) is a particular continuous function for* $x_1 \le \phi \le x_2$ and if

$$\int_{x_1}^{x_2} \eta(x) \phi(x)\, dx = 0 \qquad ...(10)$$

for every choice of continuously differentiable function h(x) for which $\eta(x_1) = \eta(x_2) = 0$ then $f(x) = 0$ indentically in $x_1 \le x \le x_2$.

Proof:

Let the lemma be not true for all x in $x_1 \le x \le x_2$. Let x′ be a point where $\phi(x') \neq 0$ and > 0. But $\phi(x)$ is continuous in $x_1 \le x \le x_2$ and in particular is continuous at x = x′. Hence there exist an interval $x'_1 \le x \le x'_0$ around x′ where $\phi(x) > 0$. For the other x's $\phi(x)$ any both vanish; then the equation $\int_{x_1}^{x_2} \eta(x)\phi(x)dx = 0$ could be written as $\int_{x_1}^{x_2'} \eta(x)\phi(x)dx = 0$.

But η(x) is at our disposal; we could choose it as greater than zero for $x_1' \le x \le x_2'$ and η(x) = 0 for other values of x. Thus $\eta(x)\ \phi(x) > 0$ for $x_1' \le x \le x_2'$; consequently integrand cannot be zero. Thus we arrive at the contradiction. Hence $\phi(x) = 0$ in $x_1 \le x \le x_2$.

Certain Remarks About Euler-Lagrange's Equation

(A) We have $(\partial f/\partial y) - (d/dx)(\partial f/\partial y') = 0$

where f = f (x, y, y′)

which can be re-written as

$$(\partial f/\partial y) - (dp/dx) = 0 \text{ where } p = (\partial f/\partial y') \qquad ...(11)$$

But $\left(\frac{dp}{dx}\right) = \left(\frac{\partial p}{\partial x}\right) + \left(\frac{\partial p}{\partial y}\right)\left(\frac{dy}{dx}\right) + \left(\frac{\partial p}{\partial y'}\right)\left(\frac{dy'}{dx}\right) = \left(\frac{\partial^2 f}{\partial x \partial y'}\right) +$

$(\partial^2 f/\partial y \partial y')(dy/dx) + (\partial^2 f/\partial y'^2)(d^2y/dx^2)$. Hence (9) give

$$(\partial f/\partial y) - [(\partial/\partial x)(\partial f/\partial y') + (\partial/\partial y)((\partial f/\partial y')(dy/dx)$$

$$+ (\partial^2 f/\partial y'^2)(d^2y/dx^2)] = 0$$

or $$(\partial^2 f/\partial y'^2)(d^2y/dx^2) + (\partial^2 f/\partial y\, \partial y')(dy/dx) - (\partial f/\partial y) + (\partial^2 f/\partial x\, \partial y') = 0 \qquad ...(12)$$

This is a second order differential equation for determining y as a function of x. The solution will contain two arbitrary constants which could be determined from the conditions:

$$y = y_1 \text{ at } x = x_1 \text{ and } y = y_2 \text{ at } x = x_2 \qquad ...(13)$$

(B) Consider $\frac{d}{dx}\left[f - y'\frac{\partial f}{\partial y'}\right] = \frac{\partial f}{\partial y'}$

i.e. $\frac{dQ}{dx} = \frac{\partial f}{\partial x}$ which $Q = f - y' \frac{\partial f}{\partial y'}$ and f is a function of x, y, y¢

$$\Rightarrow \frac{\partial f}{\partial x}\left[f - y'\frac{\partial f}{\partial y'}\right] + \frac{\partial}{\partial y}\left[f - y'\frac{\partial f}{dy'}\right]\frac{dy}{dx} + \frac{\partial}{\partial y'}\left[f - y'\frac{\partial f}{\partial y'}\right]\frac{d^2y}{dx^2} = \frac{\partial f}{\partial x}$$

$$\Rightarrow -y'\frac{\partial^2 f}{\partial x \partial y'} + \frac{\partial f}{\partial y}\frac{dy}{dx} - y'\frac{\partial^2 f}{\partial y \partial y'}\frac{dy}{dx} + \frac{\partial f}{\partial y'}\frac{d^2y}{dx^2}\frac{\partial f}{\partial y'}\frac{\partial^2 y}{\partial x^2} - y'\frac{\partial^2 f}{\partial y'^2}\frac{d^2y}{dx^2} = 0$$

$$\Rightarrow y'\left[\frac{\partial f}{\partial y} - \frac{\partial^2 f}{\partial x \partial y'} - y'\frac{\partial^2 y}{\partial y \partial y'} - y''\frac{\partial^2 f}{\partial y'^2}\right] = 0$$

$$\Rightarrow f_y - f_{xy'} - y' fyy' - y'' f_{yy'} = 0 \text{ (if } y' \neq 0) \qquad ...(15)$$

which is same as (12) and shows that if Euler's equation is satisfied equation (14) is also satisfied but if (14) is satisfied, Euler's equation may not be true.

(C) If f does not contain x explicity, i.e. $l = \int_{x_1}^{x_2} f(y, y')dx$,

then from (14), we have

$$(d/dx)\{f - y'(\partial f/\partial y')\} = 0 \qquad \{\because (df/x) = 0\}$$

$$\Rightarrow f - y'(\partial f/\partial y') = \text{constant} = C \text{ (say.)} \qquad ...(16)$$

Obviously (14) is a first order differential equation.

(D) If f does not contain y explicity, then we have $l = \int_{x_1}^{x_2} f(x, y')dx$.

$\therefore$ from the equation

$(\partial f/\partial y) - (d/dx)(\partial f/\partial y') = 0$ we obtain $(d/dx)(\partial f/\partial y') = 0$

$$\Rightarrow (\partial f/\partial y') = C \text{ (say).} \qquad ...(17)$$

(E) If we assume that the end points are not fixed.

i.e. $\eta(x_1) \neq 0$ and $\eta(x_2) \neq 0$.

Euler's equation still holds in this case, if $(\partial f/\partial y') = 0$ when $x = x_1$, x_2 This becomes the boundary condition of the problem.

(F) Euler's equation gives only a necessary condition; even if it is satisfied, there may be no extremum.

(G) The conditions to be imposed on f are such that its partial derivative are continuous and differentiable.

(H) The conditions to be imposed on the curves are that they should be continuous curves.

BRACHISTOCHRONE PROBLEM

Given two points A and B in a vertical plane, to find for the moveable particle M, the path A MB, descending along which by its own gravity and beginning to be urged from the point A, it may in the shortest time reach the point B. It is a tacit in the statement that the particle descends without friction.

Let A be the origin, then velocity of the particle of mass M at any time t is given by

$$v = (ds/dt) = \sqrt{(2gy)}$$

$$\Rightarrow dt = (ds/\sqrt{2gy})$$

$$\Rightarrow t = \int_A^B \frac{ds}{\sqrt{(2gy)}} = \int_A^B \frac{\sqrt{(1+y'^2}}{\sqrt{(2gy)}} dx$$

$$\frac{1}{\sqrt{(2g)}} \int_A^B \frac{\sqrt{1+y'^2}}{\sqrt{2}} dx = \frac{1}{\sqrt{(2g)}} \int_A^B f(y, y') dx,$$

where $f(y, y') = \dfrac{\sqrt{(1+y'^2)}}{\sqrt{y}}$

As x is absent in f, so we have $f - y'(\partial f/\partial y') A$

$$\Rightarrow \frac{\sqrt{(1+y'^2)}}{\sqrt{y}} - y'\frac{y'}{\sqrt{\{y(1+y'^2\}}} = A$$

$$\Rightarrow \frac{1}{\sqrt{\{y(1+y'^2)\}}} = \text{constant} \Rightarrow y(1+y'^2) = \text{constant} = 2c, \text{ (say)}$$

But $(dy/dx) = \tan\psi \Rightarrow y' = \tan\psi$

$\therefore y\sec^2\psi = 2c$ or $y = 2c\cos^2\psi = c(1+\cos^2\psi)$

Also $dx = \cot\psi\, dy = -2c \sin 2\psi \cot\psi\, d\psi$

$= -4c\cos^2\psi\, d\psi = -2c(1+\cos^2\psi)\, d\psi$

$$\Rightarrow x = a - 2c\left(\psi + \frac{\sin^2\psi}{2}\right) = a - 2c(\psi + \sin\psi\cos\psi)$$

Hence the required path is

$x = a - c(2\psi + \sin 2\psi)$, $y = c(1 + \cos^2\psi)$ (cycloid)

EXTENSION OF THE VARIATIONAL METHOD

Suppose the n-co-ordinates q_1, q_2, ...q_n are each functions of an independent variable t and that we required the solution of the variational problem

$$I = \int_{t_1}^{t_2} f(q_1, q_2, ..., q_n; \dot{q}_1 \dot{q}_2, ..., \dot{q}_n; t)dt$$

$$= \int_{t_1}^{t_2} f(q, \dot{q}, t) \ dt = \text{stationary value}$$

where f is of known functional from, t_1 and t_2 are fixed and each q_i is to be determined.

Let $q_i = q_i$ (t) give the stationary value to I and let $q_i = q_i (t) + \varepsilon_i \eta_i (t)$ be the neighbouring path.

$$\Rightarrow I(\varepsilon_1, \varepsilon_2, ..., \varepsilon_n) = \int_{t_1}^{t_2} f\{q_i(t) + \varepsilon_i \eta_i(t); \dot{q}_i(t) + \varepsilon_i \eta_i(t), t\} \, dt$$

$$= \int_{t_1}^{t_2} f(q_i \dot{q}_i; t)dt + \Sigma\varepsilon_i \int_{t_1}^{t_2} \left\{ \dot{\eta}_i(t) \frac{\partial f}{\partial q_i} + \dot{\eta}_i(t) \frac{\partial f}{\partial \dot{q}_i} \right\} dt + o(\varepsilon_i^2)$$

First variation $\delta I = \Sigma\varepsilon_i \int_{t_1}^{t_2} \left\{ \dot{\eta}_i(t) \frac{\partial f}{\partial q_i} + \dot{\eta}_i(t) \frac{\partial f}{\partial \dot{q}_i} \right\} dt$

$$= \Sigma\varepsilon_i \ \dot{\eta}_i(t) \frac{\partial f}{\partial \dot{q}_i} \, dt + \Sigma\varepsilon_i \int_{t_1}^{t_2} \dot{\eta}_i(t) \frac{\partial f}{\partial \dot{q}_i} \, dt$$

$$= \Sigma\varepsilon_i \int_{t_1}^{t_2} \dot{\eta}_i(t) \frac{\partial f}{\partial q_i} dt + \Sigma e_i \left[\left\{ n_i(t) \frac{\partial f}{\partial \dot{q}_i} \right\}_{t_1}^{t_2} - \int_{t_1}^{t_2} \dot{\eta}_i(t) \frac{d}{dt}\left(\frac{\partial f}{\partial \dot{q}_i} \right) dt \right]$$

$$= \Sigma\varepsilon_i \int_{t_1}^{t_2} \dot{\eta}_i(t)_i \left[\frac{\partial f}{\partial q_i} - \frac{d}{dt}\left(\frac{\partial f}{\partial \dot{q}_i} \right) \right] \qquad \{\because 0 = \eta_i(t_1) = \eta_i(t_2)\}.$$

As η_1's are arbitrary and I is stationary, so we have

$$(\partial f / \partial q_i) - (d/dt)(\partial f / \partial \dot{q}_i) \qquad (i = 1, 2, ..., n)$$

PASSAGE FROM THE HAMILTONIAN TO THE LAGRANGIAN

Suppose that we are given a function *H (q, p, t) and are told that the motion of the system satisfies the canonical equations

$$\dot{p}_i = -(\partial H / \partial qw_i) \text{ and } \dot{q}_i = (\partial H / \partial qw_i) \qquad ...(1)$$

Then we want to find a function

$L(p_1, p_2, \ldots, p_n; q_1, q_2, \ldots, q_n; t)$, *i.e.* $L(p, q, t)$

Such that the motion also satisfies the equations

$(d/dt)\ (\partial L/\partial \dot{q}_i) - (\partial L/q_i) = 0.$ ...(2)

Solve the first set of equations in (1) for the p's in terms of the q's, the $\dot{q}$'s and t.

Then write $L = \sum_{i=1}^{n} \dot{q}_1 p_i - H$ and express L as a function of the q's, the $\dot{q}$'s and t. This is the required Lagrangian.

$\therefore (\partial L/\partial q_i) = p_i \left(\text{using } L = \sum_{i=1}^{n} p_i \dot{q}_i - H \right)$

$\Rightarrow (d/dt)(\partial L/\partial \dot{q}_i) = \dot{p}_i$ and $(\partial L/\partial q_i) = -(\partial H/\partial q_i)$

$\Rightarrow \frac{d}{dt}(dL/\partial \dot{q}_i) - (\partial L/\partial q_i) = \dot{p}_i + (\partial H/\partial q_i) = 0$

i.e. L satisfies (2) assuming (1).

SOLVED EXAMPLES

Example 1:

A uniform square lamina, of mass M and side 2a, is moving freely about a diagonal with uniform angular velocity where one of the corners not in that diagonal becomes fixed, show that the new angular velocity is $\frac{1}{7}\omega$ *and that the impulse of the force on the fixed point is* $\frac{\sqrt{2}}{7} Ma\omega$.

Solution:

Initially the square ABCD is revolving about the diagonal AC with angular velocity ω; this diagonal AC is set free and the corner B is suddenly fixed. The square will begin to turn about the line BE through B parallel to AC. Let ω′ be angular velocity about BE. The impulse is at B, thus the moment of the impulse about BE zero. Therefore the moment of momentum about BE remains the same before and after the fixture,

i.e. $Mk^2\omega' + M(BG)^2\omega' = Mk^2\omega$ ...(1)

But k^2 about AC $= \frac{a^2}{3}\cos^2\frac{\pi}{4} + \frac{a^2}{4}\sin^2\frac{\pi}{4} = \frac{a^2}{3}$

and $BG = a\sqrt{2} \therefore M\frac{a^2}{3}\omega' + M2a^2\omega' = M\frac{a^2}{3}\omega$ *i.e.* $\omega' = \frac{\omega}{7}$.

After the fixture, the velocity of G, the centre of gravity of the square is $a\sqrt{2}\omega'$ at right angles to BG.

$$= M\left(a\sqrt{2}\omega' - O\right) = Ma\sqrt{2}\omega' = Ma\sqrt{2}.\frac{1}{7}\omega = \frac{\sqrt{2}}{7}Ma\omega.$$

Example 2:

A cube is turning with angular velocity ω about a diagonal, when suddenly the diagonal is let go and of the edges which does not meet this diagonal is fixed; show that the resulting angular velocity about this edge is $\left(\frac{\omega\sqrt{3}}{12}\right)$

Solution:

Let 2a be the side of the cube. Take OA, OB, OC as the axes of x,y,z, respectively. Then the direction cosines of the diagonal OP are $\frac{1}{\sqrt{3}}, \frac{1}{\sqrt{3}}, \frac{1}{\sqrt{3}}$. Let the cube revolve about the diagonal OP with angular velocity ω. Obviously the resolved parts of this angular velocity about OA, OB, OC will be $\frac{\omega}{\sqrt{3}}, \frac{\omega}{\sqrt{3}}, \frac{\omega}{\sqrt{3}}$ respectively.

Let the diagonal OP be set free and edge LA be suddenly fixed. Let ω' be the angular velocity about the edge. The distance of the edge from the centre is

Since the moment of momentum remains the same before and after the fixture, therefore we have $M\left(\frac{2a^2}{3}\omega' + 2a^2\omega'\right) = m.o + m\frac{2a^2}{3}\frac{\omega}{\sqrt{3}}$

[$\because$ For the cube about any axis through its centre, $k^2 = \frac{2a^2}{3}$]

$$\text{or } \frac{8a^2}{3}\omega' = \frac{2a^2}{3\sqrt{3}}\omega \text{ i.e. } \omega' = \frac{\omega}{4\sqrt{3}} = \frac{\sqrt{3}\omega}{12}.$$

Example 3:

A uniform cube is spinning freely with angular velocity w about a diagonal through a corner O, when suddenly the diagonal through O′ of one of the faces through O becomes fixed, show that the new angular velocity is $\frac{2\sqrt{3}}{5\sqrt{3}}\omega$.

Solution:

Let the side of cube be 2a, so the length of its diagonal

$$= \sqrt{[(4a^2 + 4a^2 + 4a^2)]} = 2a\sqrt{3}$$

and the length of the diagonal of a face $= \sqrt{[(4a^2 + 4a^2)]} = 2a\sqrt{2}.$

∴ inclination of the diagonal of the cube to the diagonal of the face

$$= \cos^{-1}\frac{2a\sqrt{2}}{2a\sqrt{3}} = \cos^{-1}\sqrt{\left(\frac{2}{3}\right)}$$

Initially the cube was revolving about its diagonal with angular velocity ω, resolved part of this angular velocity along diagonal of the face is $\omega\sqrt{\left(\frac{2}{3}\right)}$.

The distance of the centre of gravity of the cube from the diagonal of the face equal to a. Since the diagonal of the face is suddenly fixed the impulse is at this diagonal, so moment of the impulse about this diagonal is zero, therefore the moment of momentum about this diagonal of the face is the same before and after the fixture.

$$\text{i.e } m\left(\frac{2a^2}{3}\omega' + a^2\omega'\right) = m\left[\left(\frac{2a^2}{3}\omega\right)\sqrt{\left(\frac{2}{3}\right)}\right]$$

$$5\omega' = 2\sqrt{\left(\frac{2}{3}\right)}\omega \text{ or } 5\omega' = \left(\frac{2\sqrt{2}}{5\sqrt{3}}\right)\omega.$$

Example 4(a):

A heave ring of radius a is moving in its own plane which is vertical. At a certain instant when its velocity is V horizontally from left to right and the angular velocity is (V/2a) clockwise, the highest point of the ring is suddenly fixed. Prove that the ring will describe a complete revolution about the point of fixing, if $V^2 > 32$ ag

Solution:

Consider a ring of radius am moving in its own plane with velocity V. Let its angular velocity be ω, τηεν $\omega = \frac{V}{2a}$ (given).

Just after fixing at the highest point of A, let the angular velocity be ω′. Τηε velocity of the centre will be = aω′.

Obviously the impulse is at A, therefore the moment of the impulse about the axis through A (which is horizontal) is zero thus moment of momentum about A will remain unaltered just before and after the fixture at the point

A *i.e.* $mk^2\omega' + ma\omega'.a = mk^2\omega - m.\,a = mk^2\omega - mVa$

or $mk^2\omega' + ma\omega'.a = mk^2\omega - mVa$

or $(a^2\ a^2)\omega' = a^2\left(\frac{V}{2a}\right) - Va$ i.e. $\omega' = -\left(\frac{V}{4a}\right)$

When the ring is turning about A, let the radius through A make an angle θ with the vertical. Also the energy equation gives

$$\frac{1}{2}m\left\{a^2\dot{\theta}^2 + \dot{\theta}^2\right\} - \frac{1}{2}m\left(a^2 + \omega'^2 + \right) = mg\left(a - a\cos\theta\right)$$

or $\dot{\theta}^2 - \omega'^2 = -\left(\frac{g}{a}\right)(a - \cos\theta)$

or $\dot{\theta}^2 - \omega'^2 - \left(\frac{g}{a}\right)\left(1 - \cos\theta\right) = \left(\frac{V^2}{16a^2}\right) - \frac{g}{a}\left(1 - \cos\theta\right)$

Now the ring will make a complete revolution if θ is greater than or equal to zero when $\theta = \pi$.

When $\theta = \pi$, $\qquad \theta = \pi.\ \dot{\theta}^2 = \left(\frac{V^2}{16a^2}\right) - \frac{g}{a}\left(1 - \cos\pi\right) = \frac{2g}{a}$

it is positive only when $\left(\frac{V^2}{16a^2}\right) > \left(\frac{2g}{a}\right)$ i.e. $V^2 > 32ag$.

Example 4(b):

Three equal uniform rods placed in a straight line are freely jointed and move with a velocity v perpendicular to their lengths. If the middle point of the middle rod be suddenly fixed, show that the ends of the other two rods will meet in time (4 π a)/(9v), *where a is the length of each rod.*

Solution:

Let AB, BC, CD be the three rods each of length a and mass m. Freely jointed at B and C, and placed in a straight line. Let these rods move with velocity v in a direction perpendicular to their lengths. Let G_1, the middle point of the BC be suddenly fixed.

Consider the motion of the rod AB, the impulsive action acts at B with the result the moment of the impulsive force at B vanishes. Hence the angular momentum of AB about the end B remains unaltered before and after the fixture.

$$\therefore mv\left(\frac{a}{2}\right) = m\left(a/2\right)^2\omega + (m/3)(a/2)^2\omega = \left(\frac{ma^2\omega}{3}\right)$$

(where ω is the angular velocity of AB just after the fixture)

i.e. $\omega = \left(\frac{3v}{2a}\right)$

Similarly, considering the motion of the rod CD, we can prove that its angular velocity after the fixture = $\left(\frac{3v}{2a}\right)$

The end A and D of the rods will meet in E and in that position the triangle CBE will be an equilateral triangle so that $\angle ABE = \angle DCE = 120°$. Now $(d\theta/dt) = \omega$.*i.e.* $\theta = \omega t$.

$$\therefore \text{ Required time } = \frac{\theta}{\omega} = \frac{120°}{(3v/2a} = \frac{2\pi}{3}\cdot\frac{2a}{3v} = \frac{4\pi a}{9v}.$$

Example 4(c):

If an actant of ellipsoid bounded by three principal planes be rotating about the axis of "a" with angular velocity ω and if this axis suddenly becomes free when the axis of "b" becomes suddenly fixed at the same instant, show that the new angular velocity is $\left(\frac{2abw}{\pi(a^2 + c^2}\right)$.

Solution:

Let the axis of "a" be taken as the axis of x. Let ω_x, ω_y, ω_z be the angular velocities about the axes just before the fixture, then we have $\omega_x = \omega$ given $\omega_y = 0$, $\omega_z = 0$.

Let ω'_x, ω'_y, ω'_z, be the angular velocities about the same axis after the fixing of the axis b.*i.e.* y – axis.

$\therefore\ \omega'_x = 0,\ \omega'_z = 0$ and ω'_y, is to be determined.

Now as y – axis is suddenly fixed, the impulse is at y – axis and moment of the impulse about y – axis vanishes, therefore the moment of momentum about y – axis remains unaltered before and after the fixture.

i.e. $B\ (\omega'_y - \omega_y) - D\,(\omega'_z - \omega_z) - (\omega'_x - \omega_x) = 0$

where B, D, F have their usual meaning in moments of inertia.

Now substituting the values of ω_x, ω_y, ω_z, and $\omega'_x\ \omega'_z$, we get $B\ \omega'_y = -F\ \omega$ But $B = M\left[\frac{c^2 + a^2}{5}\right]$, therefore, we have

$$M\frac{c^2 + a^2}{5}\omega' y = -\omega F = -\omega \iiint p.xy dx dy dz$$

the integration being taken throughout the region bounded by

$$x > 0,\ y > 0, z > 0 \text{ and } \frac{x^2}{a^2} + \frac{y^2}{b^2} + \frac{z^2}{c^2} \leq 1,$$

and M is the mass of the octant. Let $dx = \frac{1}{2}aX^{-\frac{1}{2}}dX$

Similarly, $dy = \frac{1}{2}bY^{-1/2}$ $dZ = \frac{1}{2}cZ^{-1/2}dZ$.

making these substituting in (1), we get

$$M\frac{c^2 + a^2}{5}\omega' y$$

$$= -\omega p \iiint aX^{1/2}bY^{1/2}\frac{1}{2}aX^{-1/2}\frac{1}{2}bY^{-1/2}\frac{1}{2}cZ^{-1/2} dXdYdZ$$

$$= -\frac{a^2b^2c}{8}\omega\rho\iiint X^{\circ}Y^{\circ}Z^{-1/2}dXdYdZ \text{ where } X + Y + Z \leq 1$$

$$= -\frac{a^2b^2c}{8}\omega\rho\iiint X^{1-1}Y^{1-1}Z^{(1/2)^{-1}} dXdYdZ$$

$$= -\frac{a^2b^2c}{8}\frac{1.1.\Gamma(1)\Gamma\left(\frac{1}{2}\right)}{\frac{5}{2}\cdot\frac{3}{2}\cdot\frac{1}{2}\cdot\Gamma\left(\frac{1}{2}\right)}\omega\rho = \frac{6M}{\pi abc}\cdot$$

$$M\frac{(c^2 + a^2)}{5}\omega' y = \frac{a^2b^2c}{15}\cdot\omega\frac{6M}{\pi abc} \text{ i.e. } \omega' y = \frac{2abc\omega}{\pi(c^2 + a^2)}$$

Example 4(d):

Two like rods AB and BC, each of length 2a are freely jointed at B; AB can turn round the end A, and C can move freely on a vertical straight line through A and they are then released. Initially the rods are held in a horizontal line, C being in coincidence with A and they are then released. Show that when the rods are inclined at an angle θ to the horizontal, the angular velocity of either is

$$\sqrt{\left(\frac{3g}{a}\cdot\frac{\sin\theta}{1 + 3\cos^2\theta}\right)}$$

Solution:

Take the fixed point A as origin, the lines AX and AY as the axes. Let G_1 and G_2 be the centres of gravity of the rods AB and BC respectively. Then Co-ordinates of G_1 are (a cos θ, a sin θ). Hence the energy equation gives

$$\frac{1}{2}m\left[\frac{a^2}{3}\dot{\theta}^2 + (-a\sin\theta\dot{\theta})^2 + (a\cos\theta\dot{\theta})^2\right]$$

$$\frac{1}{2}m\left[\frac{a^2}{3}\dot{\theta}^2+(-a\sin\theta\dot{\theta})^2\right]$$

$= \text{mg a sing } \theta + \text{mg. 3 a } \sin\theta$

Or $\frac{1}{2}m\left[\frac{4}{3}a^2\dot{\theta}^2\right]+\frac{1}{2}m\left[\frac{1}{3}a^2\dot{\theta}^2+a^2\sin^2\dot{\theta}\theta^2+9a^2\cos^2\theta\dot{\theta}^2\right]$

$= 4g\, a\sin\theta$

Or $\left[\frac{4}{3}a^2\dot{\theta}^2+a^2\dot{\theta}^2+8a^2\cos^2\theta\dot{\theta}\right]=8g\, a\sin\theta$

Or $\frac{8}{3}(a^2\dot{\theta}^2+a^2\cos^2\theta\dot{\theta}^2=8ga\sin\theta$

Or $(a^2\dot{\theta}^2+3a^2\cos^2\theta\dot{\theta}^2)=3ga\sin\theta$

Or $a\dot{\theta}^2(1+3\cos^2\theta)=3g\sin\theta$ or $\dot{\theta}=\sqrt{\left(\frac{3g}{a}\frac{\sin\theta}{1+3\cos^2\theta}\right)}$.

Example 5:

A uniform straight rod of length 2a. has two small rings at its ends which can respectively slide on thin smooth horizontal and vertical wires OX and OY. The rods starts at an angle α, to the horizontal with angular velocity [{3g (1 – sin α)}/2a]$^{1/2}$ *and downwards. Show that it will strike the horizontal wire at the end of time*

$$2\sqrt{\left(\frac{a}{3g}\right)}\log\left[\left\{\cot\left(\frac{\pi}{8}-\frac{\alpha}{4}\tan\right)\right\}\frac{\pi}{8}\right].$$

Solution:

At any time t, let the rod AB be inclined at an angle θ to the horizontal al. If G be the C.G. of the rod AB, then co-ordinates of G are given by $x = a\cos\theta$, $y = a\sin\theta$, so that

$$\dot{x}=-a\sin\theta\dot{\theta},\ \dot{y}=a\cos\theta\dot{\theta},$$

$\therefore$ velocity of $G=\sqrt{\{(\dot{x}^2+\dot{y}^2)\}}$

$$=\sqrt{[(-a\sin\theta\dot{\theta})^2+(a\cos\theta\ddot{\theta}^2]}=\dot{\theta}$$

$\therefore$ kinetic energy of the rod at time t

$$=\frac{1}{2}m\left[\frac{a^2}{3}\dot{\theta}^2+a^2\dot{\theta}^2\right]=\frac{2}{3}ma^2\dot{\theta}^2 \quad ...(1)$$

Initial angular velocity is

$$=\left[\frac{3a}{2a}(1-\sin\alpha)\right]^{1/2} \quad \text{(given)}$$

$\therefore$ Initial kinetic energy $=\frac{2}{3}ma^2\cdot\frac{3g}{2a}(a-\sin\alpha)$

[from (1)] = mga (1 – sin α).

Hence energy equation gives

$$\frac{2}{3}ma^2\dot{\theta}^2 = mga(1-\sin\alpha) = mg(a\sin\alpha) - mg(a\sin\theta)$$

i.e. $\frac{2}{3}a^2\dot{\theta}^2 - ga(1-\sin\alpha) = ga\,(\sin\alpha - \sin\theta)$

i.e. $\frac{2}{3}a^2\dot{\theta}^2 = ga\,(1-\sin\theta)$

or $\dot{\theta} = \frac{d\theta}{dt} = -\sqrt{\left(\frac{3g}{2a}\right)}\sqrt{(1-\sin\theta)}$

(negative sign is taken with radical sign, since θ decreases)

or $dt = -\left(\frac{2a}{3g}\right)^{1/2}\frac{1}{\sqrt{1-\sin\theta}}d\theta.$

Hence the required time

$$= -\left(\frac{2a}{3g}\right)^{1/2}\int_{\alpha}^{0}\frac{1}{\sqrt{1-\sin\theta}}d\theta = \left(\frac{2a}{3g}\right)^{1/2\alpha}\int_{0}^{\alpha}\frac{1}{\sqrt{(1-\sin\theta)}}d\theta$$

$$= \left(\frac{2a}{3g}\right)^{1/2}\int_{a}^{\alpha}\frac{d\theta}{[\{\cos^2(\theta/2)+\sin^2(\theta/2)-2\sin(\theta/2)\cos(\theta/2)]^{1/2}}$$

$$= \left(\frac{2a}{3g}\right)^{1/2}\int_{0}^{\alpha}\frac{d\theta}{\{\cos(\theta/2)-\sin(\theta/2)\}}$$

$$= \left(\frac{2a}{3g}\right)^{1/2}\int_{0}^{\alpha}\frac{d\theta}{\sqrt{2}\left[\frac{1}{\sqrt{2}}\cos(\theta/2)-\frac{1}{\sqrt{2}}\sin(\theta/2)\right]}$$

$$= \left(\frac{2a}{3g}\right)^{1/2}\frac{1}{\sqrt{2}}\int_{0}^{\alpha}\frac{d\theta}{\sin\left(\frac{\pi}{4}-\frac{\theta}{2}\right)} = \left(\frac{a}{3g}\right)^{1/2}\int_{9}^{\alpha}\operatorname{cosec}\left(\frac{\pi}{4}-\frac{\theta}{2}\right)d\theta$$

$$= -2\sqrt{\left(\frac{a}{3g}\right)}\left[\log\tan\left(\frac{\pi}{8}-\frac{\theta}{4}\right)\right]_0^{\alpha}$$

$$= 2\sqrt{\left(\frac{a}{3g}\right)}\left[\log\tan(\pi/8) - \log\tan\left(\frac{\pi}{8}-\frac{\alpha}{4}\right)\right]$$

$$= 2\sqrt{\left(\frac{a}{3g}\right)}\frac{\tan(\pi/8)}{\tan[(\pi/8)-(\alpha/4)]}$$

$$= 2\sqrt{\left(\frac{a}{3g}\right)}\log\left[\tan\frac{\pi}{8}\cot\left(\frac{\pi}{8}-\frac{\alpha}{4}\right)\right].$$

Example 6:

A sphere rolls down a rough included plane; if x be the distance of the point of contact of the sphere from a fixed point on the plane, find the acceleration.

Solution:

We have $T = \frac{1}{2}m(\dot{x}^2 + k^2\dot{\theta}^2) = \frac{1}{2}m\left(\dot{x}^2 + \frac{2}{5}a^2\dot{\theta}^2\right)\left(\because k^2 = \frac{2}{5}a^2\right)$

$$= \frac{1}{2}m\left(\dot{x}^2 \frac{2}{5}\dot{x}^2\right) = \frac{7}{10}m\dot{x}^2; \qquad ...(1)$$

$$V = -\, mgx\sin\alpha \qquad ...(2)$$

$$\therefore\; L = T - V = \frac{7}{10}m\dot{x}^2 + mgx\sin\alpha \qquad ...(3)$$

Now $p_x = (\partial L/\partial\dot{x}) = \frac{7}{10}m\dot{x} \Rightarrow \dot{x} = (5p_x/7m)$

Thus $H = -L + p_x\dot{x} = -\frac{7}{10}m\dot{x} - mgx\sin\alpha + p_x.(5p_x/7m)$

$$= \frac{5}{14}(p_x^2/m) - mgx\sin\alpha$$

$\therefore$ One of Hamilton's equations gives

$$p_x = -(\partial H/\partial x) = mg\sin\alpha \Rightarrow \frac{7}{5}m\ddot{x} = mg\sin\alpha \Rightarrow \ddot{x} = \frac{5}{7}g\sin\alpha$$

Example 7:

A particle of mass ma moves in a force field of potential V.

(a) Write the Hamiltonian and

(b) Hamilton's equations in artesian co-ordinates.

Solution:

(a) we have

$$T = \frac{1}{2}m(\dot{x}^2 + \dot{y}^2 + \dot{z}^2)$$

$$\Rightarrow L = \frac{1}{2}m(\dot{x}^2 + \dot{y}^2 + \dot{z}^2) - V(x,y,z) \quad ...(1)$$

$$\Rightarrow \dot{x} = (p_x / m), \ \dot{y} = (p_y / m), \dot{z} = (p_z / m).$$

Thus $H = \Sigma p_x \dot{q}_z - \dot{p}_x x + \dot{p}_y y + \dot{p}_z z - \frac{1}{2}m(\dot{x}^2 + \dot{y}^2 + \dot{z}^2) + V(x,y,z)$

$$= p_x (p_x/m) + p_y (p_y/m) + (p_z (p_z/m)$$

$$- \frac{1}{2}m[p_x^2 / m^2) + (p_y^2 / m^2) + (p_z^2 / m^2)] + V(x,y,z)$$

$$= (p_x^2 / 2m) + (p_y^2 / 2m) + (p_z^2 / 2m) + V(x,y,z)$$

= total energy of the system.

(b) Hamilton's equations are:

$\dot{p}_x = -(\partial H / \partial x); \dot{p}_y = -(\partial H / \partial y); \dot{p}_z = -(\partial H / \partial z)$ and

$\dot{x} = (\dot{p}_x = -(\partial H / \partial x); \dot{p}_y \ \dot{x} = (\partial H / \partial p_x); \dot{y} = (\partial H / \partial p_y); \dot{z} = (\partial H / \partial p_z).$

$$\Rightarrow \dot{p}_x = -(\partial V / \partial x), \dot{p}_y = -(\partial V / \partial y), \dot{p}_z = -(\partial V / \partial z)$$

$$\dot{x} = (p_x / m), \dot{y} = (p_y / m), \dot{z} = (p_z / m).$$

Example 8(a):

If the Hamiltonian H is independent of time explicity, prove that it is

(a) a constant, and

(b) equal to the total energy of the system.

Solution:

$$\text{(a)} \ (dH/dt) = \sum_{i=1}^{n} (\partial H / \partial p_i)\dot{p}_i + \sum_{i=1}^{n} (\partial H / \partial q_i)\dot{q}_i \quad ...(1)$$

$$\sum_{i=1}^{n} \dot{q}_i \dot{p}_i + (-\dot{p}_i)\dot{q}_i = 0 \qquad [\because (\partial H / \partial p_i) = \dot{q}_i, (\partial H / \partial q_i) = -\dot{p}_i]$$

$\Rightarrow$ H = constant = E say.

(b) By Euler's theorem on homogeneous functions, we have

$$\sum_{i=1}^{n} \dot{q}_i (\partial T / \partial \dot{q}_i) = 2T. \quad ...(2)$$

Put

$$p_i = (\partial L/\partial \dot{q}_i) = \{\partial(T-V)/\partial \dot{q}_i\} = (\partial T/\partial \dot{q}_i) - (\partial V/\partial \dot{q}_i) = (\partial T/\partial \dot{q}_i)$$

$$\{\because (\partial V/\partial \dot{q}_i) = 0 \text{ as V is independent of } \dot{q}_i\}$$

$$\therefore (2) \Rightarrow \sum_{i=1}^{n} \dot{q}_i p_i = 2T$$

Thus $H = \sum_{i=1}^{n} p_i \dot{q}_i - L = 2T - L = 2T - (T-V) = T + V = E.$

Example 8(b):

A uniform rod of length 2a, is placed with one end in contact with a smooth horizontal table and is then allowed to fall, if α be its initial inclination to the vertical, show that its angular velocity when it is inclined at an angle θ is

$$\left[\frac{6g}{a}\cdot\frac{\cos\alpha - \cos\theta}{1+3\sin^2\theta}\right]^{1/2}$$ *find also the reaction of the table.*

Solution:

Let AB be the rod such that AB = 2a. Let G be the C.G. of the rod AB. Since there is no horizontal force on the rod AB, therefore its centre of gravity G moves down a vertical line through G Let this vertical line through G cut the horizontal table at the point O, then O is fixed point.

Taking O as the origin and the line OX and OY as the axes, the co-ordinates of G are (0, a cos θ).

i.e. $x_G = 0, y_G = a\cos\theta$

$$\Rightarrow \dot{x}_G = 0. \dot{y}_G = -a\sin\theta\dot{\theta} \Rightarrow v_{G^2} = \dot{x}_{G^2} + \dot{y}_{G^2} = a^2\sin^2\theta\dot{\theta}^2$$

Hence the energy equation gives

$$\frac{1}{2}m\left[v_G^2 + k^2\dot{\theta}^2\right] = \frac{1}{2}m\left[\frac{a^2}{3}\dot{\theta}^2 + a^2\sin^2\theta\dot{\theta}^2\right] = m\,ga\cos\alpha - mga\cos\theta$$

i.e., $$\frac{a^2\dot{\theta}^2}{6}[1+3\sin^2\theta] = ga(\cos\alpha - \cos\theta)$$

i.e. $$\dot{\theta}^2 = \frac{6g}{a}\,\frac{\cos\alpha - \cos\theta}{1+3\sin^2\theta}$$

i.e. $$\dot{\theta} = \sqrt{\left(\frac{6g}{a}\right)}\sqrt{\left(\frac{\cos\alpha - \cos\theta}{1+3\sin^2\theta}\right)} \quad ...(1)$$

To find the reaction R, take moments about G.

$$R\, a \sin \dot{\theta} = m\frac{a^2}{3}\theta\ddot{\theta} \text{ i.e., } R = \frac{m}{3} a \operatorname{cosec} \theta\ddot{\theta} \qquad ...(2)$$

To find the value of $\ddot{\theta}$, we differentiate $\dot{\theta}^2$ found above

$$2\,\dot{\theta}\ddot{\theta} = \frac{6g}{a}\frac{[(1+3\sin^2\theta)\sin\theta - (\cos\alpha - \cos\theta).6\sin\theta\cos\theta]\dot{\theta}}{(1+3\sin^2\theta)^2}$$

$$\Rightarrow \ddot{\theta} = \frac{3g}{a}\frac{\sin\theta[1+3\sin^2\theta - 6\cos\theta(\cos\alpha - \cos\theta)]}{(1+3\sin^2\theta)^2}$$

$$= \frac{3g\sin\theta}{a(1+3\sin^2\theta)^2}[1+3\sin^2\theta + \cos^2\theta - 6\cos\theta\cos\alpha]$$

$$= \frac{3g\sin\theta}{a(1+3\sin\theta)^2}[1+3(\sin^2\theta + \cos^2\theta) + 3\cos^2\theta - 2\cos\theta\cos\alpha]$$

$$= \frac{3g\sin\theta}{a(1+3\sin^2\theta)^2}[4+3(\cos^2\theta - 2\cos\theta\cos\alpha) - 3\cos^2\alpha]$$

$$= \frac{3g\sin\theta}{a(1+3\sin^2\theta)^2}[4-3\cos + 3(\cos\theta - \cos\alpha)^2].$$

Substituting this value of $\ddot{\theta}$ in (2), we get

$$R = \frac{ma}{3\sin\dot{\theta}} = \frac{3g\sin\theta}{a(1+3\sin^2\theta)^2}[4-3\cos^2\alpha) + 3(\cos\theta - \cos\alpha)^2]$$

$$= \frac{ma}{(1+3\sin^2\theta)^2}[4-3\cos^2\alpha + 3(\cos\theta - \cos\alpha)^2]$$

we observe that value of the reaction is positive for all values of θ, hence the lower end of the rod will never leave the horizontal table.

Example 8(c):

A straight uniform rod, of mass m, is placed at right angles to smooth plane of inclination α with one end in contact with it; the rod is then released. Show that when its inclination to the plane is ϕ, the reaction of the plane will be

$$mg\frac{3(1-\sin\phi)^2+1}{(3\cos^2\phi+1)^2}\cos\alpha.$$

Solution:

Let at any time t, the rod AB make an angle ϕ with the inclined plane. Then the equation of motion of the C.G. of the rod

AB $m\frac{d^2}{dt^2}(a\sin\phi) = R - mg\cos\alpha$ ...(i)

$m\frac{a^2}{3}\ddot{\phi} = -R\,a\cos\phi$ ...(ii)

Substituting the value of R from (ii) in (i),

we get $m\frac{d^2}{dt^2}(a\sin\phi) \;= -mg\cos\alpha - \frac{ma}{3}\sec\phi.\ddot{\phi}$

or $\frac{d}{dt}(a\cos\phi.\dot{\phi}) \;= -g\cos\alpha - \frac{1}{3}a\sec\phi.\ddot{\phi}$

or $\left(a\cos\phi + \frac{a}{3}\sec\phi\right)\ddot{\phi} - a\sin\phi\dot{\phi}^2 = -g\cos\alpha$

or $a\left(\frac{1}{3} + \cos^2\phi\right)\ddot{\phi} - a\sin\phi\cos\phi\dot{\phi}^2 = -g\cos\alpha\cos\phi$

Multiplying by 2 $\dot{\phi}$ throughout, we get

$2a\left(\frac{1}{3} + \cos^2\phi\right)\dot{\phi}\ddot{\phi} - 2a\sin\phi.\dot{\phi}^3 = -2g\cos\alpha\cos\phi\dot{\phi}$

or $\frac{d}{dt}\left[a\left(\frac{1}{3} + \cos^2\phi\right)\dot{\phi}^2\right] = -2g\cos\alpha\cos\phi\dot{\phi}.$

Integrating it we get a $a\left(\frac{1}{3} + \cos^2\phi\right).\dot{\phi}^2 = -2g\cos\alpha\sin\phi + C.$

Initially $\phi = (\pi/2)$ and $\dot{\phi} = 0 \;\therefore C = 2g\cos\alpha.$

Hence $a\left(\frac{1}{2} + \cos^2\phi\right).\dot{\phi}^2 = 2g\cos\alpha(1 - \sin\phi)$

or $\dot{\phi}^2 = \frac{2g\cos\alpha(1-\sin\phi)}{a\left(\frac{1}{3} + \cos^2\alpha\right)}$

Substituting the value of $\dot{\phi}^2$ in (iii), we get

$$a\left(\frac{1}{3} + \cos^2\phi\right)\ddot{\phi} - a\sin\phi\cos\phi.\frac{2g\cos\alpha(1-\sin\phi)}{a\left(\frac{1}{3} + \cos^2\phi\right)} = -g\cos\alpha\cos\phi$$

or $a\left(\frac{1}{3} + \cos^2\phi\right)^2\ddot{\phi} - 2g\cos\alpha\sin\phi\cos\phi(1 - \sin\phi)$

$= -g\cos\alpha\cos\phi\left(\frac{1}{3} + \cos^2\phi\right)$

or $a(1+3\cos^2\phi)^2\ddot{\phi} - 3g\cos\alpha\cos\phi[6\sin\phi - 6\sin^2\phi - 1 - 3\cos^2\phi)]$
$= 3g\cos\alpha\cos\phi[6\sin\phi - 3\sin^2\phi - 3(\sin^2\phi + \cos^2\phi) - 1]$
$= 3g\cos\alpha\cos\phi[-3(\sin^2\phi - 2\sin\phi + 1) - 1]$
$= 3g\cos\alpha\cos\phi[-3(1-\sin\phi)^2 - 1]$
$= -3g\cos\alpha\cos\phi[3(1-\sin\phi)^2 + 1]$

or $\ddot{\phi} = -\dfrac{3g\cos\alpha\cos\phi[3(1-\sin\phi)^2 + 1]}{a(11+3\cos^2\phi)^2}$

putting this value of $\ddot{\phi}$ in (ii), we get R = m g cos `

$\alpha \dfrac{3(1-\sin\phi)^2 + 1}{(1+3\cos^2\phi)^2}$.

Example 9:

A hemisphere of mass M and radius a is placed with its plane base on a smooth table, and a heavy rod of mass m is constrained to move in a vertical line with one end p on the curved surface of the hemisphere. If at any time t, the radius to p makes an angle θ with the vertical, show that

$$a\dot{\theta}^2[M\cos^2\theta + m\sin^2\theta] = 2mg(\cos\alpha - \cos\theta).$$

Solution:

Let pQ (= 2a) be the rod moving in a vertical line, with its one end p in contact with the curved surface of the hemisphere. Let G_1 be the C.G. of the hemisphere and G_2 the C.G. of the rod pQ. Let the vertical line along which the rod move cut the table at O, then O is a fixed point. Take O as the origin and horizontal line OA as - axis and vertical line through O as y-axis. Then the co - ordinate of G_2 are

(0, a + a cosθ) and co-ordinates of G_1 are $\left(a\sin\theta, \dfrac{3a}{8}\right)$

Hence the energy equation gives

$\frac{1}{2}k\left(-a\sin\theta\dot{\theta}\right)^2 + \frac{1}{2}M\left(a\cos\theta\dot{\theta}\right)^2 = mg\left(a\cos\alpha - a\cos\theta\right)$

or $a\dot{\theta}(M\cos^2\theta + m\sin^2\theta) = 2mg\left(\cos\alpha - \cos\theta\right)$.

Example 10:

A uniform rod, of length 2a is freely jointed at one end to a small ring, whose mass is equal to that of the rod. The ring is free to slide on smooth wire and initially is at rest and the rod is vertical and below the ring rotating with angular velocity $\sqrt{}(3g/a)$ in a vertical plane passing through the wire. When the rod is inclined at an angle θ to the vertical, show that its angular velocity is $\sqrt{\left(\frac{3g}{a}\frac{1+\cos\theta}{8-3\cos^2\theta}\right)}$ and find the velocity of the ring.

Solution:

Let A′ B′ be the initial position of the rod.

At time t, let q be the angle through which the rod A B has turned and the ring has moves a distance x from its initial position A ′ along the smooth horizontal wire.

Now taking A′ as the origin and the lines, A′A,A′B′ as axes of x and y repetitively , the co-ordinates of G are (X + a sinθ, a cos θ).

∴ Its velocity components are $\left(\dot{x}+a\cos\theta\dot{\theta}\right)$ and $\left(-a\sin\theta\dot{\theta}\right)$ and velocity of the ring is $\dot{x}$. Thus the horizontal momentum at time t is

$$=m\dot{x}+m\left(\dot{x}+a\cos\theta\dot{\theta}\right)$$

Initial horizontal momentum $=m\left[0+a\left(\frac{3g}{a}\right)^{1/2}\right]=m\sqrt{(3ag)}$

$$\{\text{since initially } \theta=0, \theta=(\frac{3g}{a}\}^{1/2} \text{ and } x=0\}$$

Since there is no horizontal force on the system, therefore the horizontal momentum remains unaltered throughout the motion.

$$\therefore m\ddot{x}+m(\dot{x}+a\cos\theta\dot{\theta})=m\sqrt{(3ag)}$$

or $2\dot{x}+a\cos\theta\dot{\theta})=\sqrt{(3ag)}$ i.e. $\dot{x}=\frac{1}{2}[\sqrt{(3ag)}-a\cos\theta\dot{\theta}]$ ...(1)

Kinetic energy at time t

$$=\frac{1}{2}m[\dot{x}^2+\frac{1}{2}m[\dot{x}+a\cos\theta.\dot{\theta})^2+(-a\sin\theta.\dot{\theta})^2+\frac{1}{3}a^2\dot{\theta}^2]$$

$$=\frac{1}{2}m[\dot{x}^2+\dot{x}^2+a^2\dot{\theta}^2+2a\dot{x}\dot{\theta}\cos\theta+\frac{1}{3}a^2\dot{\theta}^2]$$

$$=\frac{1}{2}m[2\dot{x}^2+\frac{4}{3}a^2\dot{\theta}^2+2a\dot{x}\dot{\theta}\cos\theta]$$

Initial kinetic energy

$$\frac{1}{2}m\left\{0+\frac{4}{3}a^2\cdot\frac{3g}{a}+0\right\}\left(\text{since initially } \dot{x}=0, \theta=0 \text{ and } \dot{\theta}^2=\frac{3g}{a}\right)=2mga$$

Thus the energy equation gives

$$\frac{1}{2}m[2\dot{x}^2+\frac{4}{3}a^2\dot{\theta}^2+2a\dot{x}\dot{\theta}\cos\theta]-2mga=-mg(a-a\cos\theta)$$

$$\text{or } \dot{x}^2+\frac{2}{3}a^2\dot{\theta}^2+a\dot{x}\dot{\theta}\cos\theta-2ga+ga\cos\theta)$$

$$\text{or } \dot{x}^2+\frac{2}{3}a^2\dot{\theta}^2+a\dot{x}\dot{\theta}\cos\theta=ga(1+\cos\theta)$$

Substituting the value of $\dot{x}$ from (1), we get

$$\left\{\frac{\sqrt{(3ag)}-a\cos\theta\dot{\theta}}{2}\right\}^2+\frac{2}{3}a^2\dot{\theta}^2+a\dot{\theta}\cos\theta\times$$

$$\times\left\{\frac{\sqrt{(3ag)}-a\cos\theta\dot{\theta}}{2}\right\}=ga(1+\cos\theta)$$

$$\text{or } \frac{1}{4}[3ag+a^2\cos^2\theta\dot{\theta}^2-1\sqrt{(3ag)}.a\cos\theta\dot{\theta}+\frac{2}{3}a^2\dot{\theta}^2$$

$$+\frac{1}{2}a\dot{\theta}\cos\theta\sqrt{(3ag)}-\frac{1}{2}a^2\cos^2\theta\dot{\theta}^2=ga(1+\cos\theta)$$

$$\text{or } \left(\frac{2}{3}a^2-\frac{1}{4}a^2\cos^2\theta\right)\dot{\theta}^2=\frac{ga}{4}+ga\cos\theta$$

$$\text{or } (8a^2-3a^2\cos^2\theta)\dot{\theta}^2=3ga(1+4\cos\theta) \quad \dot{\theta}=\sqrt{\left\{\frac{3g(1+4\cos\theta)}{a(8-3\cos^2\theta)}\right\}}$$

Example 11:

A uniform rod, of length 2a, hangs in a horizontal position being supported by vertical strings, each of length 1, attached to its ends, the other extremities being attached to fixed points. The rod its given an angular velocity ω about a vertical axis through its centre; find its angular velocity when it has turned through an angle, and show that it will rise through a distance $(a^2\omega^2)/(6g)$. *prove also that the time of a small oscillation about the position of equilibrium is* $2\pi\sqrt{(1/3g)}$.

Solution:

Let A'B' be the initial position of the rod, supported by two vertical strings DA', CB, such that $DA'=CB'=1$. The rod is then turned in such a way that its middle point moves in a vertical through O′. Let AB be the position of the rod when it has turned through an angle θ. Let O′O = x and φ is the angle which either string makes with the vertical.

Let the horizontal plane through AB cut CB′, DA′ in N and M respectively.

Obviously $\angle AMD = \angle BNC = \pi/2$. and $x = OO' = CB' - CN = l - l\cos\phi$ where t is the length of the vertical string

$$\therefore \dot{x} = l\sin\phi\dot{\phi}$$

From the right angle Δ CBN we have $\dot{B}N = l\sin\phi$ Also from isosceles triangle OBN, we have BN = 2a sin (θ/2)

$$\therefore l\sin\phi = 2a\sin(\theta/2)$$

so that $l\cos\ \phi\dot{\phi} = a\dot{c}os(\theta/2).\dot{\theta}$

i.e. $\dot{\phi} = \dfrac{a\cos(\theta/2}{l\cos\phi}\dot{\theta}$

Therefore form (2), we have

$$\dot{x} = l\sin\phi.\frac{a\cos(\theta/2)}{l\cos\phi}\dot{\theta}$$

$$= 2a\sin\frac{\theta}{2}a\cos\frac{\theta}{2}\frac{1}{\sqrt{l^2 - 4a^2\sin^2\theta/2}}\dot{\theta}\sin cel\sin\phi = 2a\sin(\theta/2)$$

$$= \frac{la^2\sin\theta\dot{\theta}}{[l^2 - 4a^2\sin^2(\theta/2)]\cdot}$$ Now energy equation gives

$$\frac{1}{2}m\dot{x}^2 + \frac{1}{2}mK^2\dot{\theta}^2 - \frac{1}{2}k^2\omega^2 = -mgx \qquad ...(3)$$

$$\Rightarrow \frac{1}{2}m\left(\dot{x}^2 + \frac{a^2}{3}\dot{\theta}^2\right) - \frac{1}{2}m\frac{a^2}{3}\omega^2 = -mgx$$

$$\Rightarrow \frac{1}{2}m\left[\frac{l^2a^4\sin^2\theta\dot{\theta}^2}{l^2 - 4a^2\sin^2(\theta/2)]^2} + \frac{a^2\dot{\theta}^2}{3}\right] - \frac{1}{2}m\frac{a^2}{3}\omega^2 = -mgx$$

$$\Rightarrow a^2\dot{\theta}^2\left[\frac{la^2\sin^2\theta}{[l^2 - 4a^2\sin^2(\theta/2)]^2}\right] - \frac{1}{3}a\omega^2 = 2gx \qquad ...(4)$$

This equation gives the angular velocity of the rod in any position. The rod will come to instantaneous rest when $\dot{\theta} = 0$.

Thus putting $\dot{\theta} = 0$. in (4), we get 2gx

$$= \frac{1}{3} a^2 \omega^2 \text{ i.e. } x = \frac{a^2 \omega^2}{6g}$$

For finding the time of small oscillation, we consider the motion of AB in the vertical direction.

$$\therefore 2T \cos \varphi - mg = mx$$

$$= m \frac{d^2}{dt^2}(1 - \text{loc}\phi)) = m \frac{d^2}{dt^2}[l(1 - \cos \phi)]$$

$$= m \frac{d^2}{dt^2}\left[1.2 \sin^2 \frac{\phi}{2}\right] = m \frac{d^2}{dt^2}\left[1.2. \frac{[\sin(\phi/2]^2}{(\phi/2)^2}\left(\frac{\phi}{2}\right)\right]$$

$$= m \frac{d^2}{dt^2}\left[\frac{1}{2 \cdot}\left(\frac{a\theta}{1}\right)^2\right] = m \frac{d^2}{dt^2}\left[\frac{a^2\theta^2}{2l}\right]$$

$$\left[\sin \text{cel} \sin \phi = 2a \sin \frac{\theta}{2} \therefore \frac{l \sin \phi}{\phi} \cdot \phi = 2a \frac{\sin \theta / 2}{(\theta/2)} \cdot (\theta / 2 \text{i.e.} l\phi = a\theta\right]$$

$$\therefore 2T \cos \phi - mg = \frac{ma^2}{2l} \frac{d}{dt}(2\theta\dot{\theta}) = \frac{2ma^2}{2l}[\theta\ddot{\theta} + \dot{\theta}^2] = \frac{ma^2}{l}[\theta\ddot{\theta} + \dot{\theta}^2]\Big]$$

or $2T - mg\ 0$ [since $\dot{\theta}$ $\ddot{\theta}$ and are small so they could be neglected ϕ being small $\dot{\phi} = 1$].

Now taking moments about the vertical axis through O, we have

$$m \frac{a^2}{3} \ddot{\theta} = -2T \sin \phi \text{ [perpendicular from O on BN]}$$

$$= -2T \sin \phi . a \cos \frac{\theta}{2} = -2a \cos \frac{\theta}{2 \cdot} T \sin \phi$$

$$= -2T \cos \frac{\theta}{2 \cdot} \frac{2a}{1} \sin \frac{\theta}{2} = -\frac{2a^2}{1} T \sin \theta$$

Now since $2T = mg$ i.e $T = \frac{1}{2} mg$. putting this value above, we get

$$m \frac{a^2}{3} \ddot{\theta} = -\frac{2a}{1 \cdot} \frac{2a}{1} \sin \frac{\theta}{2} = -\frac{2a^2}{1} T \sin \theta$$

or $\ddot{\theta} = -\frac{3g}{1} \sin \theta$, when θ is small.

$\therefore \ddot{\theta} = -\frac{3g}{1}\theta$, which is simple harmonic motion. Hence time of small oscillation $= 2\pi \sqrt{\left(\frac{1}{3g}\right)}$

Example 12(a):

If the earth, supposed to be a uniform sphere, had in a certain period contracted slightly so that its radius was less by (1/n) th than before, show that the length of the day would have shortend by (48/n) hours.

Solution:

Let a be the radius, m be the mass and ω the angular velocity of the earth. Supposed to be a uniform sphere, about its axis of rotation which is its diameter.

$\therefore$ The length of a day $= l_1 = 2\pi/\omega = 24$ hours.

Now when the earth contracted slightly, let a_1 be its new radius then a_1 be its new radius then $a_1 = a - a/n = a\,(1 - 1/n)$. ...(1)

If ω_1 is the angular velocity of the earth in this position, then shortened length of a day $= l_2\ 2\pi/\omega_1$.

$$\therefore \frac{\text{Decrease in the length of a day}}{\text{Length of a day}} = \frac{l_1 - l_1}{l1} = 1 - \frac{l_2}{l_1} = 1 - \frac{\omega}{\omega_1}$$

i.e. Decrease in the length of a day $= \left(1 - \dfrac{\omega}{\omega_1}\right)$

Length of a day $= \left(1 - \dfrac{\omega}{\omega_1}\right)$. 24 hours. ...(2)

The moment of the external forces (*i.e.* weight o the earth and the reaction on the axis of rotation) about the axis of rotation (*i.e.* diameter) of the earth is zero. Therefore the moment of the momentum about the axis of rotation (*i.e.* diameter) of the earth will remain constant.

$$\therefore mk^2\omega = mk_1{}^2\omega_1$$

$$\text{or} \quad \frac{1}{5}ma^2\omega = \frac{2}{5}ma^2\ \omega_1 \ \text{ or } \ \frac{\omega}{\omega_1} = \left(1 - \frac{1}{n}\right)^2,$$

$$= 1 - \frac{2}{n'}\left(\text{Neglecting } \frac{1}{n^{2'}} \text{ etc. being small quantities}\right)$$

Putting in (2), we have

the decrease in the length of a day $= \left\{1 - \left(1 - \dfrac{2}{n}\right)\right\}24 = \dfrac{48}{n}$ hours.

i.e. the length of the day is shortened by (48/n) hours.

Example 12(b):

A sphere of radius b, rolls without slipping down the cycloid $x = a(\theta + \sin\theta)$; $y = a(1 - \cos\theta)$. It starts from rest with its centre on the horizontal line $y = 2a$. Show that velocity v of its centre when at its lowest point is given by $v^2 = \frac{10}{7}g(2a - b)$.

Solution:

Initially, let the point A′ of the sphere be at the cusp A. Let C be the centre of this sphere, then the radius CA′ is a fixed line in the sphere.

Let at time t the line CA make an angle ϕ with the vertical and tangent at the point of contact P, make an angle ψ with the horizontal line OX.

Let the arc OP = s and arc OA = c.

$\therefore$ are AP = c – s.

But we are given that sphere rolls without slipping.

$\therefore$ arc AP = arc A′ P *i.e.* $c - s = b\ (\phi - \psi)$

[since $\angle A¢CP = \phi - \psi$]

Differentiating it, we get $- s = b\ (\phi - \psi)$ ...(1)

Let the centre C of the sphere move with the velocity v, then

v = velocity of P + velocity of C relative to P.

or $v = \dot{s} - b\dot{\psi} = -b(\dot{\phi} - \dot{\psi}) - b\dot{\psi} = -b\dot{\phi}$ ...(2)

Now energy equation gives $\frac{1}{2}m\left[v^2 + \frac{2b^2\dot{\phi}^2}{5}\right] = mg(2a - b)$

or $\left[v^2 + \frac{2b^2}{5}\cdot\frac{v^2}{b^2}\right] = 2g(2a - b)$ [from (2)]

or $\frac{7}{5}v^2 = 2g(2a - b)$ *i.e.* $v^2 = \frac{10g(2a - b)}{7}$

Example 12(c):

A sphere of radius a, and of radius of gyration k about an axis through its centre, rolls with linear velocity v on a horizontal plane, the direction of motion being perpendicular to a vertical face of a fixed rectangular block are perfectly rough and perfectly elastic. Show that the sphere will surmount the block if $(a^2 - ah + k^2)v^2 > 2gha^2(a^2 + k^2)$.

Solution:

Let O be the centre of the sphere and AB vertical face of the fixed block; AB = h. The sphere strikes the block at A, therefore the impulse is at the point A. Let ω and ω′ be the angular velocities of the sphere just after the impulse. After striking at A, the sphere begins to turn about A so the velocity of the centre O after impulse is at right angles to OA and is equal to aw. Since the impulse is at the point A, therefore the moment of the impulse about A is zero, hence the moment of momentum about A remains unaltered before and after the impulse.

Moment of momentum about A before impulse = $mk^2\omega + mv(a-h)$

[∴ moment of momentum = $mk^2\dot{\theta} + mvp$]

Moment of momentum about A after impulse = $mk^2\omega' + ma\omega'.a$

$\therefore mk^2 + mv(a-h) = mk^2\omega' + ma^2\omega'$ or $kk^2\omega + (a-h) = (k^2+a^2)\omega'$

or $k^2\dfrac{v}{a} + v(a-h) = (k^2+a^2)\omega'$ [since ω = v/a]

or $(k^2+a^2)\,\omega' = \dfrac{v}{a}\left(a^2 - ah + k^2\right)$

or $\omega' = \dfrac{(a^2 - ah + k^2)}{(k^2 + a^2)}$

Now the sphere will surmount the block if the sphere goes on turning about A till its lowest point L rises a height h, the height of the block. This can happen only if the sphere after impact acquires a kinetic energy, which is sufficient to take the sphere to a height h. It implies that the K.E. of the sphere must be more then that would be required for doing work of taking it upto a height h. But K.E. of the sphere after the impact

$$= \frac{1}{2}m\left(k^2\omega'^2 + a^2\omega'^2\right),$$

$$\therefore \frac{1}{2}m(k^2\omega'^2 + a^2\omega'^2) > mgh \text{ or } (k^2+a^2)\omega'^2 > 2gh$$

or $(k^2+a^2)\dfrac{v^2(a^2-ah+k^2)^2}{a^2(k^2+a^2)^2} > 2gh$ [from (1)]

or $(k^2 - ah + k^2)^2 v^2 > 2gha^2(k^2+a^2)$

Example 13:

Write the Hamiltonian and equation of motion for a simple pendulum.

Solution:

We have $T = \dfrac{1}{4}ml^2\dot{\theta}^2$ and $V = mgl(1 - \cos\theta)$,

$$\therefore L = T - V = \frac{1}{2}ml^2\dot{\theta}^2 - mgl(1-\cos\theta) \qquad ...(1)$$

$$\Rightarrow H = \Sigma p_i\dot{q}_i - L = p_\theta\dot{\theta} - L = ml^2\dot{\theta}^2 - \left\{\frac{1}{2}ml^2\theta^2 - mgl(1-\cos\theta)\right\}$$

$$= \frac{1}{2}ml^2\dot{\theta}^2 + mgl\,(1-\cos\theta) = T + V \;=\; \text{total energy.}$$

Now $p_\theta = (\partial L/\partial\dot{\theta}) = ml^2\dot{\theta}^2 = (p_\theta/ml^2)$

$$\therefore H = \frac{1}{2}ml^2(p_\theta/ml^2)^2 + mgl(1-\cos\theta) = (p_\theta^2/2ml^2) + mgl(1-\cos\theta)$$

$$\Rightarrow (\partial H/\partial p_\theta), \dot{p}_\theta = -(\partial H/\partial\theta) \Rightarrow \dot{\theta} = (p_\theta/ml^2) \text{ and } \dot{p}_\theta = -mgl\sin\theta,$$

These represent Hamilton's equations for a simple pendulum.

From above, we have $p_\theta = ml^2\dot{\theta}$, i.e. $p_e = ml^2\ddot{\theta}$

$$\therefore ml^2\ddot{\theta}, = -mgl\sin\theta \Rightarrow \ddot{\theta} + (g/l)\sin\theta = 0$$

This gives the equation of motion of the simple pendulum.

Example 14:

A uniform plane of thickness 2h rests across the top of a fixed circular cylinder of radius a whose axis is horizontal. prove that if it be set in motion, the equation of energy is

$$\frac{1}{2}[k^2 + h^2 + a^2\theta^2]\theta^2 + g[a\theta\sin\theta - (a+h)(1-\cos\theta)] = \text{const.}$$

on the assumption that the motion is of pure rolling. Hence show that if a>h, horizontal position is stable and period of small oscillation is the same as for a simple pendulum of length $\dfrac{k^2+h^2}{a-h}$.

Solution:

The figure is the vertical cross section of the system through the centre of gravity G of the plank. From G drawn GB perpendicular to the edge of the plank, so GB = h. Initially the point B of the plank coincided with the highest point A of the fixed cylinder. Let ω be the initial angular velocity. After time t, the edge of the plank make an angle θ with the horizontal. Since it is given that there is no slipping between the plank and the cylinder, therefore arc AP = BP *i.e.* BP = $a\theta$. With O as origin OX and OY as axes, the co-ordinates of G are given by

$$x = MP - PL + BD = a\sin\theta - a\theta\cos\theta + h\sin\theta$$

$$y = OM + BL + GD = a\cos\theta + a\theta\sin + h\cos\theta$$

$$\therefore \dot{x} = a\cos\theta\dot{\theta} + h\cos\theta\dot{\theta} - a(\cos\theta\dot{\theta} - \theta\sin\theta\dot{\theta})$$

$$= (a\theta\sin\theta + h\cos\theta)\dot{\theta}$$

and $\dot{y} = -a\sin\theta\dot{\theta} - h\sin\theta\dot{\theta} + a(\theta\dot{\theta} - \sin\theta\dot{\theta})$

$$= (a\theta\cos\theta - h\sin\theta)\dot{\theta}$$

$\therefore$ velocity of G at time $t = \sqrt{\dot{x}^2 + \dot{y}^2}$

$$= \theta\sqrt{[(a\theta\sin\theta + h\cos\theta)^2 + (a\theta\cos\theta - h\sin\theta)^2]}$$

$$= \dot{\theta}\sqrt{(a^2\theta^2 + h^2)}$$

K.E. at time $t = \frac{1}{2}m[k^2 + \dot{\theta}^2 + h^2\dot{\theta}^2] = \frac{1}{2}m\dot{\theta}^2[k^2 + a^2\theta^2 + h^2]$

Initial K.E. $= \frac{1}{2}m\omega^2[k^2 + h^2]$

[since in the beginning, $\theta = 0$, and $\dot{\theta} = \omega$]

Hence energy equation gives

$$\frac{1}{2}m[k^2 + a^2\theta^2 + h^2]\dot{\theta}^2 - \frac{1}{2}m(k^2 + h^2)\omega^2$$

$$= mg\,(a + h) - (a\cos\theta + a\theta\sin\theta + h\cos\theta)]$$

or $(k^2 + a^2\theta^2 + h^2)\dot{\theta}^2 - [(k^2 + h^2)\omega^2 = 2g(a + h) - (a + h)\cos\theta - a\theta\sin\theta]$

or $(k^2 + a^2\theta^2 + h^2)\dot{\theta}^2 - 2g[(a + h)(1 - \cos\theta) - a\theta\sin\theta]$

$$= (k^2 + h^2)\omega^2 = \text{const.}$$

or $\frac{1}{2}(k^2 + h^2 + a^2\theta^2)\dot{\theta}^2 + g\,[a\theta\sin\theta - [a + h)(1 - \cos\theta)]\text{const.}$

Small oscillations: We take q to be small then from the above equation,

we have $\frac{1}{2}(k^2 + h^2)\dot{\theta}^2 + g\left[\theta^2 - (a + h).\frac{2\sin^2(\theta/2)}{(\theta/2)^2}\left(\frac{\theta}{2}\right)^2\right] = \text{const}$

$$\left[\therefore \theta\sin\theta = \theta.\frac{\sin\theta}{\theta}\theta = \theta^2 \text{ approx.}\right]$$

or $\frac{1}{2}(k^2 + h^2)\dot{\theta}^2 + ga\theta^2 - 2ga(a + h).\left(\frac{\sin\frac{1}{2}\theta}{\frac{1}{2}\theta}\right)^2\frac{\theta^2}{4} = \text{const.} = 0$

or $\frac{1}{2}(k^2+h^2)\dot{\theta}^2 = g(a+h)\theta^2 - 2ga\theta^2 = -g(a-h)\theta^2$

or $\dot{\theta}^2 = -\frac{g(a-h)}{(k^2+h^2)}2\theta\dot{\theta}$

Differentiating it, we get

$$2\dot{\theta}\ddot{\theta}^2 = -\frac{g(a-h)}{(k^2+h^2)}2\theta\dot{\theta}$$

i.e. $\ddot{\theta} = -\frac{g(a-h)}{(k^2+h^2)}\theta$

which is a simple harmonic motion implying that the horizontal position of equilibrium is stable.

The length of the simple equivalent pendulum $= \frac{k^2+h^2}{a-h}$.

Example 15:

A heavy circular disc is revolving in a horizontal plane about the centre which is fixed. An insect of mass $\frac{1}{n}$ *th that of the disc walks from the centre along a radius and then flies away. Show that the final angular velocity is* $\frac{n}{n+2}$ *times the original angular velocity of the disc.*

Solution:

Let ω be the initial angular velocity of the disc. The insect flies away when it reaches the edge of the disc, let at this time ω′ be the angular velocity of the disc.

The forces acting on the system are either vertical or pass through O; thus their moments vanish about the vertical through O; so moment of , momentum about the vertical through O remains constant the motion

$$\therefore m0.\omega' + m\frac{a^2}{2}\omega' + \frac{m}{n}a^2\omega' = m.0.\omega + m\frac{a^2}{2}\omega$$

Example 16:

A circular ring of mass M and radius a, lies on a smooth horizontal plane and an insect of mass m, resting on it starts and walks round it with uniform velocity v relative to the ring. Show that the centre of the ring describes a circle with angular velocity $\frac{m}{M+2m}\cdot\frac{v}{a}\cdot$

Solution:

Let O be the centre of the ring and G the common centre of gravity of the ring and the insect resting at A.

If OG = x and GA = y, we have Mx = my

$$\Rightarrow \frac{x}{m} = \frac{y}{m} = \frac{x+y}{m+M} = \frac{a}{m+M}$$

$$\Rightarrow x = \frac{ma}{m+M'} y = \frac{Ma}{m+M}$$

where m is the mass of the insect and M the mass of ring.

So there is no horizontal force on the system throughout the motion, the horizontal momentum remains constant throughout the motion, which implies that the velocity of G (common C. G. of ring and the insect) in the horizontal plane will be zero throughout the motion because of its initial velocity being zero. In other words G remains fixed.

Hence the point O will describe a circle of radius $\frac{ma}{m+M}$ about G.

Now let ω be the angular velocity of the ring about G. Since all the forces [weight of the ring, weight of the insect] acting on the system are vertical, therefore their moments vanish about the vertical through G, hence moment of momentum about the vertical through G remains constant throughout the motion.

i.e. $(Mk^2\omega + M\,x\,\omega.\,GO) + (m\,\omega\,G\,A.\,G\,A + m\,v.\,G\,A) = 0$

for ring ← for insect →

$$\Rightarrow Ma^2\omega + M\omega x^2 + (m\omega y^2 + mv\,y) = 0$$

i.e. $\{Mk^2\omega_1 + Mx\omega_1.GO\} + m\omega_2 GA.GA = 0 + mu.GA$

for tubefor particle [tube] [practical]

$$\therefore M(a^2 - x^{2)}\omega + my^2\omega + mv(a-x) = 0(\because y = a - x)$$

$$\Rightarrow \omega[Ma^2 + Mx^2 + My^2] = -mv(a-x)$$

$$\Rightarrow \omega[Ma^2 + \frac{Mm^2a^2}{(m+M^2} + \frac{mM^2a^2}{(m+M^2}] = -mv\left[a - \frac{ma}{m+M}\right]$$

$$\Rightarrow \frac{Ma^2}{(m+M)^2}[(m+M)^2 + m^2 + mM]\omega = -\frac{amMv}{(m+M)}$$

$$\Rightarrow \frac{Ma^2}{(m+M)^2}[(m+M)^2 + m(m+M)]\omega = -\frac{amMv}{(m+M)}$$

$$\Rightarrow \frac{a(m+M)(2m+M)\omega}{m+M} = -mv \Rightarrow \omega = -\frac{m}{M+2m}.(v/a)$$

Negative sign shows that disc will turn in a direction which is opposite of the direction of v

Example 17:

A particle of mass m, moves within a rough circular tube of mass M lying on a horizontal plane and initially the tube is at rest while the particle has an angular velocity round the tube. Show that by the time the relative motion ceases, the fraction $\frac{M}{M+2m}$ *of the initial kinetic energy has been dissipated by friction.*

Solution:

Let O be the centre of the circular tube; the common C.G. of m and M where m = mass of the particle and M = mass of the tube. If OG = x and GA = y, then M x = my.

$$\Rightarrow \frac{x}{m} = \frac{y}{m} = \frac{x+y}{m+M} = \frac{a}{M+m}$$

$\Rightarrow x = ma/(M+m),\ y = Ma/M+m).$

Let u be the initial of the particle when the tube is at rest and v the velocity with which G, the common C.G. begins to move.

Now there being no force acting on the system in the horizontal plane throughout the motion, the horizontal momentum remains unchanged throughout the motion.

$$\therefore (M+m)v = mu \Rightarrow v = \frac{mu}{M+m}$$

When the system begins to move, let ω_1 and ω_2 be the angular velocities of the tube and the particle about G. Since the forces acting on the system are all vertical, therefore their moments vanish about the vertical axis through G throughout the motion, which implies that the moment of momentum about G remains the same throughout the motion.

i.e. $muy = M(a^2\omega_1 + x^2\omega_1) + my^2\omega_2$

The relative motion will cease when $\omega_1 = \omega_2$, then from (1), we have.

$$\omega_1 = \frac{muy}{M(a^2+x^2)+my^2} \qquad ...(2)$$

At this time, the kinetic energy of the system is

$$= \frac{1}{2}(M+m)v^2 + \frac{1}{2}M\,(a^2+x^2)\omega_1^2 + \frac{1}{2}my^2)\omega_1^2$$

[particle] [Tube] [K.E. of G.]

$$= \frac{1}{2}(M+m)\frac{m^2u^2}{(M+m)^2} + \frac{1}{2}(my^2+Ma^2)\times\frac{m^2u^2y^2}{(my^2+Ma^2+Mx^2)^2}$$

$$= \frac{1}{2}m^2\,u^2\left[\frac{1}{M+m} + \frac{y^2}{(my^2+Ma^2+Mx^2)}\right]$$

$$= \frac{1}{2}mu^2 - \frac{1}{2}mu^2\,\frac{M}{M+2m}$$

[by putting the values of x and y]

$$\therefore \text{ Loss of K.E.} = E\cdot = \frac{1}{2}mu^2 - \left[\frac{1}{2}mu^2 - \frac{1}{2}mu^2\frac{M}{M+2m}\right]$$

$$= \frac{1}{2}mu^2 \times \frac{M}{M+2m} = \frac{M}{M+2m} \text{ of the initial K.E.}$$

Example 18:

A small insect moves along a uniform bar; of mass equal to itself and of length 2a the ends of which are constrained to remain on the circumference of a fixed circle whose radius is $(2a/\sqrt{3})$. *If the insect starts from the middle point of the bar and moves along the bar with relative velocity V, show that the bar in time t will turn through an angle.*

$$\frac{1}{\sqrt{3}}\tan^{-1}\left(\frac{Vt}{a}\right)$$

Solution:

Let θ be the angle through which the bar AB turn in time t, the dotted line through A represents the original position. Let P be the position of insect time t.

$$\therefore CP = Vt \text{ and } OC = \left[\left(\frac{2a}{\sqrt{3}}\right)^2 - a^2\right]^{1/2} = \frac{a}{\sqrt{3}}$$

$$\therefore OP = \sqrt{[OC^2+CP^2)]} = \sqrt{\left[(\frac{1}{3}a^2+V^2t^2)\right]}$$

Since the moment of external forces [weights, reaction about vertical axis through O vanishes throughout the motion so the moment of momentum about this line remains unchanged.

$$i.e.\ 0 = (OC\dot{\theta}.OC + \frac{1}{3}a^2\dot{\theta}) + m(VOC + OP\dot{\theta}.OP)$$

$$\Rightarrow \frac{1}{3}a^2\dot{\theta} + \left(\frac{1}{3}a^2 + V^2t^2\right)\dot{\theta} + \frac{a}{\sqrt{3}}V = O$$

$$\Rightarrow a^2\dot{\theta} + V^2t^2\,\dot{\theta} + \frac{a}{\sqrt{3}}V = O$$

$$\Rightarrow \dot{\theta} + (a^2 + V^2t^2) + \frac{a}{\sqrt{3}}V = 0 \Rightarrow \theta = -\frac{Va}{\sqrt{3\left(a^2 + V^2t^2\right)}}$$

$$\therefore \theta = \frac{Va}{\sqrt{2}}\int_0^1 \frac{dt}{V^2t^2 + a^2} = -\frac{Va}{\sqrt{3}}\frac{1}{V^2}\int_0^1 \frac{dt}{t^2 + \left(a^2/V^2\right)}$$

$$= -\frac{Va}{\sqrt{3}\cdot}\frac{1}{V^2}\frac{V}{a}\left[\tan^{-1}\frac{Vt}{a}\right]_0^1 = -\frac{1}{\sqrt{3}}\tan^{-1}\left(\frac{Vt}{a}\right)$$

Negative sign shows that the rod turns in a direction opposite to that of the insect.

Example 19:

A uniform circular wire of radius a, lies on a smooth horizontal table and is movable about a fixed point O on its circumference. An insect of mass equal to that of the wire, starts from the other end of the diameter through O and crawls along the wire with a uniform velocity v relative to the wire. Show that at the end of time t, the wire has turned through an angle

$$\left(\frac{vt}{2a}\right) - \left(\frac{1}{\sqrt{3}}\right)\tan^{-1}\left[\left(\frac{1}{\sqrt{3}}\right)\tan\left(\frac{vt}{2a}\right)\right]$$

Solution:

Let in time t the diameter O A have turned through an angle ϕ from its initial position OA_0, let the insect be at P such that

$\angle ACP = \theta \therefore OP = 2a\cos(\theta/2)$.

Let O N be perpendicular on the tangent at P

$\therefore ON = OP\cos(\theta/2) = 2a\cos^2(\theta/2)$

The forces acting on the system are (i) Reaction at the point O (ii) Weight of the insect vertically downwards. (iii) Weight of the wire acting vertically down wards. Since the forces acting on the system are either vertical or pass through O, therefore moment of all forces about the vertical axis through O vanishes throughout the motion. Thus the moment of momentum of the system about this axis remains unchanged throughout the motion. The insect at P has a velocity v along the tangent at P and a velocity OP perpendicular to OP ϕ due to rotation of the circular wire.

So the moment of momentum of the insect about a vertical through

$$O = m(v.ON + OP\dot{\phi}.OP = m\left[2av\cos^2\frac{\theta}{2}. + \left(2a\cos\frac{\theta}{2}\right)^2\dot{\phi}\right]$$

$$= m\left[2av\cos^2\frac{\theta}{2} + 4a^2\cos^2\frac{\theta}{2}.\dot{\phi}\right]$$

Also the moment of momentum of the circular wire about the vertical axis though O.

= moment of momentum of the mass m moving with the C.G + moment of momentum about the C.G.

$$= m[a\dot{\phi}.a + a^2\dot{\phi}]$$

$= 2ma^2\dot{\phi}$ [C.G. describes a circle of radius a with velocity a $\dot{\phi}$]

Hence the total moment of momentum at time t

$$= m\left[2av\cos^2\frac{\theta}{2} + 4a^2\cos^2\frac{\theta}{2}.\dot{\phi}\right] + 2ma^2\dot{\phi}$$

But the moment of momentum of the system remains unchanged and initial the body was at rest, therefore we have

$$m\left[2av\cos^2\frac{\theta}{2} + 4a^2\cos^2\frac{\theta}{2}.\dot{\phi}\right] + 2ma^2\dot{\phi}$$

$$\Rightarrow 2a^2\left(1 + 2\cos^2\frac{\theta}{2}\right)\dot{\phi} = -2av\cos^2\frac{\theta}{2}$$

$$\Rightarrow 2a^2\left(1 + 2\cos^2\frac{\theta}{2}\right)\dot{\phi} = -2av\cos^2\frac{\theta}{2}$$

$$\Rightarrow \dot{\phi} = -\frac{v\cos^2(\theta/2)}{a\{1 + 2\cos^2(\theta/2)\}} \qquad ...(1)$$

Now the arc AP is covered by the insect in time moving with uniform velocity v.

$\therefore$ arc AP = vt, also arc AP = aθ $\therefore$ $a\theta = vt \Rightarrow \theta = \left(\frac{vt}{a}\right)$

putting this value of θ in (1), we get

$$=-\left(\frac{\pi}{4}\right)+\left(\frac{72}{3\sqrt{7}}\right)\tan^{-1}\left(\frac{\sqrt{7}}{3}\right)=\left(\frac{24}{\sqrt{7}}\right)\tan^{-1}\left(\frac{\sqrt{7}}{3}\right)-(\pi/4).$$

Hence the angle turned by the diameter OA in time t is given by

$$\theta=\int_o^t\frac{-v\cos^2(vt/2a)}{a[1+2\cos^2(vt/2a)}dt=-\frac{v}{a}\int_o^t\frac{\cos^2(vt/2a)}{1+2\cos^2(vt/2a)}dt$$

$$=-\frac{v}{a}\int_o^t\frac{1}{2}\left[1-\frac{1}{1+2\cos^2(vt/2a)}\right]dt$$

$$=-\frac{v}{2a}\int_o^t\left[1-\frac{\sec^2(vt/2a)}{\sec^2(vt/2a)+2}\right]dt$$

$$=-\frac{v}{2a}\int_o^t\left[1-\frac{\sec^2(vt/2a)}{3+\tan^2(vt/2a)}\right]dt$$

$$=-\frac{v}{2a}\int_o^t\left[t-\frac{2a}{v}\frac{1}{\sqrt{3}}\tan^{-1}\left(\frac{1}{\sqrt{3}}\tan\frac{vt}{2a}\right)\right]$$

$$=-\left[\frac{vt}{2a}-\frac{1}{\sqrt{3}}\tan^{-1}\left(\frac{1}{\sqrt{3}}\tan\frac{vt}{2a}\right)\right].$$

Negative sign shows that wire moves in a direction opposite to that of the insect.

Example 20:

A circular disc is moving with an angular velocity ω about an axis through its centre perpendicular to its plane. An insect alights on its edge and crawls along a curve drawn on the disc in the form of lemniscale with uniform relative angular velocity $\frac{1}{8}\omega$, the curve touching the edge of the disc. The mass of the in sect being $\frac{1}{16}$ th of that of the disc. Show that the angle turned through by the disc when the insect gets to the centre is

$$\left(\frac{24}{\sqrt{7}}\right)\tan^{-}\left[\frac{\sqrt{7}}{3}\right]-\left(\frac{\pi}{4}\right)$$

Solution:

Let the disc turn through an angle ϕin time t and the distance of insect from 0, the centre of the circle, be than r (OP = r) at an angle θ with the axis of the lemniscate, then equation of leminiscate is

$$r^2 = a^2 \cos 2\theta$$

Moment of momentum before insect alights of the disc $= M\frac{1}{2}a^2\omega,$

and moment of momentum of the system about O after time t

$$= \frac{1}{2}Ma^2\dot{\phi} + \frac{1}{16}M\left(r^2\dot{\phi} + r^2.\frac{1}{8}\omega\right)$$

Since all forces acting on the system are vertical or pass through the centre O, therefore the moment of all the forces about a vertical axis through O. vanishes, thus the moment of momentum about this axis through O remains unchanged throughout the motion.

$$\therefore M.\frac{1}{2}a^2\dot{\phi} + \frac{1}{16}M\left(r^2\dot{\phi} + \frac{1}{8}r^2\omega\right) = \frac{1}{2}Ma^2\omega$$

$$\text{or } Ma^2\dot{\phi} - \frac{1}{8}M\left(\dot{\phi} + \frac{1}{8}\omega\right)a^2\cos 2\theta = Ma^2\omega\left[\because r^2 = a^2\cos 2\theta\right]$$

$$\text{or } 64\dot{\phi} + \left(8\dot{\phi} + \omega\right)\cos 2\theta = 64\omega$$

$$\text{or } \left(64 + 8\cos 2\theta\right)\dot{\phi} = 64\omega - \omega\cos 2\theta$$

$$\text{or } \dot{\phi} = \frac{d\theta}{dt} = \frac{\left(64 - \cos 2\theta\right)\omega}{64 + 8\cos 2\theta.} \text{ or } \frac{d\phi}{d\theta}\cdot\frac{d\theta}{dt} = \frac{\left(64 - \cos 2\theta\right)\omega}{64 + 8\cos 2\theta}$$

$$\text{or } \frac{d\phi}{d\theta}\cdot\frac{1}{8}\omega = \frac{\left(64 - \cos 2\theta\right)\omega}{8\left(8 + \cos 2\theta\right)} \qquad \left[\because \frac{d\theta}{dt} = \frac{1}{8}\omega\right]$$

$$\text{or } d\phi = \frac{64 - \cos 2\theta}{8 + \cos 2\theta}d\theta.$$

If ϕ be the angle turned by the disc when the insect reaches the centre of the circle (*i.e.* when $\theta = \pi/4$), then

$$\phi = \int_0^{\pi/4} \frac{64 - \cos 2\theta}{8 + \cos 2\theta}d\theta = \int_0^{\pi/4}\left[-1 + \frac{72}{8 + \cos 2\theta}\right]d\theta$$

$$= \int_0^{\pi/4}\left[-1 + \frac{72}{8 + \left(2\cos^2\theta - 1\right)}\right]d\theta$$

$$= \int_0^{\pi/4}\left[-1 + \frac{72}{\left(7 + 2\cos^2\theta\right)}\right]d\theta = \int_0^{\pi/4}\left[-1 + \frac{72\sec^2\theta}{7\sec^2\theta + 2}\right]d\theta$$

$$=\int_0^{\pi/4}\left[-1+\frac{72\sec^2\theta}{(9+7\tan^2\theta)}\right]d\theta=\int_0^{\pi/4}\left[-1+\frac{72}{7}\frac{\sec^2\theta}{\left(\frac{9}{7}\right)+\tan^2\theta}\right]d\theta$$

$$=\left[-\theta+\frac{72}{3\sqrt{7}}\tan^{-1}\left(\frac{\sqrt{7}}{3}\tan\theta\right)\right]_0^{\pi/4}$$

$$=-\left(\frac{\pi}{4}\right)+\left(\frac{72}{3\sqrt{7}}\right)\tan^{-1}\left(\frac{\sqrt{7}}{3}\right)=\left(\frac{24}{\sqrt{7}}\right)\tan^{-1}\left(\frac{\sqrt{7}}{3}\right)-(\pi/4).$$

Example 21:

A rod OA can turn freely in a horizontal plane about the end O and lies at rest. An insect, whose mass is one third that of the rod alights on the end A and commences crawling along the rid with uniform velocity V, at the same instant the rod is set in rotation about O in such a way that the initial velocity of A is V, when the insect reaches O, prove that the rod has rotated through a right angle and that the angular velocity of the rod is then twice the initial angular velocity.

Solution:

Let OA be the initial position of the rod, when the insect is at A. Let the rod turn through an angle θ time t, also let the insect reach the point p in this time. since the insect moves along the rod will uniform velocity V so the distance moved in time t.

AP = Vt, and OP = 2a –t where 2a is length of the rod.

Since the rod moves about 0 with initial velocity V so its angular velocity in its initial position

$$=\frac{V\cdot 2a}{(a)^2}=\frac{V}{2a}\qquad\left[\therefore \text{angular velocity}=\frac{pV}{r^2}\right]$$

Moment of momentum of the system about 0, when the insect is at A

$$=\frac{1}{3}M.V.2a+M\left(a\frac{V}{2a}a+\frac{1}{3}a^2\frac{V}{2a}\right)=\frac{4}{3}MVa$$

also moment of momentum after time t

$$=\frac{1}{3}M(2a-vt)\theta(2a-Vt)+M\left(a\theta a+\frac{1}{3}a^2\dot{\theta}\right)$$

[formula $mvp+mk^2\dot{\theta}$]

[since now the angular velocity of rod is θ and that of insect due to rotation of the rod is $(2a - Vt)\theta$]

$$= \frac{M}{3}(2a - Vt)^2 + \frac{4}{3}Ma^2 \qquad \text{...(ii)}$$

The moment of all the forces about the vertical through 0 vanishes , so the moment of momentum about 0 remains unchanged throughout the motion. therefore we have from (i) and (ii)

$$\frac{M}{3}(2a - Vt)^2\theta + \frac{4}{3}Ma^2\theta = \frac{4}{3}MVa$$

or
$$[(2a - Vt)^2 + 4a^2]\theta = 4Va$$

or
$$\frac{d\theta}{dt} = \frac{4Va}{(2a - Vt)^2 + 4a^2}$$

Now the insect will reach the point 0 (moving with uniform velocity V) in time (2a/V). If θ be the angle described by the rod in this time, then

$$\int_0^{2a/v} \frac{4Va}{\left(2a - Vt^2\right) + 4a^2} dt \text{ Put } 2a\text{–}Vt = z, \therefore -Vdt = dz$$

$$i.e.\ \theta = \int_{2a}^{0} \frac{4Va}{z^2 + 2a^2} \cdot \frac{-dz}{V} = \int_{2a}^{0} \frac{4a}{z^2 + (2a^2)} dz$$

$$= 4a\int_0^{2a} \frac{dz}{z^2 + (2a)^2} = 4a \cdot \frac{1}{2a}\left[\tan^{-1}\frac{z}{2a}\right]_0^{2a} = \left(\frac{\pi}{4} - 0\right) = (\pi/2).$$

Hence the rod has turned through a right angle in this time the angular velocity of the rod $\dot{\theta}$ when the insect reaches 0 is obtained by putting $t = \left(\frac{2a}{V}\right)$ in (iii) , i.e., $\dot{\theta} = \left(\frac{V}{a}\right) = 2\left(\frac{V}{2a}\right)$ = twice the initial angular velocity

Example 22:

A uniform square plate ABCD of mass M and side 2a, lies on a smooth horizontal plane, it is struck a A by a particle of mass m moving with velocity V in the direction of AB, the particle remaining attached to the plate. Determine subsequent motion of the system and show that its angular velocity is $\frac{m}{M + 4m} \cdot \frac{3v}{2a}$

Solution:

Let O be the centre of the square ABCD of mass M and G the centre of gravity of the system.

$\therefore$ M.O.G = m.AG

or $\dfrac{OG}{m} = \dfrac{Ag}{M} = \dfrac{OG++AG}{m+M} = \dfrac{AO}{m+M}$

$\therefore OG = \dfrac{ma\sqrt{2}}{m+M}$ or $AG = \dfrac{Ma\sqrt{2}}{m+M}$

$\because AO = \dfrac{1}{2}AC = \dfrac{1}{2}\sqrt{\left[(2a)^2+(2a)^2\right]} = 2\sqrt{2.}$

The external forces acting on the system *i.e. (i)* the weight of the square (ii) the weight of the particle and (iii) the reaction of the smooth plane, are all vertical, thus there is no horizontal component of the force throughout the motion. Hence the total horizontal momentum of the system will remain constant throughout the motion. Le V′ is the velocity of the system after impact, then we have (M + m) = mV′ + M.O or V′ mV/M +m)

Again the moment of the external forces (acting vertically) about the vertical axis through G is zero, so the angular momentum of the system about this vertical axis through G remains constant throughout the motion. Let w is the angular velocity of the system after impact, then we have

$\therefore \{Mk^2\omega + M.OG\} + mAG.\omega.AG = mV.GL$

(for square) (for particle)

or $\dfrac{2}{3}Ma^2\omega + M\omega.\left(\dfrac{ma\sqrt{2}}{m+M}\right)^2 + m\omega\left(\dfrac{Ma\sqrt{2}}{m+M}\right)^2 = mV.\dfrac{Ma\sqrt{2}}{m+M}\sin 45^\circ$

$\because GL = AG\sin 45^\circ$

or $2a^2\left[\dfrac{1}{3} + \dfrac{m(m+M)}{(m+M)^2}\right]\omega = \dfrac{mVa}{m+M}$

or $\dfrac{2a^2}{3}\left(\dfrac{m+M+3m}{m+M}\right)\omega = \dfrac{mVa}{m+M}$

or $\omega = \dfrac{m}{M+4m}\cdot\dfrac{3V}{2a}$

Example 23:

A circular plate rotates about an axis through its centre perpendicular to its plane with angular velocity ω. This axis is set free and a point, in the circumference of the plate is fixed, show that the resulting angular velocity is $\dfrac{1}{3}\omega$.

Solution:

In this case centre C is set free and a point A on the circumference is suddenly fixed. Let ω' be the angular velocity just after the fixture. The impulse is at A, so the moment of the impulse about A will be zero. Hence moment of momentum about the axis through A remains the same before and after the fixture. *i.e.*

$$m\left[a^2\omega' + \frac{a^2}{2}\omega'\right] = \frac{ma^2}{2}\omega \text{ or } \omega' = \frac{1}{3}\omega.$$

Example 24:

A circular plate is turning in its own plane about a point A on its circumference. Suddenly A is freed and a point B on the circumference is fixed. Show that the plate will be reduced to rest if the arc AB is one third of the circumference.

Solution:

Consider a circular plate with centre C turning in its own plane about a point A on its circumference. A is set free and B is suddenly fixed. Let w¢ be angular velocity just after the fixture. Let . Since the impulse is at B, so the moment of the impulse about B is zero, hence moment and momentum about B will remain the same before and after fixture

$$i.e. \quad m\left[\frac{a^2}{2}\omega + a^2\omega'\right] = \left[\frac{a^2}{2}\omega + a\omega a\cos\theta\right]$$

$$\text{or} \quad \omega' = \frac{1}{3}(1 + 2\cos\theta)\omega.$$

The plate will reduced to rest if $\omega' = 0$

$$i.e. \text{ if } 1 + 2\cos\theta = 0 \text{ or } \cos\theta = -\frac{1}{2} \text{ i.e.} \theta = \frac{2\pi}{3}$$

or $a\,\theta = \frac{1}{3}(2\pi a)$ *i.e.* are AB is $\frac{1}{3}$ of the circumference.

Example 25(a):

A uniform circular disc is spinning with angular velocity ω about a diameter, when a point P on its rim is suddenly fixed. If the radius vector to P makes an angle α with this diameter, show that the angular velocities after the fixing, about the tangent and normal at P are $\frac{1}{5}\omega\sin\alpha$ and $\omega\cos\alpha$.

Solution:

Consider a circular plate with centre C, rotating with angular velocity ω

about the diameter ACB. The resolved part of ω about the normal at P *i.e.* about PC = $\omega \cos \alpha$. The resolved part of ω about the tangent at P = $\omega \sin \alpha$. Now the diameter AB is set free and the point P is suddenly fixed. Let ω_1 and ω_2 be the angular velocities about the tangent and normal at P just after the fixture. The impulse is at P, therefore moment of the impulse about P is zero. Hence the moment of momentum about the tangent at P remains unchanged after the fixture.

i.e. $m\left(\frac{a^2}{4}\right)\omega_1 + ma^2\omega_1 = m\left(\frac{a^2}{4}\right)\omega \sin a$. *i.e.* $\omega_1 = \left(\frac{1}{5}\right)\omega \sin \alpha$. Similarly the moment of momentum about the normal must be the same before and after the fixture.

$$m\left(\frac{a^2}{4}\right)\omega_2 = m\left(\frac{a^2}{4}\right)\omega \cos a \text{ i.e. } \omega_2 = \omega \cos \alpha.$$

Example 25(b):

An equilateral triangular lamina is rotating in its plane with uniform angular velocity about an axis through one vertex. If this vertex is released and one of the other vertices fixed. Show that the new angular velocity is 1/5 of its former value.

Solution:

Let the vertex C be released and the vertex B be suddenly fixed. Let ω and ω' be the angular velocities just before and after the fixture. If m is the mass of the triangular lamina, then it is equivalent to three particles each of mass m/3 placed at the mid - points of the sides.

The moment of momentum about axis through B remains unchanged *i.e.*

$$\left(\frac{m}{3}\right)a^2\omega + 0 + 0 = \left(\frac{m}{3}\right)a^2\omega' + \frac{m}{3}a^2\omega' + \left(\frac{m}{3}\right)a^2\omega'$$

$$\text{i.e.} \quad 5\omega' = \omega \text{ i.e } \omega' = \frac{1}{5}\omega.$$

Example 25(c):

A rod of length 2a is moving about one end with uniform angular velocity upon a smooth horizontal plane. Suddenly this end is set free and a point distant b from this end is fixed; find the motion, considering the case when $b >=< \frac{4a}{3}$.

Solution:

Let AB be the rod, moving with uniform angular velocity ω about the end A. The end A is set free and a point O of the rod is suddenly fixed, where AO = b. Let the angular velocity, of the rod after the fixture be ω'. The impulse is at O, thus moment of impulse about O will be zero, hence the moment of momentum about the axis through O will remain unchanged before and after the fixture *i.e.*

$$i.e.\ m\frac{a^2}{3}\omega' + m(a-b)^2\omega' = m\frac{a^2}{3}\omega + ma(a-b)\omega$$

$$\text{or}\quad \left[\frac{a^2}{3}+(a-b)^2\right]\omega' = \left[\frac{a}{3}+(a-b)\right]\omega$$

$$\text{or}\quad [a^2+3(a-b)^2\ \omega' = a(4a-3b)\omega$$

$$i.e.\ \omega' = \frac{a(4a-3b)}{a^2+3(a-b)^2}\omega.$$

(i) If $b < \frac{4a}{3}$ then $\omega' = \frac{3\left(\frac{4}{3}a-b\right)}{a^2+3(a-b)^2}$ is positive which implies that the rod will rotate in the same direction after the fixture as before.

(ii) If $b = \frac{4a}{3}$, then $\omega' = 0$, the rod is reduced the rest after the fixture.

(iii) If $b < \frac{4a}{3}$, then ω' is negative, which implies that the rod will move in opposite sense after the fixture.

Example 25(d):

A lamina in the form of an ellipse is rotating in its own plane about one of its foci with angular velocity ω. This focus is set free and other focus, at the same time is fixed, show that the ellipse now rotates about it with angular velocity $\omega\frac{2-5e^2}{2+3e^2}$

Solution:

Consider an ellipse, rotating with angular velocity ω about the focus S. If O is the midpoint of the major axis then OS = OS′ = ae, where length of the major axis AA′ = 2a. The focus S is set free and suddenly the other focus S ′ fixed.

Let the angular velocity after the fixture be ω¢ the impulse is at S ′, therefore the moment of the impulse about S′ is zero. Hence moment of momentum about this axis remains unaltered before and after the fixture

$$i.e.\ m\frac{a^2+b^2}{4}\omega' + ma^2e^2\omega' = m\frac{a^2+b^2}{4}\omega - ma^2e^2\omega$$

$$\text{or} \quad \left(\frac{a^2+b^2}{4}+a^2e^2\right)\omega' = \left(\frac{a^2+b^2}{4}-a^2e^2\right)\omega$$

$$\text{or} \quad \left(a^2+b^2+4a^2e^2\right)\omega' = \left(a^2+b^2-4a^2e^2\right)\omega$$

$$\text{or} \quad \omega' = \frac{a^2+b^2-4a^2e^2}{a^2+b^2+4a^2e^2}\cdot\omega$$

$$\text{or} \quad \omega' = \frac{a^2+a^2(1-e^2)-4a^2e^2}{a^2+a^2(1-e^2)+4a^2e^2}\omega \qquad \{\because b^2 = a^2(1-e^2)\} = \frac{2-5e^2}{2+3e^2}\omega.$$

Example 26:

An elliptic area, of eccentricity e is rotating with angular velocity ω about one latus rectum; suddenly this latus rectum is loosened and the other fixed. Show that the new angular velocity is $\frac{1-4e^2}{1+4e^2}\omega$.

Solution:

Consider the ellipse rotating with angular velocity ω, about its latus rectum L S L′.

Let this latus- rectum be set free and suddenly the other latus rectum L″S′L‴ be fixed. Let the angular velocity be ω' just after the fixture. The impulse is at L″S′L‴ . Hence the moment of the impulse about the rectum L¢¢S¢L¢¢¢ remains the same before and after the fixture.

$$i.e. \quad m\frac{a^2}{4}\omega' + ma^2e^2\omega' = m\frac{a^2}{4}\omega - ma^2e^2\omega$$

$$\text{i.e} \quad \left(a^2+4a^2e^2\right)\omega' = \left(a^2-4a^2e^2\right)\omega$$

$$i.e. \quad \omega' = \frac{1-4e^2}{1+4e^2}\omega.$$

Example 27:

An elliptic lamina is rotating about its centre on a smooth horizontal table. If $\omega_1, \omega_2, \omega_3$ be its angular velocities when the extremity of its major axis its focus and the extremity of its minor axis respectively become fixed prove that

$$\left(\frac{7}{\omega_1}\right) = \left(\frac{6}{\omega_2}\right) + \left(\frac{5}{\omega_3}\right)$$

Solution:

Initially the lamina was rotating with angular velocity ω about its O.

(i) *Let the centre O be set free and the end A of the major axis suddenly fixed.* The impulse is at A, therefore moment of the impulse about the axis through A (perpendicular to the plane of motion) is zero, therefore the moment of momentum about this axis remains the same before and after the fixture. If ω1 ισ the angular velocity just after the fixture, we have

$$\Rightarrow mk^2\omega + o = mk^2\omega_1 + ma\omega_1 \cdot AOm\frac{a^2+b^2}{4}\omega_1 + ma^2\omega_1.$$

$$\text{or } \left(a^2+b^2\right)\omega = \left[\left(a^2+b^2\right)+4a^2\right]\omega_1 \; i.e.\; \omega_1 = \frac{a^2b^2}{5a^2+b^2}\omega \qquad ...(1)$$

(ii) *Let the centre O be set free and the focus S be suddenly fixed.*

Again the impulse will be at S and the moment of the impulse about S is zero. Therefore the moment of momentum about the axis through S will remain the same before and after the fixture

$$i.e.\; m\frac{a^2+b^2}{4}\omega = m\frac{a^2+b^2}{4}\omega_2 + ma^2e^2\omega_2$$

$$\text{or } \left(a^2+b^2\right)\omega = \left(a^2+b^2+4a^2e^2\right)\omega_2$$

$$i.e.\; \omega^2 = \frac{a^2+b^2}{a^2+b^2+4a^2e^2}\omega$$

$$\text{or } \omega_2 = \frac{a^2+b^2}{a^2+b^2+4(a^2-b^2)}\omega \qquad \left[\because b^2 = a^2\left(1-e^2\right)\right] = \frac{a^2+b^2}{5a^2+3b^2}\omega$$

(iii) *Let the centre be set free and the end B of the minor axis be suddenly fixed.* In this case, we have

$$m\frac{a^2+b^2}{4}\omega = m\frac{a^2+b^2}{4}\omega_3 + mb^2\omega_3$$

$$\text{or } \left(a^2+b^2\right)\omega = \left(a^2+b^2+4b^2\right)\omega_3$$

$$i.e.\quad \omega_3 = \frac{a^2+b^2}{a^2+5b^2}\omega \qquad ...(3)$$

from (1), (2) and (3), we have

$$\frac{7}{\omega_1}-\frac{6}{\omega_2}-\frac{5}{\omega_3} = \frac{35a^2+7b^2}{\left(a^2+b^2\right)\omega}-\frac{30a^2-18b^2}{\left(a^2+b^2\right)\omega}-\frac{5a^2+25b^2}{\left(a^2+b^2\right)\omega}$$

$$= \frac{\left[35a^2 + 7b^2 - 30a^2 + 18b^2 - 5a^2 - 25b^2\right]}{\left(a^2 + b^2\right)\omega} = 0 \Rightarrow \frac{7}{\omega_1} = \frac{6}{\omega_2} = \frac{5}{\omega_3}.$$

Example 28:

Write the Hamiltonian function and equation of motion for a compound pendulum.

Solution:

We have $L = \frac{1}{2}I\dot{\theta}^2 + mgh\cos\theta \Rightarrow p_\theta = (\partial L/\partial\theta) = I\theta.$

where I mk²

$\therefore H = \Sigma p_i \dot{q}_i - L = p_\theta\dot{\theta} - L = I\dot{\theta}\dot{\theta} - \frac{1}{2}I\dot{\theta}^2 - mgh\cos\theta$

$= \frac{1}{2}I\dot{\theta}^2 - mgh\cos\theta$

$\Rightarrow \frac{1}{2}H = (p_\theta / I)^2 - mgh\cos\theta = (p_\theta^2 / 2I)\, 2\, mgh\cos\theta \{\because \dot{\theta} = (p_\theta / I)\}$

$\therefore (\partial H/\partial p_\theta) = (p_\theta / I), (\partial H/\partial\theta) = mgh\sin\theta$

Thus the Hamilton's equations for $\dot{\theta}$ and $\dot{p}_\theta$ are given by

$\dot{\theta} = (\partial H/\partial p_\theta),\ \dot{p}_\theta = -(\partial H/\partial\theta)$

i.e. $\dot{\theta} = (p_\theta / I)$ and $\dot{p}_\theta = -mgh\sin\theta$. But. $p_\theta = I\theta \Rightarrow \dot{p}_\theta = I\ddot{\theta}.$

$\therefore I\ddot{\theta} = -mgh\sin\theta \Rightarrow \ddot{\theta} + \frac{mgh}{I}\sin\theta = 0.$

This is exactly the same as obtained previously using Lagrange's equations.

Example 29(a):

Obtain Euler's equations from Hamilton's equations.

Solution:

We know that $2T = (A\omega_1^2 + B\omega_2^2 + C\omega_3^2)$,

$\Rightarrow L = T - V = \frac{1}{2}(A\omega_1^2 + B_2^2 + C\omega_3^2) - V$

Also Euler's geometrical relations give

$\omega_1 = \dot{\theta}\sin\psi\sin\theta\cos\psi;$

$\omega_2 = \dot{\theta}\cos\psi + \dot{\phi}\sin\theta\sin\psi$; and

$\omega_3 = \psi + \phi\cos\theta.$

Now H = T + V $= \frac{1}{2}(A\omega_1^2 + B\omega_2^2 + C_3^2) + V$

$$P_\phi = \frac{\partial L}{\partial \dot{\theta}} = \frac{\partial L}{\partial \omega_1}\frac{\partial \omega_1}{\partial \dot{\theta}} + \frac{\partial L}{\partial \omega_2}\frac{\partial \omega_2}{\partial \dot{\theta}} + \frac{\partial L}{\partial \omega_3}\frac{\partial \omega_3}{\partial \dot{\theta}}.$$

$A\omega_1 \sin\psi + B\omega_2 \cos\psi + C\omega_3 . 0$

$$p_\theta = \frac{\partial L}{\partial \dot{\phi}} = -A\omega_1 \sin\theta\cos\psi + B\omega_2 \sin\theta\sin\psi + C\omega_3\cos\theta \quad ...(2)$$

and $p_\psi = (\partial L/\partial\dot{\psi}) = C\omega_3$

Solving the three equations for ω_2 , ω_2, ω_3 we have

$$\omega_1 = \frac{1}{A}\left[p_\theta \sin\psi + (p_\psi\cos\theta - p_\phi)\frac{\cos\psi}{\sin\theta}\right]$$

$$\omega_2 = \frac{1}{A}\left[p_\theta \cos\psi - (p_\psi\cos\theta - p_\phi)\frac{\sin\psi}{\sin\theta}\right]; \text{ and } \omega_3 = \frac{1}{C}.p_\psi$$

Also, Hamilton's equations are $\dot{p}_\psi = -\frac{\partial H}{\partial \psi}$ and $\dot{\psi} = \frac{\partial H}{\partial p_\psi}$

Now $\dot{p}_\psi = -\frac{\partial H}{\partial \psi}$

$$\Rightarrow C\dot{\omega}_3 = -\left[\frac{\partial H}{\partial \omega_1}\frac{\partial \omega_1}{\partial \psi} + \frac{\partial H}{\partial \omega_2}\frac{\partial \omega_2}{\partial \psi} + \frac{\partial H}{\partial \omega_3}\frac{\partial \omega_3}{\partial \psi}\right] - \frac{\partial V}{\partial \psi}$$

$$= -\left[A\omega_1.\frac{1}{A}B\omega_2 + B\omega_2\left(\frac{A}{B}\omega_1\right) + C\omega_3.0\right] - \frac{\partial V}{\partial \psi}$$

$$= (A - B)\omega_1\omega_2 - \frac{\partial V}{\partial \psi}$$

$$\Rightarrow C\frac{d\omega_1}{dt} - (A - B)\omega_1\omega_2 = N\left(\because -\frac{\partial V}{\partial \phi} = N\right)$$

This is Euler's third familiar dynamical equation

Also, $\dot{\psi} = (\partial H/\partial p_\psi) = \frac{\partial H}{\partial \omega_1}\frac{\partial \omega_1}{\partial p_\psi} + \frac{\partial H}{\partial \omega_2}\frac{\partial \omega_2}{\partial p_\psi} + \frac{\partial H}{\partial \omega_3}\frac{\partial \omega_3}{\partial p_\psi}$

$= (\omega_1 \cos\psi - \omega_2 \sin\psi)\cot\theta + \omega_3 = -\phi\sin\theta\cot\theta + \omega_3$

i.e. $\dot{\psi} = -\dot{\phi}\cos\theta + \omega_3 \Rightarrow \omega_3 = \dot{\psi} + \dot{\phi}\cos\theta$

This is Euler's third geometrical equation.

One the same lines, we can deduce Euler's other equations (dynamical and geometrical).

Example 29(b):

Prove that

$$\left(\frac{\partial H}{dt}\right)=\left(\frac{\partial H}{dt}\right)$$ *where H is the Hamilton's function.*

Solution:

Let $q_1, q_2 ..., q_n$ be the generalised co-ordinates then Hamilton's equation are given by

$$\dot{p}_i = -\frac{\partial H}{dq_i} \text{ and } \dot{q}_i = \frac{\partial H}{dp_i}(i=1,2,...,n) \qquad ...(1)$$

But Hamiltonian H is a function of q's and ps'

$$\therefore \frac{\partial H}{dt} = \frac{\partial H}{\partial t} + \sum_{i=1}^{n}\frac{\partial H}{\partial q_i}\dot{q}_i + \sum_{i=1}^{n}\frac{\partial H}{\partial p_i}\dot{p}_i$$

$$= \frac{\partial H}{\partial t} + \sum_{i=1}^{n}\frac{\partial H}{\partial q_i}(-\dot{p}_i)\,\dot{q}_i + \sum_{i=1}^{n}\dot{q}_i\dot{p}_i = \frac{\partial H}{\partial t} \qquad \text{[using (1)]}$$

Example 29(c):

Use Hamilton's equations to find the equations of motion of a projectile in space.

Solution:

Let (x,y,z) be the co-ordinates of the projectile in space at time t, then we have

$$T = \frac{1}{2}m(\dot{x}^2 + \dot{y}^2 + \dot{z}^2),\ V = mgz$$

$$\therefore L = T - V = \frac{1}{2}m\,(\dot{x}^2 + \dot{y}^2 + \dot{z}^2) - mgz$$

$$\Rightarrow p_x = \frac{\partial L}{\partial \dot{x}} = m\dot{x},\ p_y = \frac{\partial L}{\partial \dot{y}} = m\dot{y}, p_z = \frac{\partial L}{\partial \dot{z}} = m\dot{z}$$

But L does not involve t explicity therefore Hamiltonian H is given by

$$H = T + V = \frac{1}{2}(\dot{x}^2 + \dot{y}^2 + \dot{z}^2) + mgz$$

$$= \frac{1}{2}m\left(\frac{p_x^2}{m^2} + \frac{p_y^2}{m^2} + \frac{p_z^2}{m^2}\right) + mgz = \frac{1}{2m}(p_x^2 + p_y^2 + p_z^2) + mgz$$

Now Hamilton's equations are given by

$$\dot{p}_x = -\frac{\partial H}{\partial x} = 0 \qquad ...(1)$$

$$\dot{x} = \frac{\partial H}{\partial p_x} = \frac{p_x}{m} \qquad ...(2)$$

$$\dot{p}_y = -\frac{\partial H}{\partial_y} = 0 \qquad ...(3)$$

$$\dot{y} = \frac{\partial H}{\partial p_y} = \frac{p_y}{m} \qquad ...(4)$$

$$\dot{p}_z = -\frac{\partial H}{\partial_z} = -mg \qquad ...(5)$$

$$\dot{z} = \frac{\partial H}{\partial p_z} = \frac{p_z}{m} \qquad ...(6)$$

Using (1) and (2), we have $\ddot{x} = 0$ ...(7)

Using (3) and (4), we have $\ddot{y} = 0$ ...(8)

Again making use of (6) and (5), we have

$$m\ddot{z} = \dot{p}_z = 0 \text{ mg or } \ddot{z} = -g. \qquad ...(9)$$

These (7, 8, 9) are the equations of motion of the projectile in space.

Example 29(d):

Using cylindrical coordinates (ρ, ϕ, z) *write the Hamiltonian and Hamilton's equations for a particle of mass m moving in a force field of potential V* (ρ, ϕ, z).

Solution:

In cylindrical coordinates, co-ordinates of any points are

$$x = \rho\cos\phi, y = \rho\sin\phi, z = z \qquad ...(1)$$

$$\therefore T = \frac{1}{2}m(\dot{x}^2 + \dot{y}^2 + \dot{z}^2) = \frac{1}{2}m(\dot{\rho}^2 + \rho^2\dot{\phi}^2 + \dot{z}^2) \qquad ...(2)$$

$$\Rightarrow L = T - V = \frac{1}{2}m(\dot{\rho}^2 + \rho^2\dot{\phi}^2 + \dot{z}^r) - V(\rho, \phi, z) \qquad ...(3)$$

$$\Rightarrow p_p = \frac{\partial L}{\partial \dot{\rho}} = m\dot{\rho}, p_\phi = \frac{\partial L}{\partial \dot{\phi}} = mp^2\dot{\phi} \text{ and } p_z = \frac{\partial L}{\partial \dot{z}} = m\dot{z}$$

Evidently, L does not involve t explicity, therefore Hamiltonian H is given by

$$h = T + V\frac{1}{2}m(\dot{\rho}^2 + \rho^2\dot{\phi}^2 + z^2) + V$$

$$= \frac{1}{2}\left[\frac{p_\rho^2}{m^2} + \frac{p_\phi^2}{m^2\rho^2} + \frac{p_z^2}{m^2}\right] + V = \frac{1}{2m}\left[p_\rho^2 + \frac{p_\phi^2}{\rho^2} + p_z^2\right] + V$$

Hence, Hamilton's are given by:

$$\dot{p}_\rho = -\frac{\partial H}{\partial \rho} = \frac{p_\phi^2}{m\rho^2} - \frac{\partial V}{\partial \rho}; \dot{\rho} = \frac{\partial H}{\partial p_\rho} = \frac{p_\rho}{m_\phi}$$

$$\dot{p}_\phi = -\frac{\partial H}{\partial \phi} = -\frac{\partial V}{\partial \phi}; \dot{\phi} = \frac{\partial H}{\partial p_\phi} = \frac{p_\phi}{m\rho^2}$$

$$\dot{p}_z = -\frac{\partial H}{\partial z} = -\frac{\partial V}{\partial z}; \dot{z} = \frac{\partial H}{\partial p_z} = \frac{p_z}{m}$$

Example 30:

Using cylindrical coordinates, write the Hamiltonian and Hamilton's equations for a particle of mass moving on the inside of a frictionless cone $x^2 + y^2 = z^2 \tan^2 \alpha$

Solution:

Like previous example, we have

$$T = \frac{1}{2}m(\dot{x}^2 + \dot{y}^2 + \dot{z}^2) = \frac{1}{2}m(\dot{\rho}^2 + \rho^2\dot{\phi}^2 + \rho^2 \cot^2 \alpha)$$

$$[\because x = \rho\cos\phi, y = \rho\sin\phi, z = \rho\cot\alpha]$$

$$= \frac{1}{2}m(\dot{\rho}^2 \operatorname{cosec}^2\alpha + \dot{\rho}^2\dot{\phi}^2) \qquad ...(1)$$

and $V = -W = -mgz = mg\rho\cot\alpha$,

[$\because$ the particle is above the vertex (origin)

$$\Rightarrow L = T - V\frac{1}{2}m(\rho^2 \operatorname{cosec}^2\alpha + \rho^2\dot{\phi}^2) - mg\rho\cot\alpha \qquad ...(2)$$

This gives, $p_\rho = \frac{\partial L}{\partial \dot{\rho}} = m\rho \operatorname{cosec}^2\alpha, p_\phi = \frac{\partial L}{\partial \dot{\phi}} = m\rho^2\dot{\phi}$ ...(3)

Again, L does not involve t explicitly, therefore Hemiltonian H is given by,

$$H = T + V\frac{1}{2}m(\dot{\rho}^2 \operatorname{cosec}^2\alpha + \rho^2\dot{\phi}^2) + mg\,\rho\cot\alpha$$

$$= \frac{1}{2}m\left[\frac{p_\rho^2}{m^2 \operatorname{cosec}^2\alpha} + \frac{p_\phi^2}{m^2\rho^2}\right] + mg\,\rho\cot\alpha + \frac{1}{2m}\left[\frac{p_\rho^2}{\operatorname{cosec}^2\alpha} + \frac{p_\phi^2}{\rho^2}\right]$$

$$+ mg\,\rho\cot\alpha$$

Thus Hamilton's equations are given by:

$$\dot{p}_\rho = -\frac{\partial H}{\partial \phi} = \frac{p_\phi^2}{m\rho^2} - mg\cot\alpha;\ \dot{\rho} = \frac{\partial H}{\partial p_\rho} = \frac{p_\rho}{m\,\mathrm{cosec}^2\alpha}.$$

$$\dot{p}_\phi = -\frac{\partial H}{\partial \phi} = 0 \qquad \dot{\phi} = \frac{\partial H}{\partial p_\phi} = \frac{p_\phi}{m\rho^2}.$$

Example 31:

Use the variational method to show that the shortest curve joining two fixed is a straight line

Solution:

We have

$$\frac{ds}{dx} = \sqrt{\left\{1 + \left(\frac{dy}{dx}\right)^2\right\}}$$ for any curve, joining two fixed points say

$A \equiv (x_1, y_1)$ and $B \equiv (x_2, y_2)$.

$$\Rightarrow s = \int_{x_1}^{x_2} \sqrt{(1 + y'^2)}\ dx = \text{length of the curve joining A to B.}$$

$$= \int_{x_1}^{x_2} f(y')\,dx \text{ where } f(y') = \sqrt{(1 + y'^2)} \qquad ...(1)$$

In equation (1), y is absent from f, therefore for s to be stationary (here minimum), we have from Euler-Lagrange's equation

$$\frac{\partial f}{\partial y'} = (\text{constant}) \Rightarrow \frac{\partial}{\partial y'}[(\sqrt{1 + y'^2})] = \text{cpmst.} \because f = (1 + y'^2)^{1/2}$$

$$\Rightarrow \frac{y'}{\sqrt{(1 + y'^2)}} = (\text{constant}) \text{ } i.e. \text{ } y' = b \text{ where b is constant.}$$

$$\Rightarrow \frac{dy}{dx} = b \text{ or } y = a + b\,x \text{ where a is a constant.} \qquad ...(2)$$

$\Rightarrow$ the shortest path joining A to B is the curve whose equation is y = a + bx and this is evidently a line.

Example 32:

Show that the area of the surface of revolution of a curve y = y (x) is

$$2\pi \int_{x_1}^{x_2} y\sqrt{(1+y'^2)}dx.$$

Hence show that for this to be a minimum,, the curve must be a catenary.

Solution:

Consider any curve, and let $A \equiv (x_1, y_1)$ and $B \equiv (x_2, y_2)$ be its extremities, then it this curve revolves about x-axis the surface of revolution is given by

$$S = \int 2\pi y\, ds = \pi \int y \frac{ds}{dx} dx = 2\pi \int_{x_1}^{x_2} y\sqrt{(1+y'^2)}dx$$

$$\Rightarrow S = 2\pi \int_{x_1}^{x_2} f(y, y')\, dx, \text{ where } f = f(y, y') = y\sqrt{(1+y'^2)}$$

Evidently, x is absent from f, so for S to be stationary (here a minimum) we have:

$$f - y'\frac{\partial f}{\partial y'} = \text{(constant) } i.e.\ y\sqrt{(1+y'^2)} - y'\frac{yy'}{\sqrt{(1+y')}} = \text{const.}$$

$$\Rightarrow \frac{y}{\sqrt{(1+y'^2)}} = c \text{ where c is constant } i.e.\ y = c\sqrt{(1+\tan^2\psi}$$

$$[\because y' = \tan\psi]$$

$\therefore$ y = c sec y and this represents catenary.

Example 33(a):

A particle of unit mass is projected so that its total energy is h in a field of force of which the potential energy is ϕ (r) at a distance r from the origin.

Deduce from the principle of energy and least action that the differential equation of the path is

$$c^2\left[r^2 + \left(\frac{dr}{d\theta}\right)^2\right] = r^4[h - \phi(r)].$$

Solution:

Let T and V be the kinetic are potential energies, then we have T + V = h $\Rightarrow$ T = h – V

= h – ϕ (r), $\because$ V = ϕ (r) given

$$\Rightarrow \frac{1}{2}v^2 = h - \phi(r) \text{ where v is velocity } \left(\because T = \frac{1}{2}mv^2 = \frac{1}{2}v^2\right)$$

$$\Rightarrow v = \sqrt{2}\{h - \phi(r)\}^{1/2}$$

Whence, the action $A = \int_{t_0}^{t_1} 2T \ dt$, by definition given earlier

$$= \int_{t_0}^{t_1} 2.\frac{1}{2}v^2 dt = \int_{t_0}^{t_1} v \, d\,s \ \text{ since } v = (ds/dt) = \sqrt{2}\int_{t_0}^{t_1}\{h - \phi(r)\}^{1/2} ds$$

$$= \sqrt{2}\int_{t_0}^{t_1}\{h - \phi(r)\}^{1/2}\left\{1 + r^2\left(\frac{d\theta}{dr}\right)^2\right\}^{1/2} dr \quad \frac{ds}{dr} = \left\{1 + r^2\left(\frac{d\theta}{dr}\right)^2\right\}^{1/2}$$

$$= \sqrt{2}\int_{t_0}^{t_1}[\{h - \phi(r)\}^{1/2}\{1 + r^2\theta'^2\}^{1/2}] \ dr \text{ where } \theta' = \frac{d\theta}{dr}$$

$$= \sqrt{2}\int_{t_0}^{t_1} f\,(\theta', r)dr, \ \text{ where } \ f(\theta', r) = \{h - \phi(r)\}^{1/2}\{1 + r^2\theta'^2\}^{1/2}$$

Evidently, θ is absent from f, therefore we must have

$$\frac{\partial f}{\partial \theta'} = c \ \text{ (const.) } \Rightarrow \frac{\partial f}{\partial \theta'}[\{h - \phi(r)^{1/2}\{1 + r^2\theta'^2\}^{1/2]} = c$$

$$\Rightarrow \{h - \phi(r)\}^{1/2}\frac{r^2}{\left\{r^2 + \left(\frac{dr}{d\theta}\right)^2\right\}^{1/2}} = c$$

$$\Rightarrow c^2\left\{r^2 + \left(\frac{dr}{d\theta}\right)^2\right\} = r^4\{h - \phi(r)\}.$$

This is the required equation.

Example 33(b):

A projectile is launched in a vertical plane with a velocity whose horizontal and vertical components are v_x and v_y respectively. Calculate the value of the integral $\int_0^{t_0} l\,dt$, *where* $t_0 = \frac{n\pi}{\omega}$.

Evaluate this integral for the varied path given by the equations

$$x = v_x t, y = V_y t - \frac{1}{2}gt^2 + \varepsilon \sin \omega t,$$

where ε is a small constant quantity.

Show that the integral $\int_0^{t_0} L \ dt$ *is greater for the varied path than for the actual path, but the result is in agreement with Hamilton's principle.*

Solution:

The path of the projectile is given by

$$x = V_x t, y = V_y t - \frac{1}{2}gt^2.$$

Also V = mgy, where V denotes the potential,

whence $L = T - V = \frac{1}{2}m(\dot{x}^2 + \dot{y}^2) - mgy$

$$= \frac{1}{2}m(y_x^2 + (v_y - gt)^2] - mg\left(v_y t - \frac{1}{2}gt^2\right)$$

$$= \frac{1}{2}m[v_x^2 + v_y^2) - 4gtv_y + 2g^2t^2] \quad ...(1)$$

$$\Rightarrow \int_0^{t_0} Ldt = \int_0^{t_0}(T - V)dt = \frac{1}{2}m\int_0^{t_0}[(v_x^2 + v_y^2) - 4gtv_y + 2g^2t^2]dt$$

$$= \frac{1}{2}m\left[(v_x^2 + v_y^2)t_0 - 2t_0^2v_y + \frac{2}{3}g^2t_0^3\right] \quad ...(1)$$

Also the varied path is given by : $x = v_x t, y = v_y t - \frac{1}{2}gt^2 + \varepsilon \sin \omega t$

Now after obtaining $\dot{x}, \dot{y}$; we have

$$T = \frac{1}{2}m[v_x^2 + (y_v - gt + \varepsilon\omega \cos \omega t)^2] \text{ and}$$

$$V = mg\left(v_y t - \frac{1}{2}gt^2 + \varepsilon \sin \omega t\right) \Rightarrow \int_0^{t_0 = n\pi/\omega} Ldt = \int_0^{t_0}(T - V)dt$$

$$\int_0^{t_0}\left[\frac{1}{2}m\{v_x^2 + v_y - gt + \varepsilon\omega \cos \omega t)^2\} - mg(v_y t - \frac{1}{2}gt^2 + \varepsilon \sin \omega t\right]dt$$

$$= \frac{1}{2}m\left[(v_x^2 + v_y^2)t_0 - 2t_0^2gv_y + \frac{2}{3}g^2t_0^3 + \frac{1}{2}\varepsilon^2\omega^2t_0\right]$$

[$\because$ other integrals vanish)

$$= \frac{1}{2}m[(v_x^2 + v_y^2)t_0 - 2t_0^2gv_y + \frac{2}{3}g^2t_0^3] + m\pi n\omega\varepsilon^2 \quad ...(2)$$

From equations (1) and (2), we see that the integral $\int_0^{t_0} L\,dt$ is greater for the varied path then for the actual path and is minimum when ε = 0.

Evidently, the minimum value in this case, the varied *path coincides with actual path.*

Example 33(c):

A particle moves in a straight line with central acceleration $\omega^2 x$ between two fixed points x_0 and x_1 in the prescribed time $t_1 - t_0$. Show that Hamilton's principal function S is

$$\frac{\omega}{2\sin\omega(t_1 - t_0)}\left[(x_1^2 + x_0^2)\cos\omega(t_1 - t_0 - 2x_1 x_0\right]$$

Solution:

Equation of motion is, $\ddot{x} = \omega^2 x$ ...(1)

Solving (1), the path is x = A cos ωt + B sin ωt ...(2)

Now, $\dot{x} = \omega[-A\sin\omega t + B\cos\omega t] \Rightarrow T = \frac{1}{2}\dot{x}^2.$

Also, we have potential V = work done against the force

$$= \int_0^{t_0} \omega^2 x dx = \frac{1}{2}\omega^2 x^2 \qquad ...(3)$$

Again, we also have

$A\cos\omega t_0 + B\sin\omega t_0 - x_0 = 0$ $(t = t_0 ; x = x_0)$

and $A\cos\omega t_1 + B\sin\omega t_1 - x_1 = 0$ $(t = t_1 ; x = x_1)$

Saving these

$$A = \frac{x_0 \sin\omega t_1 \, 0 \, x_1 \sin\omega t_0}{\sin\omega(t_1 - t_0)}, B = \frac{x_1\cos\omega t_0 - x_0\cos\omega t_1}{\sin\omega(t_1 - t_0)}$$

$$\Rightarrow B^2 - A^2 = [x_1^2\cos 2\omega t_0 + x_0^2\cos 2\omega t_1 \, 2x_1 x_0$$
$$- \cos\omega(t_1 + t_0)]\frac{1}{\sin^2\omega(t_1 - t_0)} \qquad ...(4)$$

∴ S = Hamilton's principal function

$$= \int_{t_0}^{t_1} L dt = \int_{t_0}^{t_1} (T - V)dt \quad (\because L = T - V)$$

$$= \int_{t_0}^{t_1}\left(\frac{1}{2}\dot{x}^2 - \frac{1}{2}\omega^2 x^2\right)dt \quad \text{[from above]}$$

$$= \frac{1}{2}\omega^2 \int_{t_0}^{t_1} [(-A\sin\omega t + B\cos\omega t)^2 - (A\cos\omega t + B\sin\omega t)^2]dt$$

$$= \frac{1}{2}\omega\left[(B - A)^2\sin 2\omega t + 2AB\cos 2\omega t\right]_{t_0}^{t_1}$$

$$=\frac{1}{2}\omega\left[(B^2-A^2)\cos\omega(t_1+t_0)\sin\omega(t_1-t_0)-2AB\sin\omega(t_1-t_0\sin\omega(t_1-t_0)\right]$$

Substituting for B^2-A^2 and 2AB, we obtain

$$S=\frac{\omega}{2\sin\omega(t_1-t_0)}[(x_1^2\cos 2\omega t_0+x_0^2\cos 2\omega t_1-2x_1x_0$$

$$\cos w(t_1+t_0)\}\cos w(t_1+t_0)+$$

$$\{x_1^2\sin 2\omega t_0+x_0^2\sin 2\omega t_1-2x_1x_0\sin\omega(t_1+t_0)\}\sin\omega(t_1+t_0)]$$

$$=\frac{\omega}{2\sin\omega(t_1-t_0)}[x_1^2\cos\omega(t_1-t_0)+x_0^2\cos\omega(t_1-t_0)-2x_1x_0]$$

$$=\frac{\omega}{2\sin\omega(t_1-t_0)}[x_1^2+x_0^2\cos\omega(t_1-t_0)-2x_1x_0]$$

This is the required result.

Example 34:

A particle moves in a plane curve, under the central acceleration $\omega^2 r$, between two fixed points (x_0, y_0) and (x_1, y_1) in the prescribed time t_1-t_0; prove that Hamilton's principle function S is

$$=\frac{\omega}{2\sin\omega(t_1-t_0)}[(x_1^2+y_1^2+x_0^2+y_0^2)\cos\omega(t_1-t_0)-2(x_1x_0+y_1y_0)]$$

Solution:

Equations of motion of the particle are given by

$$\dot{x}=-\omega^2 x \text{ and } \ddot{y}=-\omega^2 y \quad ...(1)$$

$$x=A\cos\omega t+B\sin\omega t;\ y=C\cos\omega t+D\sin\omega t$$

Now, kinetic energy $T=\frac{1}{2}gw^2(\dot{x}^2+\dot{y}^2)$

$$=\frac{1}{2}\omega^2[(-A\sin\omega t+B\cos\omega t)^2+(-C\sin\omega t+D\cos\omega t)^2]$$

Potential $V=\frac{1}{2}\omega^2 r^2=\frac{1}{2}\omega^2(x^2+y^2)$

Hence Hamilton's principal function S is given by

$$S=\int_{t_0}^{t_1}L\,dt=\int_{t_0}^{t_1}(T-V)\ dt \text{ since } L=T-V$$

$$=\int_{t_0}^{t_1}\left[\frac{1}{2}(\dot{x}^2+\dot{y}^2)-\frac{1}{2}\omega^2(x^2+y^2)\right]dt \quad ...(2)$$

But $\int_{t_0}^{t_1}\left(\frac{1}{2}\dot{x}^2 - \frac{1}{2}\omega^2 x^2\right)dt = \frac{\omega}{2\sin\omega(t_1 - t_0)}[(x_1^2 - x_0^2)\cos\omega(t_1 - t_0) - 2x_1x_0]$

and $\int_{t_0}^{t_1}\left(\frac{1}{2}\dot{y}^2 - \frac{1}{2}\omega^2\dot{y}^2\right)dt = \frac{\omega}{2\sin\omega(t_1 - t_0)}$

$$\times[(y_1^2 - y_0^2)\cos\omega(t_1 - t_0) - 2y_1y_0]$$

$$\therefore S = \frac{\omega}{2\sin\omega(t_1 - t_0)}$$

$$[(x_1^2 - x_0^2 + y_1^0 + y_0^2)\cos\omega(t_1 - t_0) - 2(x_1x_0 + y_1y_0]$$

Example 35(a):

Use Hamilton's principle to find the equation of motion of one - dimensional harmonic oscillator.

Solution:

The kinetic energy of harmonic oscillator is given by $T = m\dot{x}^2$ and the potential energy of the harmonic oscillator is given by

$$V = -\int F dx = \int kx dx = \frac{1}{2}kx^2.$$

$\therefore$ The Lagrangian of the system L = T – V $V = \frac{1}{2}m\dot{x}^2 - \frac{1}{2}kx^2.$

Using Hamilton's Principle, we have $\delta\int_{t_1}^{t_2} L\,dt = 0.$

$$\Rightarrow \delta\int_{t_1}^{t_2}\left(\frac{1}{2}m\dot{x}^2 - \frac{1}{2}kx^2\right)dt = 0 \Rightarrow \int_{t_1}^{t_2}\delta\left(\frac{1}{2}m\dot{x}^2 - \frac{1}{2}kx^2\right)dt = 0$$

$$\Rightarrow \delta\int_{t_1}^{t_2}(m\dot{x}\delta\dot{x} - kx\delta x)dt = 0 \text{ But } \delta\dot{x} = (d/dt)(\delta x). \qquad ...(1)$$

$$\therefore (1) \Rightarrow \int_{t_1}^{t_2} m\dot{x}\frac{d}{dt}(\delta\dot{x})\,dt - \int_{t_1}^{t_2} kx\,\delta x\,dt = 0$$

$$\text{i.e.,} [m\dot{x}\,\delta x]_{t_1}^{t_2} - \int_{t_1}^{t_2} m\frac{d}{dt}(\dot{x})\,dt - \int_{t_1}^{t_2} kx\,\delta x\,dt = 0 \qquad ...(2)$$

But $[m\dot{x}\,\delta x]_{t_1}^{t_2} = 0$ [$\because$ $\delta x = 0$ at fixed points, i.e, at instants t_1 and t_2]

$$\therefore (2) \Rightarrow \int_{t_1}^{t_2} m\frac{d}{dt}(\dot{x})\,\delta x\,dt - \int_{t_1}^{t_2} kx\,\delta x\,dt = 0 \Rightarrow \int_{t_1}^{t_2}(m\dot{x} + kx)\delta\,dt = 0$$

But δx is arbitrary, hence the above equation is only satisfied if $m\dot{x} + kx = 0$. This is the equation of motion for one dimensional harmonic oscillator.

Example 35(b):

A particle moves in the xy-plane under the influence of a central force depending only on its distance from the origin.

(a) Set up the Hamiltonian for the system.

(b) Write Hamilton's equations of motion.

Solution:

(a) Let the p potential due to the central force be V(r). Then, we have

$$T = \frac{1}{2}mv^2 = \frac{1}{2}m\,[(\text{radial velocity})^2 + (\text{transverse velocity}^2)]$$

$$= \frac{1}{2}m(\dot{r}^2 + r^2\dot{\theta}^2),$$

$$L = T - V = \frac{1}{2}m(\dot{r}^2 + r^2\dot{\theta}^2) - V(r).$$

$$\therefore p_r = (\partial L/\partial\dot{r}) = m\dot{r}, p_\theta = (\partial L/\partial\dot{r}) = mr^2\dot{\theta}$$

$$\Rightarrow \dot{r} = (p_r/m), \dot{\theta} = (p_\theta/mr^2)$$

Thus $H = \Sigma p_i\dot{q}_i - L = p_r\dot{q}_r + p_\theta\dot{q}_\theta - L$

$$= p_r.\dot{r} + p_\theta\dot{\theta} - \frac{1}{2}m(\dot{r}^2 + r^2\dot{\theta}^2) + V(r)$$

$$= p_r(p_r/m) + p_\theta(p_\theta/mr^2)$$

$$-\frac{1}{2}m\{p_r^2/m^2) + r^2.(p_\theta^2/m^2r^4)\} - V(r)\}$$

$= (p_r^2/2m) + (p_\theta^2/2m_r^2) + V(r) =$ total energy of the system.

(b) Hamilton's equations are

$$\dot{p}_i = -(\partial H/\partial q_i), \dot{q}_i = (\partial H/\partial p_i)$$

$$\Rightarrow \dot{r} = (\partial H/\partial p_r) = (p_r/m), \dot{\theta} = (\partial H/\partial p_\theta) = (p_\theta/mr^2)$$

$$\dot{p}_1 = -(\partial H/\partial r) = (p_\theta^2/mr^3) - V(r), \dot{p}_\theta = -(\partial H/\partial\theta) = 0.$$

Example 35(c):

A particle of mass m moves in a force field of potential V. Write

(a) *The Hamiltonian and*

(b) *Hamilton's equations in spherical polar co-ordinates.*

Solution:

K.E. is given by

$$T = \frac{1}{2}m(\dot{r}^2 + r^2\dot{\theta}^2 + r^2\sin^2\theta\dot{\phi}^2) \qquad ...(1)$$

$$\therefore L = T - V = \frac{1}{2}m(\dot{r}^2 + r^2\dot{\theta}^2 + r^2\sin^2\theta\dot{\phi}^2 - V \qquad ...(2)$$

We have $p_r = (\partial L/\partial\dot{r}) = m\dot{r}, p_\theta = (\partial L/\partial\dot{\theta}) = mr^2\dot{\theta},$

$p_\phi = (\partial L/\partial\dot{\phi}) = mr^2\sin^2\theta\dot{\phi}$

$$\Rightarrow \dot{r} = (p_r/m), \dot{\theta} = (p_\theta/mr^2), \dot{\phi} = p\phi/(mr^2\sin^2\theta). \qquad ...(3)$$

Now Hamiltonian is given by

$h = \Sigma p_i\dot{q}_i - L = p_r\dot{r} + p_\theta\dot{\theta} + p\phi\dot{\phi} - L$

$$= \frac{p_r^2}{2m} + \frac{p_\theta^2}{2mr^2} + \frac{p_\theta^2}{2mr^2\sin^2\theta} + V(r, \theta, \dot{\phi})$$

= total energy of the system.

(b) Hamilton's equations are given by

$\dot{q}_i = (\partial H/\partial p_i), \dot{p}_i = -(\partial H/\partial q_i)$

i.e. $\dot{r} = \frac{\partial H}{\partial p_r} = \frac{p_r}{m} \qquad \dot{p}r = -\frac{\partial H}{\partial r} = \frac{p_\theta^2}{mr^3} + \frac{p_\phi^2}{mr^3\sin^2\theta} - \frac{\partial V}{\partial r}$

$\dot{\theta} = \frac{\partial H}{\partial p_\theta} = \frac{p_\theta}{mr^2} \qquad \dot{p}\theta = -\frac{\partial H}{\partial\theta} = \frac{p_\phi^2\cos\theta}{mr^2\sin^3\theta} - \frac{\partial V}{\partial\theta}$

$\dot{\phi} = \frac{\partial H}{\partial p_\phi} = \frac{p_\phi}{mr^2\sin^r\theta} \qquad \dot{p}\phi = -\frac{\partial H}{\partial\phi} = -\frac{\partial V}{\partial\phi}$

Example 35(d):

If H is the Hamiltonian prove that if f is any function depending (df/dt) $= ((\partial f/\partial t) + [H.f]$

Solution:

We have

$(df/dt) = (\partial f/\partial t) + \{\partial f/\partial q_i)(dq_i/dt) + (\partial f/\partial p_i)(dp_i/dt)\}$

$\Rightarrow (df/dt) = (\partial f/\partial t) + \Sigma\{\partial f/\partial q_i)(\partial H/\partial q_i) - (\partial f/\partial p_i)(\partial H/\partial qi)\}$

{$\because$ By Hamilton's equations $\dot{q}i = \partial H/\partial p_i, \dot{p}_i = -(\partial H/\partial q_i)$}

$\Rightarrow (df/dt)(\partial f/\partial t) + [H, f]$ where [H, f] is the Poisson Bracket)

EXERCISES

1. A horizontal lamina of mass M can rotate freely about an axis through O, some point of it and perpendicular to it. An insect of mass m alights on A, some point of it at distance a from O and crawls along a circle having OA as diameter and returns to the point of starting Initially if the insect and the lamina were at rest, Prove that the lamina has moved through an angle relative to the ground given by $\pi\sqrt{\{Mk^2/(Mk^2.ma^2)\}]}$ where k is radius of gyration of the lamina about its axis.
2. A uniform square plate A B C D of mass M and side 2a, lies on a smooth horizontal plane; it is struck at A by particle of mass m moving with velocity V in the direction of AB, the particle remaining attached to plate. Determine the subsequent motion of the system and show that its angular velocity is $\frac{m}{M+4m}\frac{3V}{2a}\cdot$
3. Three uniform rods AB, BC, CD freely jointed at B and C lie on a smooth horizontal plane. They are originally moving as a rigid body with velocity V parallel to CB, A and B being on the same side of BC and the angles B and C being right angles, A is suddenly fixed, prove that if 2a, 2b, 2c are the lengths of rods and 3m, m, 2m their respective masses, the initial angular velocities $\omega_1 \omega_3$ of AB, CD are given by $5a\omega_1 = 20\ w = 3V$.
4. Write the Hamiltonian for the motion of a spherical pendulum and obtain Hamilton's equations. Solve them to show that Hamiltonian is a constant.
5. Derive the Euler's dynamical equations from Hamilton's equations.
6. If H is the Hamiltonian, prove that if f is any function depending on position, momentum and time, then

   $(df/dt) = (\partial f/\partial t) + [H, f]$ where [H,f] is the poisson's Bracket.
7. A uniform circular board of mass M and radius a is placed on a perfectly smooth plane and free to rotate about a vertical axis through its centre; a man of mass 'M' walks round the edge of the board whose upper surface is rough enough to prevent his slipping ; when he has walked completely round the board to his starting point, show that the board has turned through an angle

$$\frac{4\pi M'}{M+2M'}.$$

8. A right angled triangular lamina is turning about the side AC. suddenly AC is set free and BD fixed. If C be the right angle, prove that the angular velocity is 1/2 (BC/CA) of its former value.

9. A rectangular lamina whose sides are of lengths 2a and 2b is at rest on smooth horizontal plane. If one corner is caught and suddenly made to move with prescribed speed V in the plane of the lamina, then the greatest, angular velocity which can thus be imparted to the lamina is

$$\frac{3V}{4\sqrt{[(a^2+b^2)]}}$$

10. An elliptic lamina is rotating about a diameter and a point p on the edge is suddenly fixed. Prove that the velocity of the centre immediately afterwards is 1/5 of the velocity of p before it was fixed.

11. A uniform rod AB of mass 3m has a small hoop at B and swings through an angle 60° on each side of the vertical about a fixed horizontal axis through the end A. When B is passing through its lowest point the hook picks up a particle of mass m. Show that the system will afterwards swing through an angle θ on each side of the vertical given by cos θ =

$$\frac{17}{20}$$

Hint : Let ω be the angular velocity of the rod when B is in the lowest position *i.e.* when AB is vertical. Energy equation gives

$$\frac{1}{2}\cdot 3m\frac{4a^2}{3}\omega^2 = 3mg(a - a\cos 60)$$

i.e. $$\omega = \sqrt{\left(\frac{3g}{4a}\right)}$$

When mass m is picked up by the hook at B, the impulsive action is at A, let ω′ be angular velocity after impulse.

$$\therefore\ 3m\left(\frac{a^2}{3}\omega' + a^2\omega'\right) + 4m^2\omega' = 3m\left(\frac{a^2}{3}\omega' + a^2\omega'\right).$$

i.e. $$\omega' = \frac{1}{2}\omega = \frac{1}{2}\sqrt{\left(\frac{3g}{4a}\right)}$$

Let θ be the angle through which the system rises the energy equation gives

$$\frac{1}{2}\cdot 3m.\frac{4a^2}{3}\omega'^2+\frac{1}{2}\cdot m.4a^2\omega'^2=3mg(a-a\cos\theta)+mg(2a-2a\cos\theta)$$

or $8\omega^2=\frac{10g}{a}(1-\cos\theta)$ which gives $\cos\theta=\frac{1}{2}\frac{7}{0}\cdot$

12. A uniform circular disc. of radius a is rolling without slipping along a smooth horizontal plane with velocity V when the highest point becomes suddenly fixed. Prove that the disc will make a complete revolution round the point if $V^2 > 24ag$.

13. The ends of a straight uniform rod of mass m and length 2a slide on a smooth fixed horizontal circle of radius and an insect of mass m crawls along it at a constant rate V relative to the rod. If θ is the angle made by the rod at time t with a fixed direction, show that

    $\dot{\theta}\left(2c^2-\frac{5}{3}a^2+V^2t^2\right)=\text{const}$ and if $c=\frac{2a}{\sqrt{3'}}$ the angle turned through

    by the rod in time t is $\frac{1}{\sqrt{3}}\tan^{-1}\left(\frac{Vt}{a}\right)$

14. A bead of mass M can slide on a smooth straight horizontal wire and a particle of mass m is attached to the body by a light string of length l, the particle is held in contact with the wire with the string taught and is then let fall. Prove that when the string is inclined to the wire at an angle θ, the bead will have slipped a distance along the wire $\frac{mgl(1-\cos\theta)}{M+m}$ and that the angular velocity ω of the string will be given by the equation

    $(M+m\cos^2\theta)\,l\omega^2=\omega\,(M+m)\,g\sin\theta$

15. A uniform rod moves in a vertical plane, its ends being in contact with the interior of fixed smooth sphere; when it is inclined at an angle θ to the horizontal, show that the square of its angular velocity is $\frac{6cg}{3c^2+a^2}[\cos\theta-\cos\alpha]$, where a is the initial value of θ, 2a is the length of the rod, and c is the distance of its middle point from the centre of the sphere.

16. A hoop of radius a rolling on a horizontal rod with the velocity v comes onto collision with a rough inalastic kerb of height h which is perpendicular to the plane of the hoop. Show that if the hoop is to clear the kerb without the jumping then

    $$\frac{2a}{2a-h}\sqrt{\{g(a-h)\}}>v>\frac{2}{2a-h}\sqrt{(gk)}.$$

17. A sphere of radius a rolling on a rough table comes to a slit of breadth b, perpendicular to its path; if V be its velocity, show that the condition it should cross the slit without jumping is

$$v^2 > \frac{100}{7} ga(1 - \cos\alpha)\sin^2\alpha \frac{14 - 10\sin^2\alpha}{(7 - 10\sin^2\alpha)^2};$$

where b = 2a sin α and 17 gα cos α> 7V² + 10ga.

18. Discuss the principle of conservation of energy.

19. A heavy bead of mass **m** freely movable on a smooth circular wire of radius a which is made to rotate about a vertical diameter with spin ω, prove that the action,

$$A = ma^2 \int_{\theta_1}^{\theta_2} \left[\frac{2H}{ma^2} + \frac{2g}{a}\cos\theta + \omega^2 \sin^2\theta \right]^{1/2} d\theta$$

where H is the Hamiltonian and θ is the angle the radius through the bead makes with the downward vertical.

20. A particle of unit mass is projected so that its total energy is k, in a field of a force a of which the potential energy is ϕ (r) at a distance r from the origin. Find the differential equation of the path.

21. State the variational principle of Hamilton.

22. State & prove the principle of least action.

23. State & prove the principle of least action for a conservative holonomic system, is the action really least in all cases.

24. Prove that for function f (y, y′, x)

$$\frac{\partial f}{\partial y} - \frac{d}{dx}\left(\frac{\partial f}{\partial y'}\right) = 0, \text{ where } y' = \frac{dy}{dx}$$

25. Obtain Hamilton's canonical equations of motion for rigid body.

26. Define Hamiltonian function and derive Hamilton's equations of motion.

27. Apply the technique of calculus of variation to obtain Euler Lagrange's equations.

3

States of Matter

INTRODUCTION SOLIDS STATE

Many properties of solids, liquids and gases can be easily observed with the help of our sense organs. The properties which can be observed with the help of our sense organs are called macroscopic properties. The description of the behaviour of the three states of matter an terms of atomic theory is called microscopic description of matter. In chemistry, we try to explain the macroscopic behaviour of matter in terms of its microscopic description. In this chapter, we would try to explain the observable properties of different states of matter in terms of the behaviour of the constituent particles in term.

In a solid, the constituent particles atms, ions or molecules are most closely packed and have the strongest inter molecular force of attraction. As a result particles are fixed in their position and do not have any freedom of motion except that of vibration about their mean position. This give rise to the following characteristic properties to a solid.

(i) A definite shape and definite volume

(ii) Incompressibility

(iii) Rigidity

(iv) No fluidity

(v) Poor diffusibility through them

CHARACTERISTICS OF SOLIDS

Some characteristic properties of solids are given as follows :

(i) Solids maintain their volume independent of the size of shape of the container in which they are placed.

(ii) Solids are rigid and have definite shape.

(iii) Solids diffuse very slowly compared to liquids or gases. This is because the constituent particles in solids are closely paced and there is very little space to permit the movement of the particles.

(iv) Solids re nearly incompressible. The compressibility of solids is about 10th times that of the gases.

(v) Most solids melt on heating. Some undergo sublimation upon beating. Melting temperatures vary considerably from solids to solid.

(vi) Solids have very high mass-to-volume ratio (density) as compared to gases.

SOLID STATE CLASSIFICATION

Based on their structural features, the solids are classified into following three type :

(a) Crystalline solids

(b) Polycrystalline solids

(c) Amorphous solids.

These are described below.

Crystalline Solids

All crystalline solids have the following characteristics.

(i) The constituent particles in crystals are generally held by strong interatomic interionic or intermolecular forces. Particles in a solid do not possess translation motion, but they vibrate about their equilibrium positions.

(ii) Crystalline solids possess characteristic geometrical shapes.

(iii) When cut or hammered gently, crystalline solids show a clean cleavage, *i.e.*, they show fracture along a smooth surface.

(iv) Crystalline solids have sharp melting points. This indicate that there exists a long range order in crystalline solids. The long range order in crystalline solids is due to regular arrangement of the constituent particles (atoms, ions or molecules) in three dimensions.

For example in a crystal of sodium chloride, experiments (x-ray diffraction method) show that Na^+ and Cl^- ions are located at alternate sites as shown below:

Na^+	Cl^-	Na^+	Cl^-	Na^+	Cl^-
Cl^-	Na^+	Cl^-	Na^+	Cl^-	Na^+
Na^+	Cl^-	Na^+	Cl^-	Na^+	Cl^-
Cl^-	Na^+	Cl^-	Na^+	Cl^-	Na^+

The arrangement shown above is only in two dimensions. In three dimensions we find that each Na^+ ion is surrounded by a fixed number (six) of Cl^- ions and *vice-versa*. The origin of the three dimensional order in NaCl crystals is due to the strong Colombian attraction between Na^+ and Cl^- ions. A similar regular arrangement is also found in other solids.

(v) The mechanical, electrical and optical properties depend upon the direction along which they are measured, *i.e.*, crystalline solids are anisotropic in nature.

Polycrystalline Solids

In certain crystalline solids, the crystals are very find. These cannot be seen by the naked eyes. Such solids given an impression of being amorphous. Such fine crystalline solids, which appear as amorphous are termed as polycrystalline solids.

Metal powders are polycrystalline in nature. In a polycrystalline sample, there are small crystals. These crystals are randomly oriented. As a result, a sample of polycrystalline material appears to be isotropic, even when each individual crystal is anisotropic.

Amorphous Solids

Amorphous solids have some mechanical properties which are commonly associated with the word solid. But, the amorphous solids differ from the crystalline solids in many respects. So, amorphous solids are sometimes called as pseudo solids.

Some characteristic properties of the amorphous solids are;

(i) The mechanical, electrical and optical properties of amorphous solids do not depend upon the direction, *i.e.*, amorphous solids are isotropic. In this respect, the amorphous solids resemble liquids. It is for this reason that the amorphous solids are considered as supercooled liquids. There is only a short-range order in amorphous solids.

(ii) Amorphous solids do not poses sharp melting points. This indicates the absence of long-range order Amorphous soften on heating and gradually begin to flow like liquids.

(iii) Amorphous solids do not occur in characteristic geometrical shapes.

(iv) When hammered, amorphous material break in an irregular manner. When you hammer a glass sheet or a plastic sheet, these get broken in a very irregular way along many directions.

Examples. Some typical amorphous materials are Plastics, Glass, Rubber, Lamp black, pitch etc.

The solid state is next to gaseous state in its simplicity to understand, although it is also not free from complex problems. Solids are obtained by cooling liquids. There are inherent two possible ways of passing from the equilibrium liquid state into solid state namely crystallization and glass transition.

Crystallization is a transition from a state of short range order to one of long range order in the process of formation of a new phase. The concept of short range order and long range order is defined by the ratio between the distance over which the order extends and the dimensions of the elements arranged in the order.

This is brought about by slowly cooling the liquid where the molecules, due to enough thermal energy and low viscosity of the medium, get enough opportunity to arrange themselves in a geometric order. With the slow release of energy, they grow on the nucleus maintaining the shape of the crystal. It is a phase transition of the first order.

IONIC CRYSTALS

Solids in which ions are arranged in some definite geometrical pattern are called ionic crystals. These ions are bound by plane surfaces called faces. These faces intersect at an angle giving definite shape to the crystal.

Steno observed that the shape of a crystal depends on the details of crystalization but at a given temperature the angle between planes of a crystal always remain constant. This is known as Stenos law. That is why a crystal on cleavage maintains its geometry. Sometimes a substance crystallizes out in different shapes. This is known as polymorphism.

Laws of Rotational Indices

The law of rotational indices states that the intercepts of the plane of various faces of crystal on a suitable set of axes can be expressed by small integral multiples of three unit distances.

Space Lattices

The diffraction of X-rays by crystals is a phenomenon of great importance. Diffraction studies provide much of the structural information on the arrangement of atoms, ions or molecules in crystals.

Hauy suggested that the environment about any particle in space, *i.e.*, in three direction is the same as that about any other particle. Thus, a regular three-dimensional array of points, representing particles, is known as space lattice.

Unit Cell

A unit cell in the crystal is the smallest repeating unit in three dimensions, which reproduces the whole crystal by the displacement of the unit along the direction of the three axes. The crystal may be considered to consist of a very large number of unit cells, each one in direct contact with its nearest neighbours and all are similarly oriented in space. The shape and size of a unit cell are given by the spacings along the three axes (a, b, c) and the three angles (α, β and γ) between the faces.

Zero Point Energy

A crystal is found to have some vibrational and translational energy at absolute zero. This is known as the vibrational energy at absolute zero or the zero point energy of the crystal.

Symmetry Properties

The arrangement of crystal faces such that the crystal transform into image of itself is known as symmetry operation. These operations are as a result of some symmetry elements. These symmetry elements are international symbols.

An axis of rotatory inversion is the combination of a rotation axis with a centre of symmetry. It is denoted by placing a bar over the symbol.

There are 32 possible combinations of these symmetry elements. These define 32 crystallographic point groups which determine the 32 crystal classes. The symmetry operations are reversible, *i.e.*, A(BC) =

(AB)C. Only holohedral class of crystals follow the complete symmetry. Crystalline rocksalt and iron pyrites are cubic crystal. Rocksalt crystal possess the full symmetry of the cube but iron pyrites does not show all the symmetries because all the faces in iron pyrites are not equivalent.

Space Groups

The classification of crystals is based on the various groups of symmetry elements operations, such that they leave atleast one point in the crystal as invariant. Hence, they are called point groups. The possible three dimensional lattices symmetry operations of infinite figures are called space groups. It has been worked out that the internal symmetry of crystal may be classified according to 230 space groups, *i.e.*, there are only 230 ways of repeating a particular pattern.

Diffraction Methods

In 1912 Von Lave suggested that a crystal can act as a three-dimensional diffraction grating for X-rays of wave length ranging 0.001-50Å. X-rays are produced by striking atoms in a solid target by accelerated electrons. Thus, the electrons in the inner orbit of the target are displaced by bombardment and the energy liberated by the transition of electrons from a higher energy level to the vacancy in the inner orbit is given out in the form of an x-ray photon. These transitions produce characteristic lines in the x-ray spectrum. A copper target gives an intense line at 1.54Å. In addition to the above radiation, a continuous range of X-ray wave lengths are obtained. These are called white X-rays. Thus, using a nickel filter radiations of wave length 1.54Å can be picked up which are widely used for polymers.

Beams of electrons and neutrons has also been diffracted by crystals. Neutron diffraction determines the position of protons in a structure. X-rays scattered by electrons do not generally lead to a precise position of protons. Electrons do not penetrate very far into solids, but their use in gases has been found very valuable in finding the structure of gas molecules. Reflections from the surface reveal facts about their structure.

Hugens-Fresnel suggested that every point reached by the electromagnetic vibrations may be regarded as a radiation centre. Bragg suggested that it would be convenient to assume that the X-rays are reflected from the planes in the crystal. Woulfe-Bragg found a simple relation between A the wave length of the X-rays, d the distance between planes in a crystal and Q the angle of reflection

$$n\lambda = 2\ d \sin\theta \qquad ...(1)$$

Let the plane ABC be perpendicular to the incident beam of parallel X-rays and the plane DEF is perpendicular to the reflected rays

It can be seen that the ray BPE travels farther from AOD by a distance equal to LP + PM. The angle of reflection is equal to the angle of incidence θ and LP & PM is equal to the integral number of wave lengths.

$$\therefore \qquad LP + PM = n\lambda \qquad ...(2)$$

$$LP = PM = d \sin\theta \qquad ...(3)$$

$$\therefore \qquad n\ \lambda = 2d \sin\theta \qquad ...(4)$$

When n = 1, it is called the first order reflection and n = 2, it is called the second order reflection, etc. Equation 8.4 can be written as

$$\lambda = 2\left(\frac{d}{n}\right)\sin\theta \qquad ...(5)$$

$$= 2\ d_{hkl} \sin\theta$$

where d is the distance between the planes having hkl as miller indices.

The second order reflections from 100, 110, 111 planes are labelled as 200, 220, 222 respectively. In a cubic crystal distance between planes with Miller indices hk*l* is given by

$$d_{khl} = \frac{a}{(h^2 + k^2 + l^2)^{1/2}}$$

when a is the length of the side of the unit cell.

Thus, the spacings between the planes in a primitive cubic crystals may have values a, $\frac{a}{\sqrt{2}}, \frac{a}{\sqrt{3}}, \frac{a}{\sqrt{2}}, \frac{a}{\sqrt{5}}, \frac{a}{\sqrt{6}}, \frac{a}{\sqrt{8}}$ but not $\frac{a}{\sqrt{7}}$ as $h^2\ k^2 + l^2$ cannot have a value.

In the case of face centred cubic lattice, the spacings which will appear are $\frac{a}{\sqrt{3}}, \frac{a}{\sqrt{4}}, \frac{a}{\sqrt{8}}, \frac{a}{\sqrt{11}}, \frac{a}{\sqrt{12}}, \frac{a}{\sqrt{16}}, \frac{a}{\sqrt{19}}, \frac{a}{\sqrt{20}}$, etc.

In the case of body centred cubic lattice the inter planer spacings are $\frac{a}{\sqrt{2}}, \frac{a}{\sqrt{4}}, \frac{a}{\sqrt{6}}, \frac{a}{\sqrt{8}}$, etc.

Thus, knowing the missing X-ray lines it would be possible to predict the type of cubic lattice of a given crystal.

If X is known and the direction of θ in which the intensity of the scattered radiations is maximum is determined experimentally, d can be calculated and a can be known from the density of the cubic crystal.

Example:

The length of the side of a unit cell of potassium is 5.34Å. Calculate the distance between 200, 110 and 222 planes.

$$d = \frac{5.34}{\left(h^2 + k^2 + l^2\right)^{1/2}}$$

For 222 planes d is $\frac{5.34}{(4)^{1/2}} = 2.67\text{Å}$

For 110 planes d is $\frac{5.34}{(2)^{1/2}} = 3.77\text{Å}$

For 222 planes d is $\frac{5.34}{(12)^{1/2}} = 1.54\text{Å}$

Macromolecules : In macromolecules, there are three basic types of ordering; the small period, the helicals when repeat distance is along the chain and the large period.

The small period characterizes the dimensions of the smallest structural elements out of which all polycrystalline bodies are made in their various combinations.

Regularly built macromolecular chains with side substitutes assume in space helical conformation with constant pitch. The helical pitch conforms to the periodicity along the chain with a specific size of the order of several angstroms.

A special type of ordering inherent only to polymer bodies is characterized by the alternate regions of long-range and short-range orders and in the limit; amorphous and crystalline regions are obtained in oriented polymers. The large periods have the order of hundred of angstroms. Three methods are used with crystals.

Laue's Method

White x-rays are used in this method. The x-ray pattern obtained represents a system of spots from various systems of planes in a crystal. This is not used for polymers because polymer single crystals of large size are usually not obtained.

Bragg's Method

In this method a monochromatic radiation is used and the sample is rotated or tilted when it is photographed. The photographic film is arranged on the surface of the cylinder around a sample which is placed along the cylinder axis. Diffraction maxima appear in the coarse of rotation and then θ can be measured.

Debye Scherrer Method

Samples in the form of powder or a polycrystalline body are examined. The incident monochromatic X-ray is diffracted on the planes of a crystallite of a polycrystalline sample that satisfy the Woulfe-Bragg equation by their orientation with respect to incident ray. Diffracted rays will spread along a cone with the vertex angle equal to 2θ. The intersection of these cones with the flat photographic film placed perpendicular to the incident ray behind the sample produces a system of concentric rings of growing radii, each of which contains all the reflections with the same angle θ. The reflections are recorded on circular photographic film. If coarse crystals are used, the powder pattern produces rings of spots, each spot produced by the different crystal planes. If the crystals are very fine, continuous arcs are obtained on the film.

X-ray diffraction analysis is used in obtaining the structure of polymeric bodies and its change as a result of heat, mechanical and other effects, on the phase transformation of a body and to determine the conformation of chains. They also give average dimensions of crystal and their distribution according to dimensions. They make it possible to detect the origination of the smallest (10-100 A) cracks in polymeric bodies.

Primitive Cubic Lattice

In a primitive or simple cubic cell, the cations are located only at the corners of the unit cell and its centres occupy the anions, *e.g.*, Cs Cl, *i.e.*, it contains one ion pair per unit cell. Each cation is therefore, surrounded by eight anions and vice versa. This arrangement is known as cubic coordination. This type of structure is exhibited by cesium and thalium chloride, bromide and iodides and by a host of other compounds. The packing factor is 0.52.

$$\text{P.F.} = \frac{\frac{4}{3}\pi r^3}{a^3} = \frac{4\pi}{3}\frac{(r)^3}{8r^3} = 0.52 \qquad ...(1)$$

CRYSTAL STRUCTURE

Bragg during his study of X-ray diffraction derived a relation between the distance between planes in a crystal and the angle at which the reflected radiations have a maximum intensity.

$$\lambda = 2\left(\frac{d}{n}\right)\sin\theta = 2\, d_{hukl}\sin\theta \qquad ...(1)$$

where n is the order of reflection usually set equal to 1, d is the distance between planes, θ is the angle and λ is the wave length employed.

In a cybic system $$d_{hkl} = \frac{a}{\left(h^2 + k^2 + l^2\right)^{1/2}} \qquad ...(2)$$

$$= \frac{2a\sin\theta}{\left(h^2 + k'' + l^2\right)^{1/2}} \qquad ...(3)$$

thus, knowing h, k, l, θ and d_{hkl}, a can be calculated.

The Cohesive Energy of Crystals

Born Mayer derived the equation for crystal energy per mole

$$U = \Delta H_{cryst} = \frac{Z + Z - C^2 NA}{r_o}\left(1 - \frac{\rho}{r_o}\right) \qquad ...(4)$$

$$= \frac{Z^2\varepsilon^2 A}{r_o}\left(1 - \frac{\rho}{r_o}\right)$$

where ZC is the net charge on ions, r_o is the equilibrium inter nuclear separation, A the Madelung's constant, and p the repulsion energy constant and is expressed in terms of volume compressibility β. It has an average value of 0.311×10^{-8} cm for alkali halides.

In the case of NaCl the value of r_o is 2.81 A, A is 1.75 and ρ is 0.311×10^{-8}cm

$$\Delta H_{NaCl} = U_{NaCl} = -\frac{(4.8 \times 10^{-10})^2 \times 6.02 \times 10^{23} \times 1.75}{2.81 \times 10^{-8}\ \text{cm}}\left(1 - \frac{0.311 \times 10^{-8}\ \text{cm}}{2.81 \times 10^{-8}\ \text{cm}}\right)$$

$$= 7.68 \times 10^{12}\ \text{crg/mole}$$

$$= -184\ \text{K cal/mole}$$

which is in good agreement with the experimental value of –183.3 Kcal/mole.

Compressibility

The compressibility β is given by Born and Landes equation

$$\frac{1}{\beta} = -\frac{mnu_0}{9\upsilon_0} = -\frac{nu_0}{18r_0^3} \quad ...(5)$$

$$\therefore \quad n = 1 + \frac{18r_0}{\beta AZ^2\varepsilon^2} \quad ...(6)$$

The lowest potential energy per molecule is $U_0 = \frac{AZ^2\varepsilon^2}{r_0}\left(1 - \frac{1}{n}\right)$

where n is the repulsive force constant ε the charge, U_0 is molecular volume equal to $2r_0^3$.

Ionic Radii

In the absolute sense of the word it is not a constant parameter. It is not possible to find the exact ionic radii as ions are not rigid bodies, but for all practical purposes if we take half the distance of approach between two similar ions, as the ionic radii, it will give a very useful data to build up the concept of the ionic structure in the compounds. The sum of the ionic radii of both ions in compound gives idea of the approach distance between two ions, which in turn gives the idea of their geometrical arrangement. The radius of an ion remains the same in various crystals as the repulsive force increases very sharply as the internuclear distance becomes smaller than a certain value.

Four anions are arranged around a cation. This is the limiting condition of stability as the anions are touching the cations.

$$XY^2 + YZ^2 = XZ^2 \quad ...(7)$$

$$(2r_a)^2 + (2r_a)^2 = (2r_a + 2r_c)^2$$

or $$r_a + r_c = \sqrt{2}\, r_a \quad ...(8)$$

or $$r_c = r_a(\sqrt{2} - 1) = r_a \,.\, 414$$

$$\frac{r_c}{r_a} = 0.414 \quad ...(9)$$

In a three-dimensional arrangement it will represent an octahedral arrangement with two additional ions one above and one below the cation. By increasing or decreasing the radius ratio various arrangements are obtained. If the ratio is decreased, the anions would tend to arrange tetrahedrally. If the ratio is increased above 0.73 a cubic arrangement is obtained CsCl. Generally in ionic solids, cations tend to get surrounded by the largest possible number of anions and therefore, larger the radius ratio larger is the coordination number of the cation. By employing value of ionic radii for cations and anions, it is possible to predict the coordination number of cations in compounds. Radius ratio is indeed a useful concept for prediction of structures of ionic compounds.

Properties of Ionic Solids

The ionic bond is a strong bond, its cohesive energy is large. It is difficult to separate the ions from the condensed aggregate. Their melting points and boiling points are high but have low vapour pressure. There is rupture of bonds on melting although, the ions are grouped together even in melt so much so that they have strong ionic interaction. The structure of solid is very compact as the ions are closely packed.

MOLECULAR CRYSTALS

Substances that exist as aggregates of discrete molecules in the crystalline state are known as molecular crystals. Their melting points are low and are efficiently packed. They are held by Van der Waals forces.

They are either cubic close packed or hexagonally close packed. The interstitial sites or holes are larger in the case of HCP than C.C.P. The tetrahedral hole can accommodate a sphere of radius 0.41 times that of larger sphere while the tetrahedral hole can accommodate a sphere of radius 0.23 times that of larger sphere.

The solid inert gases being spherical in shape form a simple CCP structure. While N_2 and O_2 which are non-sphereoids give a face centred packing CCP. In such crystals there are two types of interatomic distances, one between the atom and its molecular partner and the other between atoms not covalently bonded to one another.

These are known as Van der Waal radii. In iodine the internuclear distance is 2.68Å, while the distance between two non-bonded atoms is 4.54Å. The Van der waals radius of iodine if about 2.3Å. Sulphur, selenium and tellurium have similar structure of long spiral chain packed together with their axis parallel. Phosphorous, arsenic and antimony are dimorphic. They assume both close packed tetrahedral structure and layer type of arrangement. In BCl_3 or BF_3, the halogen atoms form an HCP structure with boron atoms in the holes.

The cohesion energy of a molecular crystal is defined as the enthalpy of the reaction

$$M(g) \xrightarrow{\Delta H} M\ (\text{Crystal}) \qquad ...(1)$$

or $(H_{fus} + H_{vap})$. ...(2)

If the close-packed molecules possess a dipole moment then dipole-dipole interaction will take place and this, will contribute to the cohesion energy.

COVALENT SOLIDS

The atoms in a covalent solid occupy such position that the attractive force due to two protons acting on each electron completely counter balances the force of repulsion between two electrons. They equally share electrons with their neighbour. These bonds extend in fixed directions resulting into a giant interlocking structure, *e.g.*, diamond, silicon, germanium, grey tin, etc. The covalent crystals have high energies (35-150 Kcal/mole), short interatomic distances (1.01-1.6 A), relatively constant angles between successive bonds. These bonds are very strong (high melting point and large heat of sublimation). Their directional character does not lead to maximum filling of space. They are bad conductors of electricity and are chemically inert.

Internal Symmetry

Wells has shown that the symmetrical arrangements of atoms around a given atom is indicative of the type offers governing their interaction. A compound containing A and B have stoichiometric ratio AB_3. The arrangement given in (a) indicates that it is a molecular compound (Fe_3) while the arrangement (b) shows the presence of AB_2 & AB_4 groups, *i.e.*, the number of bonds in which A is engaged with B is either two or four ($AuCl^-_3$). The arrangement in C indicates that the compound exist as a dimer ($AlCl_3$). While d is a cyclic trimer (SO_3 or S_3 O_9) c & f represent a chain structure.

All these indicate that a detailed consideration of the spatial arrangement of a system can give a powerful insight of the nature of bonding. Cu, NaCl and diamond all belong to cubic system but NaCl has a heteronuclear linking and in contrast to Cu, diamond has a tetrahedral structure.

METALS

Those crystalline solids which exhibit high thermal and electrical conductivity, a metallic lusture, a large coefficient of thermal expansion and low ionization energy are known as metals. The structure and properties of metals.

Metallic elements display body-centred, face centred or HCP structures. In metals the radii are found to be functions of coordination number and very often all the neighbours are not equidistant. Goldschmidt deduced that the relative order for different coordination numbers are:

Table 1 : Structures of some polymorphic metals

	Structure	*Se angstroms*	*Temperature °K*
α Ca	CCP	3.94	291
γ Ca	BCC	3.87	770
α Fe	BCC	2.48	293
α Fe	CCP	2.57	1189
δ Fe	BCC	2.53	1667
α Sr	CCP	4.29	298
β Sr	HCP	4.31	521
γ Sr	BCC	4.19	887
Li	BCC	3.03	293
Li	HCP	3.11	351

Hume-Rothery have shown that metals will form alloys provided the following conditions are satisfied.

1. The metallic radii should differ less than 15%.
2. The crystal structures of the metals should be identical.
3. The metals must be chemically similar.

Thus, copper and zinc form alloys with five different structures.

They also form *interstitial compounds* with non-metals like carbide, boride and nitride which have a very high melting point and extreme hardness. They also form hydrides with CCP structure. They are regarded as close-packed structures of metals with smaller non-metallic elements located in interstitial sites or holes which are either octahedral .or tetrahedral. Let us examine the case of lithium on the basis of atomic orbital (a.o.) and molecular orbital (m.o.) bonding. There are as many m.o.'s as there are a.o's (with each m.o. holding two electrons with opposite spins) and since 2s a.o. of each lithium atom has only one electron, it follows that half the m.o.'s are fully occupied and the other half are vacant. In other words, the 2s-band in lithium is only half-filled. This conclusion remains valid if we consider a three-dimensional array of Li atoms.

This picture of metallic bonding readily explains metallic behaviour. Firstly, since the m.o.'s are spread over the entire crystal, electrons in these m.o.'s are completely delocalized. Secondly, the half-filled nature of the electric field, thus, permitting electrical conduction. The same two factors also account for thermal conductivity. The delocalized electrons absorb and re-emit light, rendering the metal surface shiny and lusturous. Finally, since the bonding electrons are not localized at any particular atom, the lattice can be deformed easily, thus, explaining the malleability and ductility of metals.

Our analysis of lithium band structure applies equally well to other alkali metals. If we apply the same analysis to magnesium which has two electrons in 3s level, the 3s-band would be predicted to be completely filled, thus, making magnesium a non-metal. However, the width of energy bands depends upon crystal structure and the crystal structure of magnesium is such that the 3p-band overlaps with the 2s-band. This means that electrical and thermal conduction can take place and magnesium exhibits metallic behaviour.

This example raises the obvious question, "Under what conditions band overlap can be expected?" The answer is also quite obvious. If the orbital interaction (that is orbital overlap) increases, there is greater lowering of energy and also greater overlap of bands involving different orbitals. Orbital irteraction increases with decrease in internuclear distance. One way to shorten the internuclear distance is by applying pressure. Thus, many elements (even non-metallic ones) under high pressure undergo phase transitions and become metallic (conducting).

In fact, calculations show (hat at a pressure of a few hundred thousand atmospheres, the hydrogen atoms should form a cubic close-packed structure which should conduct electricity. Another way that changes in internuclear separation can occur is when the crystal structure of a solid changes.

Thus, grey tin which has relatively open structure (density = 7.29 gm/cc) is a metal. As long as the crystal structure of a metal is not disturbed, its band structure remains unaltered. This is why foreign atoms of the same valency and similar size can readily replace atoms in a metal in arbitrary proportions (forming alloys) without significantly altering the crystal structure or band structure.

BAND MODEL

We shall try to understand it by means of a small ball in a well with steps at different heights. If the ball is rolled on the sides of the well it can either rest at the bottom of well (ground state) or on one of the steps (excited states). If there are more than one balls, the correct situation will be that we start from the bottom up and the number of balls which can be accommodated at each level depends on some parameters that characterize the steps. An atom may be thought of in a well with steps at different heights.

If there are non-interacting atoms, *e.g.*, argon, there would be many such wells separated by interatomic distance whereas in large amount of cases atoms do interact. In that case we have a large number of smaller steps or bands very close to each other between consecutive level. In each band there are as many steps as the atoms. In a diamond gem of one carat there are about 10^{23} atoms, hence, this will be the order of magnitude of steps with different energies.

It is not possible to accommodate the atoms outside the well. If the broadening of band is profound, the forbidden zone between two bands disappears and a wider band is obtained. Let us imagine some small balls in a container. These will fill the box with a lid in flat condition up to a certain level. If we tilt the container in one direction the balls will roll and occupy positions higher up with respect to the bottom of the box. If the box is completely full, tilting does not produce any moment in the balls. If these balls are electrons, at the bottom of the box, the lower energy limit and the lid, the upper energy limit, it can be seen that if the band is not full, the electrons will have a net movement in the direction of the applied field, *i.e.*, they will show electrical conduction.

We can still imagine the balls moving in a full box if they are churned but as many balls will move in one direction as in other. If the band is full, the electrons are not localized in one position, they can move but as many will go in one direction as in other. The net apparent result will be no motion, *i.e.*, no conduction, *e.g.*, diamond.

IMPERFECTIONS IN SOLIDS

In crystals, the state of complete order and of lowest energy is found at absolute zero of temperature. At any temperature above 0°K there will be some departure from complete order. In principles, any deviation from a completely ordered arrangement in a crystal constitutes native disorder. In addition to native disorder, crystals may have disorder due to the presence of impurities. The term imperfection or defect is generally employed to denote a departure from a perfectly periodic array of atoms in a crystal, immaterial of whether the imperfection is intrinsic or extrinsic in origin. Many a property of a solid such as the electrical conductivity, optical spectra or mechanical strength cannot be explained on the basis of its crystal structure alone. Imperfections may not only change the properties of a crystal, but may also give rise to certain new properties of interest.

The important types of native disorder are those associated with atomic and electronic imperfections. In addition, orientational disorder and notional disorder are also found in certain solids. A typical example of orientational disorder is found in ferromagnetic solids where all the magnetic moments are alligned in magnetic field at low temperatures. At higher temperature, the solid becomes paramagnetic due to the disorientation of the magnetic moments.

Motional disorder is exemplified by ammonium halides, where the ammonium ions rotate in phase at low temperatures. At higher temperatures the ions rotate out of phase. In this section, we shall be limiting our discussion mainly to the two more important types of native disorder arising from electronic and atomic imperfections.

Electronic Imperfections

It may appear strange at first sight to consider electrons in solids as imperfections, but we will soon see that such a classification is helpful in understanding several phenomena. In perfect covalent or ionic crystals at 0°K, electrons are present in the fully occupied lowest energy states. The distribution of the valence electrons depends on the nature of the chemical bonds in the case of covalent crystals, while they are mainly concentrated about the electronegative component in ionic crystals. In both these cases electrons do not move under an applied electric field. Above 0°K, some of the electrons may occupy higher energy states depending on the temperature and the energy distribution in the allowed states. Thus, in crystals of pure silicon some electrons are released thermally from the covalent bonds at temperatures above 0°K. These electrons which are free to move in the crystal would be responsible for the electrical conductivity in silicon. The electron deficient bond produced by the removal of an electron is referred to as a positive hole. Holes also give rise to electrical conductivity, but the direction of p motion of the holes in an electric field will be opposite to that of the electrons. Excited electrons and holes in solids are considered to be electronic imperfections. The electrons and holes are generally designated by e and h respectively and their concentrations by n and p.

The concentrations of holes and electrons will be equal in pure crystals like silicon at equilibrium. Electrons and holes can be preferentially produced in covalent crystals like germanium or silicon by adding suitable impurities. Impurities with excess valence electrons (P or As) give rise to extra electrons in germanium and silicon; impurities like Ga or Al with one electron less than Ge or Si produce holes.

In most ionic solids electrons are localized on a particular ion (unlike in metals where electrons are delocalized). The concept of electrons and holes can be employed in ionic crystals just as covalent solids. This is best understood by considering an ionic crystal where cations of two different valence states can be present. Consider a solid containing mainly ferrous ions.

If we introduce a small number of Fe^{+3} ions in this solid, it is equivalent to introducing holes. This is because, Fe^{+3} ions have one electron less than the preponderant Fe^{+2} ions in the solids. In such a solid, the electrons and holes spend appreciable time at cation sites before 'hopping' to the adjacent sites. The hopping process may be visualized in terms of the change, Fe^{+2} Fe^{+2} Fe^{+3} ® Fe^{+2} Fe^{+3} Fe^{+2} Fe^{+2}. The hopping process is responsible for the electrical conduction in such solids.

Atomic Imperfections

If imperfections in a crystal are caused by departure from the periodic arrangement in the vicinity of an atom or a group of atoms, the imperfections are called point defects. If the deviations from periodicity extend over microscopic regions of the crystal, they are referred to as lattice imperfections. Lattice imperfections may extend along lines (line defects) or surfaces (plane defects). Line defects are also called dislocations.

POINT DEFECTS

When an atom is missing from its normal lattice site a lattice vacancy (Schottky defect) is created. In the case of alkali halide crystals there are electron traps known as V and F centres. Heating the sodium chloride crystal in vapour of sodium generates F centres, while in the presence of chlorine vapour a υ centre is generated. In stoichiometric ionic crystals, a vacancy of one ion is accompanied by the creation of a vacancy of the oppositely charged ion in order to maintain etectroneutrality. Thus, equal numbers of cation and anion vacancies are found in alkali halides. In NaCl, there are approximately 10^6 Schottky pairs per cc at room temperature; in one cc there are about 10^{22} ions and therefore, there will be one Schottky defect for 10^{16} ions. Presence of a large number of Schottky defects in a crystal lowers its density markedly. The pyknometric density of vanadium monoxide, VO, is 5.6 g/cc while the x-ray density is 6.5 g/cc. Obviously, there is about 15% disorder of the Schottky kind in VO.

Atoms (or ions) which occupy the normally vacant interstitial sites in a crystal are called interstitials. The important factor determining the formation of interstitials is the size of the atom. There is another kind of point defect called the antistructure disorder, where atoms occupy sites which in the ideal crystal are assigned to atoms of a different kind. In a crystal containing atoms X and Y, some of the X atoms occupy Y sites and an equal number of Y atoms occupy X sites, if there is antistructure disorder in the crystal. Antistructure disorder is commonly found in alloys, particularly at higher temperatures, but is rare in ionic crystals since it would bright like ions next to each other.

In addition to the three basic types of point defects (Schottky, Interstitial and Antistructure) discussed above, three hybrid type of defects can arise from combinations of any two of the basic types as shown below. Of these three combinations the Frenkel defects are more common in ionic solids. In pure alkali halides one does not encounter Frenkel defects since the ions cannot get into the interstitial sites. Frenkel defects are however, found in silver halides where the Ag^+ ions are considerably smaller than the halide ions. Unlike Schottky defects, Frenkel defects do not change the density of the solid. In many real systems various types of point imperfections may occur simultaneously. Thus, in NiAl, Schottky and antistructure disorder are known to be present; in AgBr Schottky and Frenkel defects coexist.

There is a large variety of non-stoichiometric inorganic solids which contain an excess or deficiency of one of the components. Such solids showing deviations from the ideal stoichiometric composition form an important group of defect solids. There are some solids which are difficult to prepare in the stoichiometric composition; thus, the ideal composition in compounds like FeO is metastable (normally we get $FeO_{0.95}O$). The most stable compositions of praseodymium and terbium oxides under ordinary conditions are Pr_6O_{11} and Tb_4O_7 respectively. Similarly, many of the II-VI group compounds such as ZnO and CdS are difficult to prepare in their stoichiometric compositions. In fact, compounds become more interesting to a solid state chemist when there is non-stochiometry. In certain solids, gross variations in composition are also accompanied by changes in crystal structure, while in some others marked deviations from the stoichiometric composition may occur without any change in crystal structure. Thus, VO and TiO can be prepared over a range of compositions (VO_x or TiO_x; x = 0.6 –1.3) with NaCI structure.

Zinc oxide loses oxygen reversibly at high temperatures and turns yellow in colour. The excess metal is accommodated interstitially, giving rise to electrons trapped in the neighbourhood. The enhanced electrical, conductivity of the non-stoichiometric ZnO arises from these electrons (n-type conduction). In FeO, the non-stoichiometry arises from an excess of anions and the electrical conductivity is due to positive holes (p-type conduction). In such oxides the deviation from stoichiometry is accommodated by the change in the valence state of the cation; the cation valency increases if there is anion excess. These two behaviours of oxides can be understood in terms of the basic chemical reactions of the type given below:

Oxygen (anion) excess: 1/2 O_2(g) = O^{-2}(s) + cation vacancy + 2 h (p-type)

Oxygen (anion) deficiency: O^{-2}(S) = 1/2 O_2g) + anion vacancy + 2e (n-type)

Anion vacancies in alkali halides are readily produced by heating the alkali halide crystals t in an atmosphere of the alkali metal vapour. When the metal atoms deposit on the surface of the alkali halide crystal, halide ions diffuse to the surface and combine with the metal atoms. The electrons produced by the ionization of the alkali metal atoms then diffuse into the crystal and combine with a negative ion vacancy, electrons trapped in anion vacancies (referred to as F centers) give rise to interesting optical and electrical properties. Thus, excess potassium in KCl makes the crystal appear violet, and excess lithium in LiCI makes it pink. Alkali halides with holes trapped in cation vacancies (V centers) can be produced by heating the alkali halides in an atmosphere of the halogen. Such colour centers can be produced by the irradiation of alkali halides by high energy radiation like x-rays or γ-rays.

Till now the discussion of point defects has been limited to systems where there were no foreign atoms or impurities. Foreign atoms can occupy either interstitial or substitutional sites in a host crystal. The electronic structure of the impurity is of importance in the case of substitutional solid solutions while the size of the impurity atom determines the formation of interstitial solid solutions. Many of the impurity systems provide new and interesting properties of great importance; the most well-known of such solid solutions are those of Group III or Group V impurities with Group IV elements like Ge or Si. Group III elements like Ca or Al and Group V elements like P or As enter Ge or Si lattice substitutionally.

The Group V elements will have one excess valence electron after forming the four covalent bonds normally formed by Group IV elements. The excess electrons may be released to move through the host structure by thermal energies to give rise to n-type conduction. A Group III element which has only three valence electrons creates one electron deficient bond giving rise to a hole which can move in the crystal carrying an effective positive charge.

The electrical conductivity will therefore, be of p-type. In solid solutions of Group IV elements with either Group V or III elements, the overall charge neutrality is maintained. The impurities in these cases are ionized: imperfections that ionize releasing electrons (*e.g.*, P or As in Ge or Si), are called donors while those which ionize by accepting electrons (*e.g.*, Ga or Al in Ge or Si) are called acceptors.

The magnitude of energy required to ionize impurity atoms is generally small and in germanium at ordinary temperatures (0.025 eV) nearly all the donor and acceptor levels are ionized. The ionization energy decreases with the increase in impurity concentration and if the concentration is sufficiently high, a large fraction of ionized donors and acceptors is found even at low temperatures. The extra electron of the donor atom may be visualized as moving in orbits in the field of a positive charge. The picture is somewhat similar to that of the hydrogen atom, the main difference being that the electron and the positive charge are in a medium of high dielectric constant. Further, the radius of the orbit is much larger covering several atomic distances.

A common method of introducing defects in ionic solids is by adding impurity ions. If the impurity ions are in different valence state from that of the host ions, vacancies are created. Literature abounds in examples where interstitials or vacancies are produced by addition of impurities to ionic solids. For example, addition of $CdCl_2$ to AgCl or $SrCl_2$ to Nad yields solid solutions where the divalent cations occupy the Ag^+ or Na^+ sites and produce cation vacancies equal in number to that of the divalent ions. If one adds a molecule of a rare earth fluoride, MF_3 to CaF_2, the type of the concentration of defects produced will depend on the valence state of the rare earth ion in the host lattice. If the rare earth ion is in the +3 state, one F^- interstitial will be formed beside a new cation site; if the rare earth ion is in the +2 state there will be no interstitial F^- ion. The valence state and the defect structure in such systems is best studied by spectroscopic methods.

ELECTRICAL PROPERTIES

Based on their electrical conductivity, solids can be broadly classified into three types: metals, insulators and semiconductors. Conductivity of solids varies anywhere from 10^8 ohm^{-1} cm^{-1} in metals to 10^{-22} ohm^{-1} cm^{-1} in insulators. Electrical conductivity of solids may arise through the motion of electrons and holes (electronic conductivity) or other charged imperfections and ions (ionic conductivity). Substances like pure alkali halides, where the conduction is through ions only are generally insulators since the mobilities of ions are very much lower than those of holes or electrons. However, presence of vacancies or other imperfections could markedly increase the conductivity of ionic solids.

At laboratory temperatures, conductivity of metals is nearly independent of impurities and lattice defects. The electron concentration is mainly determined by the nature of the metal (number of valence electrons in the metal) and the mobility by the lattice vibrations. At low temperatures, however, the lattice vibrations are not limiting and the conductivity should be infinite. However, this is not true because of the presence of lattice imperfections and impurities. It is thus, convenient to use the resistance ratio p 300°K/ρ. 4.2°K as a measure of purity of the metal.

LINE DEFECTS

Although the concept of dislocations was first introduced primarily to explain deformation in metals, it is now established that dislocations play an important role in crystal growth, crystal transformations, electrical conductivity and other solid state phenomena. Dislocations in crystals have been observed directly with the aid of electron microscopes.

The dislocation consists of a line of atoms each of which is short of one coordinating atom than prescribed by the crystal structure. The dislocation extends along a direction perpendicular to the plane of the paper and is designated by the symbol $\perp$. The dislocation line marks the edge of the plane of atoms inserted part-way into the crystal. Edge dislocations are also called Taylor dislocations or Taylor-Orowan dislocations.

Another kind of dislocation called the screw dislocation described by Burgers (also known as Burgers dislocation) consist of a line of atoms with distorted coordination polyhedra.

This is caused by a displacement of atoms in one part of the crystal with respect to the rest of the crystal, the displacement direction being parallel to the dislocation line. The row of atoms marks the termination of the displacement. The polyhedron is not regular in the disturbed region and the distortic decreases progressively with increasing distance from the dislocation line.

(a) *Structure of an edge dislocation :* The deformation Is the result of inserting an extra plane of atoms on the upper half of the V-axis. Atoms in the upper half-crystal are compressed and those in the lower half are extended.

(b) Model to show termination of an edge dislocation in Sb of the surface of the crystal (H.C.Gatos, M.l.T.)

(c) Electron microscope picture of a thin crystal of platinum phytholocyanine (direct transmission) showing crystal planes and the presence of an edge dislocation. Magnification (300.000). (J.W. Menter, Tube Investments Research Laboratories).

An edge dislocation can be produced by applying a shear to the crystal and the net displacement of atoms in one part of the crystal relative to the atoms in the other part of the crystal determines the magnitude of the dislocation. Since atoms in acrystal occupy equilibrium positions, the net displacement is always equal to one of the equilibrium distances between atoms. If this distance (Burgers vector) is equal to the unit lattice translation, the dislocation is referred to as a unit dislocation. If the Burger's vector is shorter than a unit translation, the dislocation is called a partial dislocation.

PLANE DEFECTS

Many of the solid state materials are polycrystalline in nature. Polycrystalline solids consists of a number of small randomly-oriented interlocking crystals (grains). The grain boundary generally has a complex structure which in some way conforms to the shapes and orientations of the adjacent grains. Since a grain boundary forms a discontinuity in the periodicity of the lattice, it may be considered to be another kind of imperfection.

The disturbed region in the boundary layer may consist of a few atom layers and acts as a bridge between the adjacent grains. These disturbances are generally large due to random orientation of the initial grains. Since grain boundaries may contain a number of imperfections like impurities, interstitials, etc. (expelled by the grains during their growth), it often becomes difficult to understand the electrical, mechanical and other properties of polycrystalline solids.

Another important kind of plane defect found in crystals arises from stacking faults. In the cubic close-packed structures, we have the sequence ... ABCABC ..., while in the hexagonal case we have the sequence ... ABAB ... A fault in a stacking sequence as in ABABBCBCBC or ABCABCBCABCABC, occurs when the packing is cubic instead of hexagonal or vice versa. The sequence in the planes before and after the stacking fault would, however, be normal.

PROPERTIES OF SOLIDS

There is a close relationship between the physical properties of a solid and its structure and chemical composition. We shall now examine a few of the important properties of solids to illustrate the wide variety of possibilities. Some of these properties are exploited for various new innovations in electronic and magnetic devices such as transistors, computers, telephones, etc. The information explosion in inorganic and organic solid state materials has been so great in recent years that it would be difficult to list representative examples of all classes of materials in an elementary chemistry text of this sort.

Magnetic Properties

Materials can be divided into different classes depending on their response tq, magnetic fields. Diamagnetic materials (like NaCI or ZrO_2) are weakly repelled by magnetic fields.

In paramagnetic materials there are permanent magnetic dipoles due to the presence of atoms, ions or molecules with unpaired electrons (*e.g.*, O_2, NO, Na atoms, Ti_2O_3, VO_2) and these materials are attracted by magnetic fields; they, however, lose their magnetism in the absence of a magnetic field. Solids like TiO_2 which should be diamagnetic often exhibit paramagnetism due to the presence of slight non-stoichiometry.

Unlike paramagnetic materials, ferromagnetic materials show permanent magnetism even after the magnetic field is removed (*e.g.*, Fe, Cu_2Mn, Sn, EuO, CrO_2). The explanation for the large magnetization in these materials is that there are domains of magnetization (consisting of about a million atoms or ions), all of which cooperatively direct their magnetic moments (spans) in the same direction. Once such a material is magnetized (with all domains oriented), it remains permanently that way.

Iron, cobalt, and nickel are the only three elements which show ferromagnetism at room temperature. Manganese has most of the properties required for ferromagnetism, but the metal atoms are too close; however, addition of copper to manganese enhances the spacing and the alloy is ferromagnetic.

We have just learnt that a spontaneous alignment of magnetic moments in the same direction gives rise to ferromagnetism. If the alignment of moments is in a compensatory way so as to give zero net moment, then we get antiferromagnetism in the material (*e.g.*, MnO). Ferrimagnetism is found when the moments are aligned in parallel and antiparallel directions in unequal numbers resulting in a net moment (*e.g.*, Fe_3O_4, ferrites of the formula $M^{+2}Fe_2O_4$ (M = Mg; Cu, 3n, etc.).

All these magnetically ordered solids transform to the paramagnetic state at a higher temperature due to the randomization of spins. For example V_2O_3 and NiO transform from the anti-ferromagnetic phase to the paramagnetic phase at 150°K and 523°K respectively: Ferromagnetic CrO_2 becomes paramagnetic at 320°K; a similar transition in ferrimagnetic Fe_3O_4 is at 850°K. Magnetic properties of a few transitional metal oxides.

Table 2 : Magnetic properties of typical transition metal oxides.

TiO(p)	VO(p)	CrO(?)	MnO(af)	FeO(af)	CoO(af)	NiO(af)CuO(p)
Ti_2O_3(p)	V_2O_3(af)	Cr_2O_3(af)	Mn_2O_3(af)	Fe_2O_3(afw)	CO_3O_4(fe)	–Cu_2O(d)
TiO_2(d)	VO_2(p)		MnO_2(af)	Fe_3O_4(af)		
	V_2O_5(d)		Mn_3O_4(p)			

p = paramagnetic; ? = not known; af = antiferromagnetic; afw = antiferromagnetic with parasitic ferromagnetism; fe = temmagnetic; f = ferromagnetic.

Dielectric Properties

In insulators, electrons are closely bound to individual atoms or ions and they do not generally migrate under an applied electric field. However, dipoles are created by a shift in the charge resulting in polarization.

(i) These dipoles may align themselves in an ordered manner such that there is net dipole moment in the crystals or (ii) they may align themselves in such a manner that the dipole moments cancel each other. (iii) It is also possible that there are no polar species (dipoles) in the crystal, but only ions are present. Crystals where the situation (i) is found, exhibit piezoelectricity. When such a crystal is deformed by mechanical stress, electricity is produced due to the asymmetric displacement of ions or conversely if an electric field is applied to the crystal, there will be atomic displacements causing mechanical strain. Thus, a piezoelectric crystal acts as a mechanical-electrical transducer. Some of the polar crystals, when heated, produce a small electric current or pyroelectricity due to the asymmetric variation in interatomic distances.

In some of the piezoelectric crystals, the dipoles are permanently lined up (spontaneously polarized) even in the absence of an electric field and the direction of polarization can be changed by applying an electric field. This phenomenon is called ferroelectricity by analogy with ferromagnetism. Barium titanate ($BaTO_3$), sodium-potassium tartrate (Rochelle salt), sodium nitrite ($NaNO_2$) and potassium dihydrogen phosphate (KH_2PO_4) are typical ferroelectric solids. The occurrence of ferroelectricity undoubtedly depends on the structure and only certain geometric arrangements in the crystal can exhibit this phenomenon. If the dipoles in alternate polyhedra point up and down, there will be no net dipole moment and the crystal is said to be antiferroelectric. Lead zirconate ($PbZrO_3$) is a typical antiferroelectric. Both ferroelectric and antiferroelectric crystals transform at some temperature to the paraelectric phase where the dipoles are randomly oriented.

NON-CRYSTALLINE SOLID STATE

The liquid phase is the state of substances at temperature above their melting points. This includes all amorphous substances (silicate glass, rosin, etc.) Since silicate glass has no crystalline lattice, all amorphous solids are termed as glassy or glasses. Therefore, the two phase states of solids are crystalline and amorphous (glass).

The glass transition is the passage of a molten liquid into the solid state without changing the phase, *i.e.*, with the retention of short range order. When the liquid is cooled at a fast rate, its coefficient of viscosity increases and its energy of thermal motion decreases. The molecules remain in the random disorder characteristic of a liquid. Since it involves the breaking and reforming of strong covalent bonds, there is only a short-range order and a long-range disorder ultimately at a certain temperature at which B the coefficient of viscosity approaches 10^{13} poises which correspond to viscosity coefficient of solids. So the super-cooled liquid solidifies and the substance passes into amorphous (glassy) state.

There are two types of substances that can form glasses.

Glass

Substances, whose liquids consist of long molecules, that become very much entangled, form glasses. Organic polymers and polymeric form of sulphur, selenium, belong to this class.

Open Structure Crystals

These substances, which have directed covalent bonds and low coordination number, crystallizes in an open structure. SiO_2, H_2O, oxides of B, Si, Ge, P and As belong to this class. H_2O glasses are formed on extremely rapid cooling or upon condensation at 100°C. Alcohols, glycols and some other hydroxy substance also forms glasses.

In the case of SiO_2, each silicon atom is surrounded tetrahedrally by four oxygen atoms, each of which is bonded to two silicon atoms. The bond length of Si-O is 1.62 A and the ratio of the distance between comers of tetrahedron to the distance from a comer to its centre is 1.64. In the melt Si–O bonds are continually forming and breaking. When the melt is rapidly cooled the particular bond pattern present at that instant is frozen. The resulting glass is still a giant molecule but a rather disordered one. The result is a solid with the rigidity of a crystal but no regular order. In glassy state the tetrahedron order is distorted. The positional disorder existing within each distorted tetrahedron prevents the prediction and assignment of fixed position to the atom.

NON-CRYSTALLINE POLYMERS

Polymers undergo continuous transition from an ideal order to an absolutely disordered state. Very many polymers do not exhibit crystalline state and the crystalline order is also not found in the melt and the solutions of polymers because the thermal motion destroys the crystalline lattice. In the case of rapidly cooling the polymer solution, either a completely amorphous polymer or a body where crystalline and amonaphous region coexist, can be obtained. All the above mentioned cases form a group of non-crystalline bodies eronously called amorphous polymers.

Kargin suggested a definite degree of ordering within the limits of a non-crystalline state. He proposed a bundle model containing a set of macromolecules which are packed almost parallel and have large longitudinal dimensions. Let W(R) be the relative probability W of finding neighbouring atoms at a distance R from a certain fixed atom. For a crystalline substance this function shows constant distances at which the neighbouring atoms are arranged, exhibiting a long range order in the body. For non-crystalline polymers, it completely changes. The first high maximum conforms to the probability of an ordered arrangement of the first particle layer relative to the fixed particle showing a short range order. Such an arrangement results in ordered clusters on associates. Since the height of subsequent maxim is considerably lower, W(R) –1 and hence, the long range order is absent.

In the case of non-crystalline polymers the X-ray patterns give d_1 as 4.5-5Å and d_2 as 10Å. The presence of these maxima show that either macromolecules or segments of macromolecules are packed almost parallel to one another. Therefore, the main feature of non-crystalline polymers is a short-range order up to 10Å and absence of a long-range order.

The employment of electron microscope has shown that in crystalline polymers folding chains form the crystalline lamella, but in amorphous polymers parallel arrangement of chains occur up to 30-100Å, showing a short-range order. There is great deviation from a crystallographic order in the domain. It is assumed that these domains are connected with another by tie chains. Such domain exhibit transition from an amorphous to a crystalline structure. The non-crystalline polymers display a truely amorphous region which are formed by chain ends.

Plastics

Polymers according to their mechanical properties have been classified as elastomers and plastomers. Elastomers are rubbery elastic materials. Plastomers or plastics are macromolecular materials which flow on deformation. The deformation is permanent in plastics but in elastomers the body returns to its initial state instantaneously upon the removal of stress. Normally, in elastomers the chains are very flexible at the room temperature but they are stiff in the case of plastics. This depends on several factors namely, the degree of polymerization, cohesive forces acting between the chains, flexibility of the chains, their branching and cross linking and the degree of crystallinity. The effect of crystalline structure is discussed below.

Polymer Crystals

Polymers are products of high molecular mass and their decomposition temperatures are far below their boiling points. Hence, they exist only in liquid or solid state. Although most of the polymers are amorphous, but the crystalline polymers known as isotactic polymers have high melting point and produce fibres of high tensile strength and excellent mechanical properties. A polymer has two types of structural elements viz. monomer units and chains. The long range order is a prerequisite for the crystallization of polymer, not only in the arrangement of units but of chains as well as in three dimensions. This can happen in three ways:

1. Long-range order of arrangement of both chains and monomeric units in three dimensions.
2. Long-range order of arrangement of chains, but the monomeric units are arranged disorderly.
3. Long-range order of the arrangement of units but short range order of chain arrangement.

The high degree of ordering can be achieved in polymers in two ways, *i.e.*, by crystallization or by chain orientation without orientation of monomeric units. These correspond to crystalline and amorphous (glassy) state. It is possible to have a perfect short-range order in chains but the long-range order in arrangement of chains and monomeric units is not very perfect. Therefore, the examples of well ordered liquids and defective crystals are well known.

Crystallization of Polymers

Many polymers are incapable of crystallization under any condition. Their crystallizability is determined by their chemical constitution, *i.e.*, their packing, energy of intermolecular reaction, chain regularity, flexibility, etc.,

Regularity of Polymer Chain

A long-range order must exist in the chain in three dimensions. Sometimes irregularity in the chain may not obstruct crystallization but random and atactic copolymers cannot be crystallized under any circumstances. Natta by using certain catalysts succeeded in synthesizing several highly crystalline vinyl polymers, *e.g.*, polypropylene, polybutylene and polystyrene.

A mixture of the polymers of various degree of orientation was formed and crystalline fraction was separated from amorphous fraction by means of solvents as crystalline polymers are less soluble than amorphous polymers.

Second Order Transition Temperature

A first order transition is one at which a definite change in volume occurs. A second order transition is one at which a change occurs only in the rate of change of volume with temperature. Below T_g, polymers are rigid, glass like material and above T_g they are soft elastometric materials. The second order transitions are caused by thermal agitation releasing portions of the resin chain to rotate around its bonds.

These second order transitions can be measured conveniently by deflection temperature DT tests.

Flexibility of Polymer Chain

Polymers with sufficiently flexible chain can crystallize, but liquid chain polymers crystallize with difficulty because the thermal motion of the momeric, units disturb the established order and the crystal breaks. To crystallize a polymer, it is cooled to a temperature where the motion of its units does not hinder their orientation but excessive cooling may also hinder the rearrangement of the units. Crystallization of each polymer is possible only within a definite temperature interval, *i.e.*, between the glass temperature and flow temperature, *e.g.*, the glass temperature of isotactic polystyrene is about 100°C. Hence, it will only crystallize between 100°-220°C which is its melt temperature.

Packing of Molecules

Polymers obey the principle of close packing and macromolecules must be packed as closely as possible in the crystalline lattice. There are several possibilities for the formation of close packings of polymer chains. The first crystalline structure is built by close packing of spheres which is exhibited by globular proteins. The formation of such crystalline structures are possible because all the spheres are equal in size, *e.g.*, edestion catalase, iodobenzoyl glycogen.

The second possibility of close packing is the packing of helical macromolecules. The convexities of one helix fit into the concavity of other, *e.g.*, amylose, poly (hexamethylure adipamide), polypeptides. The third is the packing of long extended chains, *e.g.*, polyamide, polyurethanes, etc. The essential requirement is that the side chain substituents must not hinder regular arrangement of adjacent chains.

Close packing of chain molecules can only be accomplished with flexible chains capable of shifting in parts. Introduction of polar groups gives rise to two opposite effects. On one hand, intramolecular attraction increases, favouring close packing, while on the other hand, the thermodynamic flexibility (potential energy difference between two positions of the chain) of the chain decreases. The chain becomes more rigid. The rate of close packing depends on these two factors. However, with polar groups like –OH or –CN, the attraction effect prevails inspite the fact that the chains are rigid. Since rigid chains closely pack in the extended state, such polymers are highly oriented, *e.g.*, poly (vinylalchol), polyacrylo butyrate. The chains of isotactic polypropylene are highly flexible and form crystalline lattice, but rigid cellulose chains cannot form such lattices under any condition.

Polymer Crystal Cells

The crystalline cell of polymers consists of parts of chains located at definite distances a, b, and c and angles α, β and γ. They do not differ at all from cells formed by low molecular weight compounds and obey all the rules of symmetry.

Polyethylene crystals form orthorhombic unit cells with a = 7.40Å, b = 4.93Å and c = 2.53Å and density equal to 1.0 g/cm^3. All the carbon atoms in the polyethylene cell are arranged on the same plane forming a zigzag plane. In the polyethylene unit cell, chains are arranged along four edges and in the centre of the cell.

Polymers like sulphur or phosphorus are characterized by polymorphism. Polypropylene may form crystals belonging to monoclinic, hexagonal and triclinic types of symmetry. They are typical first order phase transitions. These transitions between various polymorphous formations take place either upon a change in temperature or under the action of load which brings a change in the parameters of a crystalline cell. As a matter of fact most of the polymers can be crystallized under stretched conditions. The morphology of a crystalline polymer is determined by the mutual arrangement of the unit cells within the limits of the crystalline state of a body.

Single Crystals

Single or ideal crystals are formed by the parallel transfer of unit cells along edges at distances equal to periods. Fischer first of all ascertained the possibility of high molecular compounds to form monocrystals. Polymer single crystals like other crystals have been found to contain certain defects originating in the packing of chains.

Lamellar Crystals

Polymer crystals are grown by crystallizing from a one per cent solution of polymer under gradual cooling or isothermal aging under the temperature of dilution. The dimensions, shape and regularity of structure of a single crystal depend on the chemical structure of a chain and the physical constraints like temperature, concentration of solution, the cooling rate, the nature of the solvent, etc. The polymer single crystals are monolayer flat plane Lamella which are often rhomboid of 100Å thick and the size of the plate up 1 mp. The axes a and & of a crystalline cell correspond to long and short diagonal of the rhomboid, while axis c along which the polymer chains are directed is perpendicular to the crystal plane. Since the length of the polymer is very big while thickness of the crystal is of the order of about 100Å, the long chain can only fit in the crystal by turning on its surface by 180°. This is known as chain fold conformation. To have an idea in a crystal of 120Å thickness, the fold shall contain 100 carbon atoms. A molecule having a weight of 105 shall fold about 70 times. The thickness of a lamella, the dimensions of crystallite and degree of defects of the boundary layers of a crystal are affected by the crystallization conditions. Crystals with folded chain conformation can be obtained when super cooling is in the range 15° – 20°C.

Crystallization under high hydrostatic pressure 5-10 × 10^3 atm give extended chain crystals. Polyethylene annealed under a pressure of 700 atm give extended chain crystals. When the melt is cooled under great deformation crystals of different morphological forms can be obtained. Polyhedrons known as hedrites and axialites or avoids are obtained by raising the concentration of the solutions.

Fibrillar Crystals

Fibrillar crystals are obtained by the high rate of evaporation of a solvent from a moderately concentrated solution or at cooling of melt. There are two views about their formation. One section believes that aggregation of rolled lamella is responsible for them while other hold that lamella degenerate in the formation of fibrillar crystals.

In stereospecific polymerization, fibrillar crystals are directly found. The formation is determined by the ratio of the rates of growth of a chain and its crystallization. By changing this ratio crystal either with extended or folded chains are obtained.

Globular Crystals

In globular crystals, macromolecules in coiled conformation form the lattice points. These are found in biopolymers, as a very high degree of uniformity of macromolecules is a pre-requisite for it, *e.g.*, tobacco mosaic virus synthetic polymers do not form them.

Spherulites

Spherulites are obtained when polymers crystallize from melts or concentrated solutions. They are made of crystallites which grow radially from one common centre. Macromolecular chains make at least an angle of 60° with the radius, *i.e.*, they are arranged tangentially to the spherulite radius Spherulites possess anisotropic properties. Their indices of refraction of light are different in radial or tangential directions. The orientation of the crystallographic axes continuously changes along the angular coordinates, A spherulite is called positive when the refractive index in the radial direction is greater than in the tangential one; otherwise it will be a negative spherulite.

In radial Spherulites one of the axis of the crystalline lattice retains a constant direction along all the radial directions. Ring type Spherulites are built of lamellar crystals whose orientation continuously changes along the spherulite radius. Depending on the degree of cooling, the same polymer may form different types of spherulitis. With low degrees of super cooling ring type Spherulites are obtained while radial Spherulites are obtained by high ones.

The formation of Spherulites passes through several stages. First the crystal nuclei are formed which are scattered throughout the volume of the sample and then independent growth of discrete fibrillar or lamellar crystalline structure grow at the same time in all directions. Then encountering one another they grow into regions filling the entire volumes of the body. Then their boundaries are distorted and they assume the form of a polyhedron.

The structural units in spherulite are bundles of macromolecules made out of chains in an extended conformation. This is possible when several macromolecules try to crystallize simultaneously from both the ends. Such interstructural bonds combine to form separate crystallites inside fibrillar or lamellar crystals.

Oriented State of Polymers

Polymers because of great anisotropy of the dimensions of macromolecules exist in oriented state. In the extreme case all the macromolecules will be oriented parallel to each other, but in reality this can happen on the average as the flexible chains do retain some fold conformations. In the oriented state the difference in amorphous and crystalline polymer vanishes.

Peterlin suggested that in oriented polymers microfibrils of diameter 100-200Å clearly divide crystalline blocks of chains with fold conformation. These chains are divided by defective layers and a large number of intra- and smaller number of interfibrillar chains. The microfibrils continue into larger fibrils to create the macrostructure of oriented bodies of the size 1000Å. These polymers are important for producing the fibres and films. In the first case an uniaxial orientation takes place while in the second a planar orientation takes place.

Liquid Crystals

Certain substances that contain rod-shaped molecules can exist between the crystalline and liquid state. The forces between these molecules are greatest when they-lie parallel and the crystals of such substances consist of parallel arrays of molecules stacked side by side when the crystal is heated, a point is reached at which the thermal energy is sufficient to break down the crystal lattice but not the parallel arrangement of the molecules. This results into a liquid crystal when the molecules are free to move provided they all remain parallel. At higher temperature the liquid crystal melts to a liquid. Such crystals are called *nematic.*

There are other liquid crystals where the molecules are not only parallel but their ends are aligned. Such crystals are known as *sematic.* In this case the crystals contain layers where the parallel molecules are free to move about in its layer but not in other layers. These crystals form drops in which the surface forms sharp steps with flat risers in between.

The intermolecular attraction in liquid crystals do not act equally in all directions with the result that all the degrees of freedom are not liberated. In a nematic liquid crystal the molecule can rotate only about its long axis and two rotational degrees of freedom are lost. In sematic liquid crystal the molecules are free to move only in the two directions and some translation degree of freedom is lost.

ELASTOMERS

Elastomers are non-crystalline high polymers that have a three dimensional space network structure.

They are built up like a fishing net, the siring of each mesh is replaced by a flexible chains of atoms with only weak forces between separate chains. These chains are prevented from melting by crystalline or by covalent interbonded portions. The distance between the head and tail of the chain is greatest when it is stretched out straight and less when it coils itself. It is therefore, elastic at the room temperature. Unlike steel, stretched elastomer is found to shrink on heating. It is another example of ordered and disordered arrangement.

UNIT CELL OR LATTICE

A unit cell is a convenient block, carved out from a crystal of infinite extension, bound by three pairs of parallel planes. The movement of this block along the three axes reproduces the whole crystal. Liquids on cooling give out energy, resulting in the decrease of translational energy of molecules, so much so, that they come under the influence of the field of secondary valency forces. The energy at this stage is only sufficient for rotational and vibrational motions of molecules, atoms, or ions, about a position of maximum potential energy, resulting in a solid. The lattice retains a perfect order with all the atoms occupying α-sites even at high pressure over a wide range of temperature.

The lattice can expand without loosing its order provided the kinetic energy of the molecules and the repulsive force between them decreases monotonically with increasing volume at constant temperature.

This increase in volume will create more disorder because of the movement of molecules to β-sites. The initial increase in disorder will be responsible for the increase in pressure due to the increase in repulsive force but when the volume becomes large enough, it would be followed by decrease in pressure as it would permit easy movement of molecules to β-sites. Thus, this model suggest a phase transition at constant pressure, leading to a critical temperature like gases, above which a liquid-solid transition does not occur. This phenomenon has not yet been observed even at highest pressures attainable.

Crystalline Solid

The molecules, under the influence of cohesive force, will try to orient themselves in an ordered manner in space, provided the rate of cooling is so adjusted that the molecules get enough time to orient and the density of the liquid is not too high.

Thus, a crystalline structure will result. Since there is only a single spatial arrangement of minimum potential energy for all the molecules in a compound, therefore, in the transition from liquid to solid state, the same spatial arrangement is repeated, *i.e.*, we obtain a crystalline lattice of the molecules.

Lattices and Structures

A regular array of points in space is called a lattice. If identical atoms or molecules occupy fixed position in a lattice, a crystal structure is obtained.

In liquid state, the chains of sulphur are broken at higher temperature. When the molten mass is poured in cold water, it gives plastic sulphur. Sulphur above 250°C on chilling gives thin rods which can be stretched 20 times the length at the room temperature. The resulting fibre Sϕ consists of Sγ(S_8 rings) and S4 consists of long chain molecules.

IONIC CRYSTALS

We have been dealing throughout with the structures having the balls of the same size. The majority of compounds consist of different elements. We shall now try to build up structures consisting of balls of different sizes. The simplest will be those structures which have balls of two different sizes.

Radius Ratio

Radius ratios of the two different balls play an important role in deciding the shape of the ultimate structure. It is useful to consider that the smaller ball occupies holes in a closed packed lattice of the larger balls. A closed packed lattice contains triangular holes (Z = 3), tetrahedral holes (Z = 4) and octahedral holes (Z = 6). Consideration of the geometry of the holes leads to the observation that for given larger spheres of unit radius, small spheres that can fill the holes should have the radius.

TYPES OF PLASTIC STATE

The liquids on cooling, try to attain a minimum potential energy and thus, form crystal. If it so happens that after some time the process of crystallization stops and super cooling starts, we get a series of crystalline structures included in the super cooled structures. Such a state is known as plastic state. Thus, polymers whose chains are stiff at room temperature are called plastic.

Macromolecules

Macromolecules are molecules built from a large number of atoms. They can be of natural origin like cellulose, proteins, etc., or they may be produced synthetically like poly (ethylene) or silicones. All macromolecules consist of at least one chain of atoms bonded together, not necessarily by covalent bond, but sometimes by coordinate bonds or electron deficient bonds.

This is known 'as main-chain which consists of constitutional units and of at least two end groups. A polymer is a substance composed of molecules characterized by the multiple repetition of one or more species of atoms linked to each other in amounts sufficient to provide a set of properties that do not vary markedly with the addition or removal of one or a few constitutional units (IUPAC).

Constitutional Units

A constitutional unit is the smallest unit whose repetition completely describes the main chain structure. Hence, it is known as the constitutional repeating unit.

Monomeric Unit

The monomeric unit (or mer. or base unit) is the largest constitutional unit contributed by a monomer during a polymerization process.

End Groups

End groups are the groups that occur at the end of a macromolecular chain. Linear chains have two end groups but the star shaped polymer with four arms has four end groups. The end groups of a polymer yield information on the mechanism of synthesis and this information may be used to determine the molar mass or the degree of branching.

Polymer Homologous Series

If the molecules have identical constitutional repeating units but differ only in the number of such units per macromolecule, the macromolecules are known as belonging to polymer homologous series.

The number of monomeric units joined together in a macromolecular chain is known as the degree of polymerization of a molecule. In exceptional cases the degree of polymerization is identical with the chain link number, which gives the number of atoms joined sequentially together in a chain.

Polymers with a small number of constitutional units and unspecified end groups are known as *oligomers*. Small refers to 2-20 constitutional units in synthetic polymers and several hundred units in nucleotide chemistry. Telomers are oligomers with end groups consisting of fragments of chain transfer agents. Telchelic polymers are oligomers with known functional end groups.

DEFECTS OR DEFORMATIONS

In actual practice it is not possible to arrange all the balls in positions required by specific structure. These balls slip past each other to produce what are known as defects. These defects may be point defects or linear deformations.

Linear Defects: Dislocations

The shifting of the balls from atomic sites in the crystal without separation of the atomic planes, will give rise to line defects, known as dislocations. The line defect, from above and below the slip, generates a screw pattern The distortions may be pure edge, pure screw or any possible combination of edge and screw components.

Hetero Chains

Styrene on heating changes from a clear liquid to a solid. Berthelot called this phenomenon as polymerization. He also observed that poly (styrene) on being heated at higher temperatures converts to styrene. This leads to miscelle theory of polymers. Staudinger classified colloids into colloidal dispersions miceller colloids (associated colloids) and colloidal molecules (macromolecular).

Steward and Von Weimam suggested that colloids are general possible states of material and not specific materials. Staudinger concluded that organic colloid consists of many atoms joined together by covalent bonds, *i.e.*, true macromolecules.

LIQUID STATE

The liquid state lies between the gaseous and the solid state in the sense that there is neither the ordered arrangement of constituents nor the complete disorder as in gases. It is generally observed that some of the properties of liquids closely resemble those of gases while some of the properties approach those of the solids. In the light of the kinetic molecular theory, liquids may be regarded as a continuation of gases into the region of small volumes and high intermolecular attraction. The cohesive forces in a liquid have been stronger than those in gases even at high pressures, and are sufficiently high to keep the molecules confined to a definite volume. The positions of the molecules in liquids arc not rigidly fixed. These forces are not strong enough to entirely eliminate the movements of the molecules in the liquid. Hence, the molecules in the liquid state have much shorter mean free path as compared to gas molecules.

Properties of liquids can be explained on the basis of their following characteristics :

(a) Appreciable forces of attractions exist between the molecules of a liquid. These are about 10^6 times as strong as in gases. The forces amongst the constituents of a liquid disallow them from separating spontaneously from each other, but they are not strong enough to hold the molecules in fixed positions. It is possible to explain surface tension, viscosity, fairly high heat of vaporization and vapour pressure of liquids in terms of these attractive forces.

(b) The molecules in a liquid are in a state of random motion although the extent of randomness is appreciably much smaller in comparison to gases. The molecules are relatively close together. Most of the space in the liquid gets occupied by its molecules and only a small fraction of the space is available to them for their free movement. This explains the higher density, incompressibility and slow diffusion of liquids in comparison to gases.

(c) The average kinetic energy of the molecules in a liquid is proportional to the absolute temperature. Increase in temperature would increase the proportion of the energized molecules, lowers the attractive forces between the molecules, and consequently increases the vapour pressure of the liquid.

From me above arguments, it is evident that most of the characteristic properties of the liquid arise due to the nature and magnitude of the intermolecular forces between the molecules. The important properties dealt with in this chapter are :

(i) Vapour pressure,

(ii) Surface tension, and

(iii) Viscosity.

LIQUID-VAPOUR EQUILIBRIUM-VAPOUR PRESSURE

Molecules of a liquid, like those of a gas, exhibit a distribution of energies. The molecules that are having sufficient energy to overcome intermolecular attractions are able to escape from the bulk into the gas phase (or vapour phase) provided they are close to the surface. This happens during evaporation. But the average kinetic energy of molecules in vapour state is more than that in liquid state. Therefore the temperature of the liquid falls on evaporation. The rate of evaporation of a liquid depends upon :

(a) The temperature of the liquid,

(b) Attractive forces in the liquid,

(c) Surface area, and

(d) Pressure above the liquid.

If we examine the evaporation of a liquid in an enclosed space (at constant temperature), it is found that some of the molecules of the liquid will spontaneously pass from the surface into the space above it. But the space is closed. Therefore, the molecules in the vapour state are unable to escape from the container. The molecules in the vapour phase collide with each other and with the sides of the container. Some of the molecules on collision pass on their energy to other molecules and come back to the liquid phase.

This phenomenon is termed as condensation. The rate of condensation of molecules in the vapour phase is proportional to the concentration of molecules in the vapour phase. A stage is ultimately reached when the rate of condensation becomes equal to the rate of evaporation. This situation corresponds to that of a dynamic equilibrium between the liquid and its vapour, *e.g.*,

$$H_2O(l) \underset{\text{condensation}}{\overset{\text{evaporation}}{\rightleftharpoons}} H_2O(g)$$

Thus a state of equilibrium may be defined as one in which two opposing processes occur simultaneously at the same rate. At this point there is no further change in the number of molecules in either the liquid or in the vapour state.

Molecules of a liquid in the vapour state exert some pressure. At equilibrium, this pressure is characteristic of the liquid at a given temperature and is called the vapour pressure of the liquid. The vapour pressure of a liquid is therefore defined as the pressure exerted by the vapours that are in equilibrium with the liquid at a given temperature. The vapour pressure has been found to depend on

(i) The nature of the intermolecular forces in the liquid, and

(ii) The temperature of the liquid.

(i) *Nature of the Liquid :* If in a liquid the attractive forces between the molecules are strong then the tendency of the molecules to escape from the surface of the liquid would be low. Hence, such a liquid will have low vapour pressure. However, liquids with low intermolecular forces would be having a high escaping tendency and consequently high equilibrium vapour pressure.

(ii) *Effect of Temperature :* As the temperature of a liquid gets increased the average kinetic energy of the molecules increases. The increased kinetic energy partly overcomes the attractive forces between the molecules and thus raises the escaping tendency of the molecules. This causes an increase in the equilibrium vapour pressure. The values of vapour pressure when plotted against temperature, a curve is obtained. The curve reveals that the vapour pressure increases exponentially with the increase of temperature. The temperature at which the vapour pressure of a liquid becomes equal to the atmospheric pressure is termed as the boiling point of the liquid. *The boiling point of a liquid is the temperature at which its vapour pressure is equal to the external pressure.* When the external pressure is 1 atmosphere, the term normal boiling point is used.

EQUATION OF STATE FOR LIQUIDS

In gases the distance between the molecules is large as compared to that in solids or liquids. All gases behave nearly alike at very low pressures (strictly speaking $P \to 0$) so that their behaviour can be adequately explained by an equation of state, $PV = nRT$. Solids and liquids are considered as *condensed phases*. Because of relatively small distances between the molecules, the effects of intermolecular forces are quite predominant in solids and liquids.

The magnitude of these forces tend to vary with me nature of the liquid or the solid. Thus, a general equation of state for condensed phases would not be possible.

An equation of state for solids and liquids can be deduced in terms of coefficients of thermal expansion (a) and compressibility (k). These are defined as

$$\alpha = \frac{1}{V}\left(\frac{\partial V}{\partial T}\right)_P \quad ...(1)$$

$$k = -\frac{1}{V}\left(\frac{\partial V}{\partial T}\right)_T \quad ...(2)$$

where α refers to the isobaric relative increase $\left(\frac{\partial V}{V}\right)$ in volume per degree rise in temperature while k isothermal relative decrease $\left(-\frac{dV}{V}\right)$

per unit rise in pressure, a and k depend on the nature of the condensed phases and are different for different substances.

For small temperature intervals equation (1) can be written as follows :

$$\frac{dV}{V} = \alpha\ dT$$

Integrating the above equation within limits, we get

$$\int_{V_0}^{V_t} \frac{dV}{V} = \int_0^T dT$$

$$\text{In} \frac{V_t}{V_0} = \alpha t$$

or $$V_t = V_0\ e^{\alpha t}$$

where V_t and V_0 refer to the volumes at t° and 0°C respectively. Expanding $e^{\alpha t}$ in the power series, we get

$$e^{\alpha t} = 1 + \alpha t \left(\frac{(\alpha t)^2}{2!} \right) + \ldots$$

As α is very small, square and other higher order terms may be neglected. Therefore, we can write

$$V_t = V_0(1 + \alpha t) \qquad \ldots(3)$$

Equation (3) relates the volume of a substance with temperature, α is constant over a limited temperature range and is approximately constant for gases. In case of liquids or solids, α is having its own characteristic value, α is having a positive value for gases and solids while for most of the liquids it is positive with some exception, *e.g.*, water between 0 – 4°C has negative value of α.

Similarly, integrating equation (2) we obtain

$$\int_{V_0^0}^{V0} \frac{dV}{V} = -\int_1^P kdP$$

$$\text{In} \frac{V_0}{V_0^0} = -\ k(P-1)$$

or $$V_0 = V_0^{\ 0}\ e^{-k(P-1)}$$

where V_0^0 refers to the volume at 1 atmosphere and 0°C.

On expanding $e^{-k(P-1)}$ in the power series and neglecting square and other higher terms, we obtain

$$V^0 = V_0^0 \, [1 - k(P - 1)]$$

Like α, k is always positive and is the same for all gases but each liquid or solid has its own characteristic value of k. k is constant over a fairly large range of pressure. Equation (4) predicts that the volume of a condensed phase gets decreased linearly with increase of pressure.

This behaviour is contrary to the behaviour of gases in which volume is inversely proportional to pressure. Combining equations (3) and (4) and eliminating V_0, we obtain

$$V_t = V_0^0 \, (1 + \alpha t) \, [1 - k(P - 1) \qquad ...(5)$$

The constants a and k in equation (5) depend upon the nature of the condensed phase. Table 3 gives the values of α and k for some common liquids and solids.

Table 3 : Coefficient of Thermal Expansion and Compressibility at 293 K.

Solid	*Coefficients*		*Liquid*	*Coefficients*	
	$\alpha \times 10^4\ k^{-1}$	$k \times 10^6\ atm^{-1}$		$a \times 10^{-4}\ K^{-1}$	$k \times 10^6\ a^{-1}$
Silver	0.582	1.00	C_6H_6	12.4	94.0
Copper	0.491	0.78	CH_3OH	12.0	120.0
Graphite	0.242	3.00	C_2H_5OH	11.2	110.1
Platinum	0.263	0.38	CCl_4	12.4	103.2
NaCI	1.212	4.20	H_2O	2.0	45.3

STRUCTURES OF LIQUIDS

In a perfect crystal a complete ordered arrangement of the atoms, ions or molecules constituting the crystal exists. The intermolecular forces are strong enough to hold them together. The constituents vibrate about their mean positions but cannot execute Translational motion. Hence both the short range as well as the long range orders exist in crystals. The molecules in a gas have complete random motion and the intermolecular forces between the molecules are small and are effective only at short distances. There is no possibility of any kind of ordered structure in gases. In liquids, on the other hand, the situation somewhat lies in between the two. The cohesive forces in liquids are stronger than those in gases. These forces, however, are not strong enough to disallow considerably the translational motion of the individual molecules. In terms of the arrangement of the constituents a liquid is having *short range order* but lacks long range order.

When a crystal melts there is, in general, an increase in volume of about 10% or about 3% in intermolecular spacings. The molecules in the liquid state still remain in the vicinity of the other molecules surrounding them. The intermolecular forces tend to decrease as the distance between the molecules increases. In other words, the short range forces exist while the long range forces are negligible. The thermal motion tends to introduce a disorder in the structure but the motion is not as random as in a gas. Some ordered arrangement still persists. The ordered arrangement in the crystal is not completely destroyed on melting. As the temperature gets further increased, the thermal motions of the molecules increase the kinetic energy, decrease the order until at the boiling point it completely vanishes.

The sharpness of the melting point could be explained by a two dimensional model for solids, liquids and gases. Introduction of a small region of disorder into a crystal brings about the disturbance of the long range order and destroys the crystalline arrangement. This explains the abrupt variation in the properties between solids and liquids. J.D. Bemal postulated that an atom is surrounded by five atoms instead of normal six atoms as is the requirement for a closest packed structure for solids. The remaining atoms (circles in the figure) are drawn in the utmost possible ordered arrangement. One point of abnormal coordination number is sufficient to bring about a long range disorder. That is, when the normal motion causes a disorder in one region, it spreads in all directions destroying the entire regular structure. Liquids, therefore, may be regarded to have structure like that of a crystal with a difference that the ordered arrangement extends over a short region instead of over the whole mass. This is termed as the short range order or long range disorder. It is to be noted that the short range order in a liquid structure is continuously changing because of the thermal motions.

X-ray Diffraction of Liquids : It becomes possible to calculate the distribution of atoms or molecules in a liquid by the analysis of the intensity of the scattered X-rays. The diffraction pattern of a liquid resembles a powder photograph with a difference that in place of sharp lines a few broad bands are observed. This reveals the existence of a regular order. From the intensities of these bands, it is possible to construct the radial distribution function which is interpreted in terms of the average number of atoms around a central atom at a certain distance. As the temperature of the liquid gets raised, the bands become less pronounced and the pattern resembles those of a gas revealing a more disordered arrangement at higher temperatures.

Vacancy Model for a Liquid : This model for the structure of liquids was given by Eyring (1933) who assumed the liquids more or less similar to a gas. In the liquid, most of the space gets occupied by the molecules and only a small fraction (about 3% of the total volume at ordinary temperature and pressure) of the total volume is free or void. This void space in which the molecules can move is known as the *free volume*. The vapours have only a few molecules moving randomly and thus has large free volume. With a rise in temperature, the concentration of the molecules in the vapour phase increases bringing about an increase in the vacancies of the liquid.

The density of the vapour increases while that of the liquid decreases until they become equal at the critical temperature. At the critical temperature the liquid and vapour become indistinguishable. Eyring further regarded that the free volume gets distributed randomly and the vacancies are of approximately molecular size. Molecules adjacent to a vacancy would be expected to posses gas-like properties and show maximum disorder. But a molecule away from the vacancy would possess solid-like properties and possessing a more ordered arrangement. Based on this model, they were able to show a fairly good agreement between the observed and predicted values of the properties.

Radial Distribution Function : The particles of a liquid are held together by intermolecular forces, but their kinetic energies are comparable to their potential energies. As a result, the whole structure is very mobile. The mathematical description of the average locations of the particles is expressed in terms of the radial distribution function g, this function is defined so gr^2 dr is the probability that a molecule will be found in the range dr at a distance r from another.

Calculation of g

Because the radial distribution function can be calculated by making assumptions about the intermolecular forces, it can be used to test theories of liquid structure. However, even a fluid of hard spheres without attractive interactions (a collection of ball-bearings in a container) gives a function that oscillates near the origin and one of the factors influencing, and sometimes dominating, the structure of a liquid is the geometrical problem of stacking together reasonably hard spheres. Indeed, a liquid of hard spheres shows a more pronounced oscillatory behaviour at a given temperature than any other type of liquid. The attractive part of the potential modifies this basic structure but sometimes only quite weakly.

Monte Carlo Methods

In the Monte Carlo method, the particles in the box are moved (usually one at a time) through small but otherwise random distances, and the total potential energy is calculated using one of the intermolecular potentials. Whether or not this new configuration is accepted is then judged from the following rules :

(1) If the potential energy is not greater than before the change, then the configuration is accepted.

(2) If the potential energy is greater than before the change, then it is accepted or rejected with the probability of acceptance in proportion to the value of $e^{-\Delta V_N/kT}$, where ΔV_N is the change in total potential energy of the N particles in the box.

It can be shown that this procedure ensures that at equilibrium the probability of occurrence of any configuration is proportional to the Boltzmann factor $e^{-\Delta V_N/kT}$.The configurations generated in this way can then be analyzed for die value of g simply by counting the number of pairs of particles with a separation r and averaging the result over the whole collection of configurations.

SURFACE TENSION

A molecule in the interior of a liquid is completely surrounded by other molecules and hence, on an average, it is attracted equally in all directions. However, a molecule on the surface of a liquid is subjected to a resultant inward pull because there is a greater number of molecules per unit volume in the liquid than in the vapour. Because of this inward pull, the surface of a liquid tends to contract to the smallest possible area and behaves as if it were in a state of tension. Since the natural tendency of a liquid is to decrease its surface area, any increase in the latter can only be accomplished at the expense of work. Consider a liquid film contained within a rectangular wire frame PQRS.

The side RS, of length *l* is movable. If a force 'f' is required to move RS against the force of surface tension acting in the film along RS, then the work, W, in moving the wire from RS to AB is given by

$$W = f.x \qquad ...(1)$$

This force f will have to be balanced by the force of surface tension along RS. If γ is the force acting per cm along RS due to the surface tension and since there are two surfaces to the film, the force due to surface tension would be $2\gamma l$, then

Work done in increasing the surface area $= 2\gamma.l.x$ ...(2)

or $$W = 2\gamma.l..x = f.x \qquad ...(3)$$

Hence, $$\gamma = \frac{f}{2l} \qquad ...(4)$$

Therefore, surface tension may be defined as *the force in dyne acting along the surface of a liquid at right angles to any line one cm dn length.*

From equation (3), we have

$$w = \gamma(2lx) = \gamma.\Delta A$$

where ΔA is the increase in surface area of the film. It follows that

$$\gamma = \frac{W}{\Delta A} \qquad ...(5)$$

Therefore, surface tension is the work in ergs required to generate 1 sq. cm of surface area

METHODS OF MEASURING SURFACE TENSION

The methods more frequently used in the determination of surface tension of liquids are described below.

Single Capillary Rise Method

Consider a fine capillary tube of a uniform radius 'r' immersed in a vessel containing a liquid which wets glass. The liquid will rise in the capillary till the force of surface tension f_1 acting upward is equal to the force f_2 due to the column of liquid acting downward.

Let 'γ' be the surface tension of the liquid expressed in dynes per cm of the inner circumference. This force is effective in pulling the liquid upward. Since this force is acting around the circumference, the force acting upward will be given by

$$f_1 = 2\pi r.\gamma \cos\theta \qquad ...(1)$$

This upward force is balanced by the weight of the liquid column.

Weight of the liquid column,

$$f_2 = \pi r^2 h \rho g \qquad ...(2)$$

where 'ρ' is the density of the liquid and 'g' is the Gaseous State due to gravity.

At equilibrium both the forces become equal and the height of the liquid remains steady.

Hence, $2\pi r.\gamma \cos\theta = \pi^2.h.\rho.g$

$$\therefore \qquad \gamma = \frac{r h \rho g}{2 \cos\theta} \qquad ...(3)$$

The angle θ is often called an angle of contact. For most liquids which wet glass, θ is nearly equal to zero.

$$\therefore \qquad \cos\theta = 1$$

$$\theta \qquad \gamma = \frac{r h \rho g}{2} \qquad ...(4)$$

For liquids which do not wet glass, *e.g.*, mercury, the angle of contact θ has to be taken into consideration.

In order to calculate γ by means of equation (3) or (4) r has to be determined by a travelling microscope, A by means of a cathetometer and ρ by a specific gravity bottle.

Double Capillary Rise Method

This method is useful when the given liquid is available in a small quantity.

If two capillary tubes of radii r_1 and r_2 ($r_2 > r_1$), are immersed in the same liquid whose surface tension is to be determined then the liquid rises to different heights in the tubes. Let the height to which the liquid rises in the two capallaries be h_1 and h_2 cm respectively.

Assuming that θ, the angle of contact, is negligible, it follows from equation (4)

$$\text{Surface tension} = \gamma = \frac{r_1 h_1 \rho g}{2}$$

Similarly, for second capillary tube with radius r_2

$$\gamma = \frac{r_2 h_2 \rho g}{2}$$

$$\therefore \quad \frac{\gamma}{r_1} = \frac{1}{2} h_1 \rho g$$

$$\text{and} \quad \frac{\gamma}{r_2} = \frac{1}{2} h_2 \rho g$$

$$\therefore \quad \gamma \left[\frac{1}{r_1} - \frac{1}{r_2} \right] = \frac{1}{2} (h_1 - h_2) \rho g$$

Hence if r_1, r_2 and the difference between two heights are known, γ can be calculated.

Drop Weight Method

The principle of this method consists in determining the weight of a single drop of a liquid which falls under its own weight through a capillary tube of uniform radius. At the instant the drop is about to fall the force f_1 due to surface tension pulling the drop upwards will be $2\pi r\gamma \cos\theta$. This force is counterbalanced by the weight of the drop f_2 acting downwards. At equilibrium both these forces become equal and hence $2\pi r\, \gamma \cos\theta = mg$, where 'm' is the mass of the drop.

Consider two different liquids having surface tensions γ_1 and γ_2 flowing through the same capillary tube. On assuming that their angles of contacts are equal, we get

$$\frac{2\pi r \gamma_1 \cos\theta}{2\pi r \gamma_2 \cos\theta} = \frac{m_1 g}{m_2 g}$$

$$\therefore \quad \frac{\gamma_1}{\gamma_2} = \frac{m_1}{m_2} \qquad \text{...(5)}$$

Hence, if we determine the masses of a single drop of each of the liquids, and knowing the surface tension of any one liquid that of the other can be calculated.

Drop Number Method

Since it is very difficult to determine the weight of a single drop, the drop weight method is not very convenient. In the drop number method, the number of drops formed from a fixed volume of a liquid is determined by using an instrument called the Traube Stalagmometer.

The lower portion of the stalagmometer consists of a fine capillary tube through which the liquids can flow. The upper portion is connected by rubber tubings to a Y-shaped glass tube.

At the start, the screw clamp is closed and the given liquid is sucked into the stalagmometer to a level above mark A by opening the pinch cock.

The pinch cock is then closed and the screw clamp gradually opened, so that number of drops flowing out is between 15 and 20 per minute. Once this drop rate is achieved the screw clamp is left undisturbed.

The liquid is then again sucked in by opening the pinch cock without allowing any air bubble to remain in the capillary. The number of drops formed while the liquid flows from the mark A to B is then determined. Let this number be denoted by n_1.

If $2\pi r\gamma_1 \cos\theta$ is the force due to surface tension on a single drop, then the force on n, drops will be $2\pi r\gamma_1 n_1 \cos\theta$. This latter force will have to be balanced by the weight of n_1 drops which is $M_1 g$ where M_1 is the mass of n_1 drops.

$$\therefore \quad 2\pi r n_1 \gamma_1 \cos\theta = M_1 g$$

$$\therefore \quad 2\pi r n_1 \gamma_1 \cos\theta = V.\rho_1.g \qquad ...(6)$$

where V = volume of the liquid between the marks A and B,

ρ_1 = density of the liquid.

The stalagmometer is then cleaned and dried and the same procedure repeated with another liquid of surface tension γ_2. Let the number of drops obtained with this liquid be denoted by n_2.

$$\therefore \quad 2\pi r\, n_2 \gamma_2 \cos\theta = M_2 g$$

$$\therefore \quad 2\pi r\, n_2 \gamma_2 \cos\theta = V.\rho_2.g \qquad ...(7)$$

From equations (6) and (7) we have,

$$\frac{2\pi r_1\, n_1 \gamma_1 \cos\theta}{2\pi r_2\, n_2 \gamma_2 \cos\theta} = \frac{v.\rho_1.g}{v.\rho_2.g}$$

$$\therefore \quad \frac{n_1}{n_2} = \frac{\rho_1}{\rho_2} \times \frac{\gamma_2}{\gamma_1}$$

$$\therefore \quad \gamma_2 = \frac{n_1}{n_2} \cdot \frac{\rho_2}{\rho_1} . \gamma_1 \qquad ...(8)$$

Hence, if n_1 and n_2 are determined and so also the densities of the two liquids, then the surface tension of any one liquid can be calculated if that of the other liquid is known.

In addition to these methods, there are 'maximum bubble pressure' and' Torsion balance' methods which are also frequently used. The description of these methods is outside the purview of this book.

VISCOSITY

Liquids possess another property known as viscosity which implies resistance to flow. It is developed in liquids because of the shearing effect of moving one layer of liquid past another. Suppose each plane is moving to the right with velocities v_0, v_1, v_2 etc. such that each succeeding velocity is greater than the preceding by an amount dv. This type of flow is called *laminar flow* or *streamlined flow as distinct from turbulent or vortex flow* in which parallelism of the layers is not maintained.

In laminar flow the force 'f' required to maintain a steady velocity difference 'dv' between any two parallel planes is directly proportional to the area of plane A and the velocity difference between the two planes dv. It is also inversely proportional to the distance between the planes.

Mathematically, it can be expressed as

$$\therefore \quad f \propto A.\frac{dv}{dy}$$

$$\therefore \quad f = \eta.A.\frac{dv}{dy} \qquad ...(1)$$

where, η is called the coefficient of viscosity of the liquid.

$$\therefore \quad \eta = \frac{f}{A \times \frac{dv}{dy}} \qquad ...(2)$$

When $A = 1$ sq.cm and $dv = 1\text{cm sec}^{-1}$,

$dy = 1\text{cm}$ then η = force in dynes

$$\text{units } \eta = \frac{\text{dynes}}{\text{cm}^2 \times \frac{\text{cm/sec}}{\text{cm}}} = \text{dynes.sec.cm}^{-1}$$

This is the absolute unit of viscosity coefficient. The practical unit in the C.G.S. system is known as poise.

$$\therefore \quad 1 \text{ poise} \equiv 1 \text{ dynes . sec cm}^{-1}.$$

Other units in use are centipoise (0.01 *poise*) and *millipoise* (0.001 *poise*).

Therefore, the coefficient of viscosity (η) 'Can be defined' as the force in dynes that must be exerted between two parallel planes 1 cm^2 in area and 1 cm apart in order to maintain a laminar velocity difference of 1 cm sec of one layer past another.

METHODS OF DETERMINING VISCOSITY COEFFICIENT

The method commonly used for determining viscosity coefficient is based on Poiseuilles equation,

$$\eta = \frac{\pi \, Pr^4 \, t}{8LV} \qquad ...(1)$$

where V = volume (in cc) of a liquid of viscosity η that flows through a capillary tube of radius r cm and length L cm in time t secs under a pressure head of P dynes/cm^2.

Instead of determining η by a direct application of the Poiseuille equation, the one commonly used is the one devised by Ostwald in which the viscosity of one liquid is compared with that of a liquid of known viscosity. (*e.g.*, H_2O).

Use of Ostwald Viscometer

A known volume (say V_0 cc) of the given liquid is introduced into bulb B of the viscometer. The liquid is then gently sucked into the bulb A above the mark 'x' and the time (t_2) required for the liquid to flow through the capillary from 'x' to 'y' is accurately determined using a stop watch. A mean of three observations is recorded. The viscometer is then cleaned and dried and the same volume (V_0 cc) of water, of known viscosity, is introduced into bulb B. This is again sucked into bulb A gently and the time (t_1) required for water to flow between marks 'x' and 'y' is determined as before.

From equation (1), we have,

For water $\qquad \eta_1 = \dfrac{\pi P_1 . r^4 . t_1}{8LV}$

For the given liquid $\qquad \eta_2 = \dfrac{\pi P_2 . r^4 . t_2}{8LV}$

Note that V = volume corresponding to that between the marks

'x' and 'y'

$$\therefore \qquad \frac{\eta_2}{\eta_1} = \frac{P_2 . t_2}{P_1 . t_1} \qquad ...(2)$$

$$\text{But } P = \frac{\text{Force}}{\text{Area}} = \frac{mg}{A} = \frac{V_0 . \rho g}{A}$$

$$\therefore \qquad \frac{P_2}{P_1} = \frac{V_0 . \rho_2 \, g/A}{V_0 . \rho_1 \, g/A} = \frac{\rho_2}{\rho_1}$$

From equation (2), we get

$$\frac{\eta_2}{\eta_1} = \frac{\rho_2 . t_2}{\rho_1 . t_1}$$

$$\therefore \qquad \eta_2 = \frac{t_2}{t_1} \times \frac{\rho_2}{\rho_1} \times \eta_1 \qquad ...(3)$$

Since all quantities on the R.H.S. of equation (3) are known, η_2 can be calculated. This is called the *absolute viscosity* of the given liquid. Then η_2/η_1 is the relative viscosity of liquid 2 w.r.t. liquid 1. If liquid 1 is water, then instead of using the term "relative viscosity", for the ratio η_2/η_1 we use the term "*specific viscosity*".

Determination of η by the 'Falling Sphere' Method

This method is specially useful for liquids having a high coefficient of viscosity.

Principle : A spherical body of a radius r and density p falling under gravity through a liquid medium of density ρ is acted upon by the gravitational force f_1 which is given by

$$f_1 = \frac{4}{3} \pi r^2 (\rho - \rho')g \qquad ...(4)$$

This force which tends to accelerate the body falling through the liquid is opposed by the frictional forces within the liquid which increase as the velocity of the falling body increases. Ultimately, a uniform rate of fall is attained at which the frictional forces become equal to the gravitational force and thereafter the body continues to fall with a constant velocity v called the *terminal velocity*. Stokes showed that this frictional force f_2 is given by

$$f_2 = 6\pi r \eta v \qquad ...(5)$$

where r = radius of sphere

v = terminal velocity

η = coefficient of viscosity of the liquid.

Equating the two forces, we get

$$\frac{4}{3}\pi r^3(\rho - \rho')g = 6\pi r\eta \upsilon$$

$$\therefore \qquad \eta = \frac{2r^2(\rho - \rho')g}{9v} \qquad ...(6)$$

Procedure : The viscometer consists of a vertical tube filled with the given liquid and placed in a thermostat at a particular temperature. A steel ball of density ρ and of such diameter to give a slow rate of fall is then dropped through the neck of the tube. Its time of fall between the marks 'x' and 'y' is ascertained accurately by means of a stop-watch. Another liquid of known viscosity and density is now placed in the cylinder and the time required for the same steel ball to fall between the marks 'x' and 'y' is determined. Hence from equation (6) it follows that

$$\frac{\eta_1}{\eta_2} = \frac{(\rho - p_1')}{(\rho - \rho_2')} \times \frac{v_2}{v_1}$$

But $\quad v = \dfrac{\text{distance between 'x' and 'y'}}{\text{time of fall}}$

$$\therefore \qquad \frac{\eta_1}{\eta_2} = \frac{(\rho - p_1')}{(\rho - \rho_2')} \times \frac{t_1}{t_2} \qquad ...(7)$$

EFFECT OF TEMPERATURE ON VISCOSITY

It has been found that the viscosity of a liquid decreases markedly with rise in temperature. The variation of viscosity with temperature is best expressed by the equation.

$$\eta = A.e^{E/RT} \qquad ...(1)$$

where A and E are constants for a given liquid.

Taking logarithms

$$ln\ \eta = ln\ A + \frac{E}{RT}$$

$$\therefore \quad \log_{10}\eta = \log_{10} A + \frac{0.4343\ E}{RT} \qquad ...(2)$$

Hence, a plot of $\log_{10}\eta$ versus 1/T should be linear. This has actually been borne out for a large number of liquids.

The reason why viscosity decreases with temperature is that before a molecule can take part in liquid flow it must acquire a sufficient energy to push aside the molecules surrounding it. As the temperature increases the number of surrounding molecules increases in proportion to the Boltzmann factor $e^{-E/RT}$ and hence the resistance to flow *e.g.*, viscosity may be expected to decrease according to the factor as $e^{-E/RT}$ equation (1).

LIQUID—LIQUID EQUILIBRIUM

(a) *Introduction :* Two liquids may be completely miscible, partially or completely immiscible with each other. Water and alcohol form an example of the first type, mercury and water constitute an illustration of the third type. Ether and water belonging to the second of the above categories, when mixed together form to separate layers, one being an aqueous solution of ether and the other, ethereal solution of water. The two solutions thus formed are known as conjugate solutions. It is with such cases of partial miscibility of liquids that we shall be concerned, because they represent this type of equilibria faithfully. Complete miscibility yields only one phase, while we are going to consider the cases of two components in two phases and so they are left out from our considerations. Complete immiscibility, however, yields two phases no doubt, but each phase is constituted of the pure component (only one) and as such loses interest in this connection.

(b) *Classification of L-L System :* L-L systems can be classified into three different classes:

(i) The liquid pairs completely miscible in all proportions, *e.g.*, water-alcohol, benzene-nitrobenzene etc.

(ii) The liquid pairs partially miscible with one another, *e.g.*, phenol-water, nicotine-water etc.

(iii) The liquid pairs completely immiscible with one another, *e.g.*, water-benzene, alcohol- nitrobenzene etc.

Completely Miscible Liquid Pairs

There are certain pairs of liquids which dissolve in each other in all proportions. The general rule about solubility has been that closer the chemical similarity of the two liquids, the greater will be their miscibility. The expression, *"Similia Similibus Solvunter"* (like dissolved by like) is fully justified. It is evident that the partial pressure of the components can be determined by the composition of the liquid and its temperature. This statement is present in *Duhem-Margales* equation for two components A and B, according to which

$$\frac{d \log P_A}{d \log x_A} = \frac{d \log P_B}{d \log x_B} \qquad \text{...(1)}$$

The terms x and p denote the mole fraction and partial pressure of the respective components.

Derivation of Duhem-Margules Equation

Suppose there is a binary solution which is having n_B moles of A and n_A mole of B. Applying *Gibbs-Duhem Equation*, we get at equilibrium:

$$n_A d\mu_A + n_B d\mu_B = 0 \qquad \text{...(2)}$$

were μ_A and μ_B denote the chemical potentials of A and B.

On dividing (2) by $(n_A + n_B)$, we obtain

$$\frac{n_A}{n_A + n_B}.d\mu_A + \frac{n_B}{n_A + n_B}.d\mu_B = 0 \qquad \text{...(3)}$$

where, x_A and x_B denote the mole fractions of constituents A and B respectively.

If the liquid mixture of A and B has been in equilibrium with their vapour phase at constant temperature T, the chemical potentials would be the same in both the phases for each component. Since,

$$\mu = RT \log f_i + \mu_i^0 \qquad \text{(where } f_1 = \text{fugacity)}$$

or if the vapours are to behave ideally, we can write

$$\mu_i = RT \log p_i + \mu_i^0$$

where P denotes partial pressure of ith component in the vapour phase.

Thus $\quad \mu = RT\ d \log p_i.$

On substituting it in equation (3), we get

$$x_A\ RT\ d \log p_A + x_B\ RT\ d \log p_B = 0$$

or $$\frac{x_A\, d \log p_A}{d x_A} + \frac{x_B\, d \log p_B}{d x_B} = 0$$

$\because$ $$x_A + x_B = 1,\ dx_A = -dx_B.$$

$\therefore$ $$\frac{x_A\, d \log p_A}{d x_A} + \frac{x_B\, d \log p_B}{d x_B} = 0$$

$$\frac{x_A\, d\log p_A}{d x_A} + \frac{x_B\, d\log p_B}{d x_B}$$

or $$\frac{d\log p_A}{d\log x_A} = \frac{d\log p_B}{d\log x_B} \qquad ...(4)$$

Equation (4) is known as *Duhem-Margules Equation.* It gives rise to a correlation between the composition in the liquid phase and the partial vapour pressures in the gaseous phase.

Konowaloffs Rule

It is also possible to write the *Duhem-Margules Equation* (4) in the form

$$\frac{x_A}{p_A}\cdot\frac{dp_A}{dx_A} = \frac{x_B}{p_B}\cdot\frac{dp_B}{dx_B} \qquad ...(5)$$

or since $$dx_A = -dx_B$$

$$\left[x_A = \frac{n_A}{n_A + n_B},\ x_B = \frac{n_B}{n_A + n_B},\right.$$

where n denotes number of moles of the components

$$x_A + x_B = \frac{n_A}{n_A + n_B} + \frac{n_B}{n_A + n_B} = 1$$

or $$x_A = -x_B$$

or $$\left. dx_A = -dx_B \right]$$

Hence, (4) becomes as follows:

$$\frac{x_A}{p_A}\cdot\frac{dp_A}{dx_A} + \frac{x_B}{p_B}\cdot\frac{dp_B}{dx_B} = 0$$

Hence. $$\frac{dP}{dx_A} = \frac{dp_A}{dx_A} + \frac{dp_B}{dx_A}\left(1 - \frac{x_B\cdot p_A}{x_A\cdot p_B}\right), \qquad ...(6)$$

where P = total pressure.

The factor $\frac{dp_B}{dx_A}$ must be negative, because it is equal to $\frac{-dp_B}{dx_B}$.

If $\frac{dp}{dx_A}$ is positive, then we have

$$x_B\, p_A > x_A\, p_B$$

or $$\frac{p_A}{p_B} > \frac{x_A}{x_B} \qquad ...(7)$$

If $\frac{dp}{dx_A}$ is negative *i.e.* $\frac{dp}{dx_B}$ is positive, then we have

$$x_B\, P_A < x_A\, p_B$$

or $$\frac{p_A}{p_B} < \frac{x_A}{x_B} \qquad ...(8)$$

Inequalities (7) and (8) have been the mathematical expressions of *Konowaloff's rule* that the vapour is richer in the component whose addition to the liquid mixture causes an increase of total vapour pressures.

Ideal Solutions of Liquids

An ideal solution may be defined as one in which the partial pressure of each component is proportional to the molar concentration in the mixture. This is also called *Raoult's law*. It may be mathematically stated in the form

$$p_A = p_A{}^0 \cdot x_A,$$
$$p_B = p_B{}^0 \cdot x_B,$$

where,

p_A, p_B = partial pressure of the respective components A and B

$p_A{}^0, p_B{}^0$ = vapour pressure of the two components

x_A, x_B = molar fraction of the two components.

Thus this law may be stated that the partial vapour pressure of any volatile constituent of a solution would be equal to its vapour pressure in the pure state multiplied by the mole fraction of two constituents in the solution.

Vapour Pressure—Composition Curve

Suppose there is a liquid pair A and B which are volatile and completely miscible and form an ideal solution, in which their mole fractions are x_A and x_B. Then their partial vapour pressures are as follows :

$$P_A = x_A \cdot p_A{}^0$$

and $$p_B = x_B \cdot p_B{}^0 \qquad ...(9)$$

where, $p_A{}^0$ and $p_B{}^0$ denote their vapour pressures in pure state.

The total vapour pressure (P) of the solution may be given by

$$P = P_A + P_B = x_A\, p_A{}^0 + x_B p_B{}^0 \qquad ...(10)$$

$$P = (1 - x_B)p_A{}^0 + x_B p_B{}^0 = P_A{}^0 + (p_B{}^0 - p_A{}^0)x_B \qquad ...(11)$$

The relations (9), (10) and (11) have been such that if p_A, p_B or P are made to plot against mole fraction of any component.

The dotted lines are giving plots of partial vapour pressures of A and B. It is also possible to derive the composition of the vapour by combining Dalton's law with Raoult's law. Let $P_A = x_A'P$. where x_A' denotes the composition of the vapour phase in equilibrium with the condensed phase. The following relationship can be derived for the vapour composition (only component A and B in system);

$$P = \frac{p_A^0 \cdot p_B^0}{p_A^0 - x_A^0\left(p_A^0 - p_B^0\right)} \qquad ...(12)$$

If the pressure of each constituent is plotted against its corresponding mole fraction in the liquid phase, then the plot would be a straight line which is passing through the origin.

The vapour pressure of the pure components A and B. The line obtained by joining A and B would be the vapour pressure curve of the mixture of A and B of all proportions. Thus any point x on AB denotes the vapour pressure of the mixture say 0.6 mole fraction of B.

It is evident that the vapour pressure x would be given by:

y = 0.4 × (vapour pressure of pure A) + 0.6 × (vapour pressure of pure B)

= (0.4 × AC) + (0.6 × BD).

The dotted lines BC and AD would give the partial pressures of the two components in the given various mixtures.

Ideal solutions get formed by chemically similar components, *e.g.*, ethyl bromide and ethyl iodide or n-hexane and n-heptane etc.

MEASURABLE PROPERTIES OF GASES

The behaviour of gases can be described in terms of certain parameters. These are described below.

Mass and Amount

The mass of a gas can be easily determined by weighting the container containing the gas and then emptying the container by taking out the gas and weighting the empty container again the difference between the two weights give the mass of the gas.

In chemistry, according to the IUPAC recommendations, the amount of a substance is expressed in terms of the number of moles (n). The number of moles can be obtained from the mass of the gas-by using the relationship,

No. of moles (n)

$$= \frac{\text{Mass of the gas in gram (m)}}{\text{Molar mass of the gas in gram per mole (M)}}$$

Volume (V)

Volume of the as is equal to the volume of its equation. Volumes can be expressed in litre (L), millilitre (mL), or cubic centimetre (cm^3). The SI unit of volume is cubic metre (m^3). In common practice, however smaller units such as cubic decimetre (dm^3) and cubic centimetre (cm^3) are more frequently employed. These are related to each other as,

$$1 \text{ m}^3 = 10^3 \text{ dm}^3 = 10^6 \text{ cm}^3$$

The commonly used unit, litre (L) has now been redefined so that,

$$1 \text{ L} = 1 \text{ dm}^3$$

and $1 \text{ mL} = 1 \text{ dm}^3$

The volume (V) of any gas depends upon its amount, temperature and pressure. Mathematically, volume of a gas is a function of the amount, temperature, and pressure, *i.e.*,

$$V = f \text{ (amount, temperature, pressure)}$$

or $V = f(n, T, P)$

The instrument used for the measurement of the pressure of a gas is called a manometer. It simply consists of a U-shaped tube containing mercury usually. One limb of the tube is longer than the other. Two types of manometers are used. As pressure is force per unit area, the pressure obtained in terms of the height of the mercury column can be converted into force per unit area as follows:

Suppose height of the mercury column $= h$ cm

Area of cross-section of the tube $= A cm^2$

$\therefore$ Volume of the mercury column $= A \times h cm^3$

If density of mercury at room temp $= \rho$ g cm^{-3}

then mass of the mercury column $= A \times h \times \rho$ gram

$\therefore$ weight of the mercury column $= (A \times h \times \rho) \times g$

where g if, the Gaseous State due to gravity.

This weight of the mercury column is the force acting on A cm^2. Hence.

$$\text{Pressure (P)} = \frac{\text{Force}}{\text{Area}} = \frac{A \times h \times \rho \times g}{A} = h\rho g$$

A *standard or normal atmospheric pressure* is defined as the pressure exerted by a mercury column of exactly 76 cm at 0°C. This is the pressure exerted by the atmosphere at the sea level.

The smaller unit commonly employed in expressing the pressures of gas is mm or torr (after the name of Torricelli, who invented the barometer). Hence

1 atm = 76 cm = 760 mm or 760 torr

The SI unit of pressure is pascal (Pa) which is defined as the pressure exerted by a force of 1 newton on an area of 1 m^2.

The two units are related as

1 atm = 101325 Pa or N m^{-2}

$= 1.01325 \times 10^5$ Pa or Nm^{-2}

For most of the purposes, the approximate value used is

1 atm = 10^5 Pa or Nm^{-2}

A bigger unit, called kilopascals (written as kPa) is also sometimes used. Thus

1 atm = 10^2 kPa

Measurement of Temperature : Temperature is a measure of the extent of hotness or coldness of a body. The measurement of temperature is based upon the principle that substances expand on heating. The most common substance whose expansion is made use of in the measurement of temperatures is 'mercury'. There are three different scales on which the temperatures are measured. These are

(i) Centigrade or Celsius scale (after the name of Anders Celsius)

(ii) Fahrenheit scale (after the name of Daniel Fahrenheit, a German instrument maker)

(iii) Kelvin scale (after the name of Lord Kelvin)

The Celsius scale is based upon taking the freezing point of water as 0°C and the boiling point of water as 100°C at normal atmospheric pressure (*i.e.* at sea level) and then dividing the range into 100 equal parts so that each part represents 1°C.

The Fahrenheit scale is based upon taking the freezing point of water as 32°F and the boiling point of water as 212°F and dividing the range into 180 equal parts. Thus, A large number of chemicals that go to form the basis of bigger molecules required for the synthesis of high polymers exist in gaseous state at the room temperature and atmospheric pressure, *e.g.*, acetylene, ethylene, propylene, 1-Butene, etc. In order to follow the mechanism of the formation of these bigger molecules and their subsequent polymerization, it is essential that we may not only study the characteristics and behaviour of these gases, but also the physical laws that govern their behaviour.

One of the main objectives of modern chemistry is to relate the properties of a large collection of molecules which constitute the matter in bulk (macroscopic properties) to the properties of individual atoms and molecules (microscopic properties). Inspite of the fact that the particles of a gas are always in random motion, a relation between microscopic and macroscopic behaviour is simpler in gases than in solids or liquids. The molecules of a gas, even at ordinary temperature and pressure, move with considerable kinetic energy, but they are least affected by the forces of attraction and repulsion.

The gases do not possess volume of their own but have attributes like touch and sound. They possess properties like mass, compressibility, refractive index, density and even electrical conductivity at very low pressure when the molecules get charged. Their behaviour has been found to depend on the population density of the molecules in a specified space.

TYPES OF GASES

There are two types of gases viz. *uniform* and *non-uniform*.

A gas need not normally have its mass, pressure, volume and temperature the same at all the places. Generally, it is a mixture of several elements as air is a mixture of oxygen, nitrogen, water vapour, carbon dioxide, etc. Its composition vary from place to place. Even in a chemically pure substance, *e.g.*, nitrogen, it may be considered that either the mass or volume or temperature or pressure or all of them may vary from place to place. Such gases are called non-uniform gases.

It is always possible to find out a small volume element where all the above measurable properties are constant. Such gas element are called uniform gases.

PROPERTIES OF GASES

There are two classes of properties viz. extensive and intensive.

The properties that depend on the amount of the gas like volume, mass, etc., are called extensive properties while in a uniform gas the properties that are independent of the amount of the gas, *e.g.*, pressure, temperature are called intensive properties.

It will not be possible to define a uniform gas. It is one that has the value of each of its intensive property the same irrespective of the region of the gas.

STATE OF A GAS

Each condition of a gas, *i.e.*, mass, volume, temperature, pressure, etc., which is completely distinguishable from any other condition is called as the state of the gas. The above measurable properties are called the state variable of the gaseous system. The state of a system is a function of its state variables.

Mass

The gas is composed of innumerable molecules each having a mass of the order of 10^{-24}g. So all these molecules collectively contribute towards the mass of the gas.

Occupies Space

The space occupied by all molecules of the gas is very small as compared to the space occupied by the gas. It is roughly of the order of 1000th part of the space occupied by the gas.

The remaining space at any instant is vacant, and as such it occupies a volume. In this context it would not be wrong to approximate the molecule as a point in the whole space. The volume is measured as millilitres.

Motion

At any temperature above absolute zero, the molecules of a gas are in motion. At room temperature the molecules of air move at a speed of 5×10^4 metres per second (1100 miles per hour), so they have fluidity.

Diffusivity

The process by which a gas moves from a high pressure through a porous wall is known as diffusion. The molecules of a gas are moving at a fast speed and do not occupy any appreciable volume by themselves, hence, they will not oppose expansion. So they will have very high diffusivity.

Isotropic

The molecules in a gas, occupying insignificant space, are moving randomly with high speed and as such do not have any effective attractive or repuslive force between each other with the difference that if one molecule moves in a stipulated direction then it must correlate with the other, moving simultaneously in the opposite direction. Thus, they will occupy the space of the container and as such will not have a volume of their own and can be easily added or separated if no change of composition is to be achieved. They are mutually soluble as there is too much vacant space in between them. So its movement appears to be equal in all the directions. Since they have the same movement in all directions, they are known as isotropic.

Stationary State

Any change in the motion of one molecule is correlated with the compensatory change in the other molecule. Therefore, unless external factor is introduced the state of the gas remains stationary.

Stationary Energy

The molecules of the gas collide with the walls of the container and rebound. These collisions are assumed to be perfectly elastic and as such exert a pressure on the walls of the container but do not lose energy on rebounding from it. Therefore, at a given temperature the total energy of the system remains constant.

LAWS OF NEWTON

The olecules of the gas change direction on striking against the wall of the container. In all these cases they are guided by the Newton's laws of motion. These laws provide equation of state of uniform gases at low pressures.

The behaviour of a uniform pure gas like hydrogen, oxygen, nitrogen, carbon dioxide, isopentane, etc., is governed by pressure and temperature as intensive variable and mass and volume as extensive variable. This can be mathematically expressed by the equation:

$$F(m, V, P, t) = 0 \qquad ...(1)$$

Since V and m have extensive nature

$$V = V(m, P, t) \qquad ...(2)$$

$$= mV(1, P, t) \qquad ...(3)$$

Eqn. 3 represents a function depending on P and t as variables and refers to a unit mass which can be written as

$$V = m\overline{\upsilon}\ (P, t) \qquad ...(4)$$

where $\overline{\upsilon}$ is the volume for a unit mass and is known as specific volume of the gas.

CHARLES AND GAY LUSSACS LAW

Charles and Gay Lussac in 1802 studied the effect of temperature on volume at constant pressure and concluded that all gases have the same coefficient of expansion at the same temperature provided the pressure and amount of gas are kept constant.

$$V = V_0 + \left(\frac{\partial \upsilon}{\partial t}\right)^t \text{ (n, P constant)} \qquad ...(1)$$

The intercept at the vertical axis is Vg which is the volume at t = 0° and slope of the line is $\frac{\partial \upsilon}{\partial t}$ which is α.

$$V_t = V_0 + V_0\ \alpha t$$

$$= V_0\left(1 + \frac{t}{273.16}\right) = V_0\left(\frac{t + 273.16}{273.16}\right) = V_0\ \frac{T}{T_0}$$

$$\frac{V_1}{T} = \frac{V_0}{T_0} \qquad ...(2)$$

Since both VQ and a are positive.

$$\frac{V}{V_0} = \frac{T}{T_0} = K_1(t) \quad ...(3)$$

where K_1 is a constant. By plotting K(t) vs T, a straight line was obtained showing a linear relation. It is mathematically represented as:

$$Kt\ (t) = RT \quad ...(4)$$

where R is a new universal constant.

The equation of state can now be written as:

$$P\overline{V} = K_1\ (T) = RT \quad ...(5)$$

where $\overline{V}$ is wv, *i.e.*, volume per gram mole of the gas known as Molal volume.

AVOGADRO'S HYPOTHESIS

In 1811, Amedeo Avogadro suggested that equal volumes of all gases contain equal number of molecules under the same conditions. This means that volume is proportional to the number of molecules.

Cannizzaro in 1860 on the basis of the above proved that water is H_2O and not HO as was believed earlier. Avogadro defined a new term mole of a substance. The number of grams equal to its molecular weight is known as Mole. One mole of the gas at the same temperature and pressure will contain the same number of molecules given by (6.023 × 1023). This figure is called Avogadro's number and applies to one mole of any compound. Avogadro's hypothesis in equation form can be written as:

$$V = K_2(t, P)n \quad ...(1)$$

where K_2 is a constant.

EQUATION OF STATE

Bernauli independently arrived at a similar equation of state. Let V_1, P_1 and T_1 be the volume, pressure and temperature of a mass of gas respectively. Let us do some work.

(1) At constant temperature, change the pressure P_1 to P_2, so that the volume V\1 becomes V_x

$$V_x = V_1 \frac{P_1}{P_2} \text{ (Boyle's law)} \quad ...(1)$$

(2) At constant pressure P_2, change the temperature T_1 to T_2, so that volume changes from V_x to V_2

$$V_x = V_2 \frac{V_1}{V_2} \quad \text{(Charle's law)} \qquad ...(2)$$

$$V_1 \frac{P_1}{P_2} = V_2 \frac{T_1}{T_2}$$

or
$$\frac{P_1 V_1}{T_1} = \frac{P_2 V_2}{T_2} = \frac{PV}{T} = \text{constant} \qquad ...(3)$$

Again
$$d\upsilon = \left(\frac{\partial \upsilon}{\partial P}\right)_T P + \left(\frac{\partial \upsilon}{\partial t}\right)_P T \qquad ...(4)$$

Now
$$\left(\frac{\partial \upsilon}{\partial P}\right)_T = -\frac{K_1}{P_2} \text{ (Boyles law)} \qquad ...(5)$$

$$\left(\frac{\partial \upsilon}{\partial T}\right)_P = K_2 \quad \text{(Charles law)} \qquad ...(6)$$

$$dv = \frac{K_1}{P_2} dp + K_2 \; dT \qquad ...(7)$$

$$= \frac{V}{P} dp + \frac{V}{T} dT$$

or
$$\frac{v}{V} = \frac{P}{P} + \frac{T}{T}$$

on integration

$$\frac{PV}{T} = \text{constant} \qquad ...(8)$$

The equation of state for an ideal gas can be written as

$$PV = nRT \qquad ...(9)$$

Expressing the equation in terms of density and mass,

$$P \frac{nRT}{V} = \frac{m}{M} \frac{RT}{V} = \frac{\rho RT}{M} \qquad ...(10)$$

where n, m, M and ρ are the number of moles, mass of the gas, molecular mass and density of the gas respectively.

The gas that follows the above equation is called an Ideal gas.

Gas Constant

The ideal universal gas constant R has a value of $\frac{P\overline{V}}{T}$ for one mole of a gas and is independent of the nature of the gas. One mole of a gas occupies a volume of 22.414 litres at 1 atmosphere pressure and 273°K.

$$= \frac{P\overline{V}}{T} = \frac{1 \text{ atom} \times 22.414 \text{ litres}}{1 \text{ mole} \times 273° \text{K}}$$

$$= 0.0821 \text{ litre atm mole}^{-1} \text{ °K}^{-1} \qquad ...(11)$$

R has the dimensions of energy per degree per mole.

Values of ideal gas constant R in different units

$$R = 0.0821 \text{ l atm K}^{o-1} \text{ mole}^{-1}$$
$$= 82.057 \text{ ml atm K}^{on} \text{ mole}^{-1}$$
$$= 62.3 \text{ l} - \text{mm Hg K}^{o-1} \text{ mole}^{-1}$$
$$= 8.3143 \text{ J°K}^{-1} \text{ mole}^{-1}$$
$$= 1.987 \text{ carl °K}^{-1} \text{ mole}^{o-1}$$

Boltzmann denoted the gas constant per molecule as K.

$$K = \frac{R}{W} = \frac{8.314 \times 10^7 \text{ erg deg}^{-1} \text{ mole}^{-1}}{6.023 \times 10^{23}}$$

$$= 1.3806 \times 10^{-16} \text{ erg deg}^{-1} \text{ mol}^{-1} \qquad ...(12)$$

The average kinetic energy

$$\overline{E} = \frac{3}{2} KT$$

$$= 6.17 \times 10^{-14} \text{ erg at 25°C} \qquad ...(13)$$

MIXTURE OF GASES, DALTON'S LAW OF PARTIAL PRESSURES

So far we have considered the relation between P, V, T, and n for a gaseous system in which only one component is present. If several substances are present in a gaseous mixture, we can still use the gas laws by taking into account the presence of different substances. The relationship between P, V, T and n in the mixture of gases can be illustrated by the experiment.

Suppose the sample of nitrogen is pumped into the first vessel and its pressure is found to be P_1 cm Hg; a sample of oxygen is pumped into the second vessel and its pressure is found to be P_2 cm Hg. If both the samples are pumped into the third vessel, the pressure is observed to be the sum of P_1 and P_2. This shows that each component independently contributes to the total pressure provided they do not react.

This behaviour of a mixture of two gases was first expressed in a generalized form in 1807 by John Dalton and is known as Dalton's law of partial pressures. This law states that the total pressure exerted by a mixture of non-reacting gases is equal to the sum of the pressures which each component would exert if placed separately into the container:

$$P_t = P_1 + P_2 + P_3 + \dots \text{ (T, V = constant)} \qquad \dots(1)$$

where P_t is the total pressure and P_1; P_2, P_3, are partial pressures of components 1, 2, 3 ..., etc. In a concise form, we can write the above expression as,

$$P_t = \sum_i P_i \text{ (V and T constant)} \qquad \dots(2)$$

where the symbol $\sum_i$ stands for the summation over all the components present in the mixture.

The partial pressure P_i of component; is defined as the pressure that the gas would exert if it was present alone in the same volume and at the same temperature. Thus, if a container of volume contains n_1 moles of gas 1, n_2 moles of gas 2, etc., then the partial pressures of gases 1, 2, etc. are

$$P_1 = \frac{n_1 RT}{V}, P_2 = \frac{n_2 RT}{V}$$

The total pressure of the system can be written as

$$P_t = \sum_i P_i \frac{RT}{V} \sum_i n_i = n_t \frac{n_2 RT}{V} \qquad \dots(3)$$

where n_t is the total number of moles of gas present in the mixture, $n_t = n_1 + n_2 + n_3 + \dots = \sum_i n_i$. This equation is an even more general form of the ideal gas law, valid for mixtures as well as for pure gases; it combines in one equation, the laws of Boyle, Charles, Avogadro and Dalton.

For mixtures of gases, it is sometimes convenient to express the law of partial pressures in terms of mole fractions of the various components of the mixture. The mole fractions, of a component i is defined by $x_i = n_i/n_t$, where n_i is the number of moles of component i and n_t is the total number of moles. The partial pressure P_i of component is then given by

$$P_i = x_i P_t \qquad ...(4)$$

Remember that $\sum_i x_i = 1$, so that $\sum_i P_i = P_t \sum_i x_i = P_t$

Equation 32 represents Raoults law.

A law similar to Dalton's law is Amagat's law of partial volumes. According to this law, $V_t = \upsilon_1 + \upsilon_2 + \upsilon_3...$ at constant temperature and pressure. Here V_t represents the total volume and $\upsilon_1, \upsilon_2...$ the partial volumes. Thus, when two or more gases are mixed at constant volume and temperature their pressures are additive; when they are mixed at constant pressures and temperature, their volumes are additive.

$$V_1 = n_i \frac{RT}{P} \qquad ..(5)$$

or

$$W = \frac{m_1}{P_1} \frac{RT}{V} \qquad ...(6)$$

The application of Dalton's law is illustrated by the following examples.

Example:

12.59 × 10⁻³ moles per litre of N^ when heated to 25°C exerts a pressure of 0.3942 atmos. The gas was found to contain a mixture of N_2O_2 and NO_2. Calculate the average molecular weight and partial pressures due to each.

Solution:

$$M_{Av} = \frac{(12.59 \times 10^{-3})\, 92.2 \times 8.2 \times 10^{-2} \times 298}{0.3942}$$

$$= 71.78 \text{ g mole}^{-1}.$$

$$^{46x}NO_2 + {}^{92}N_2O_4 = 71.78 \text{ g mole}^{-1}$$

$$46 - {}^{46x}N_2O_4 + {}^{92x}N_2O_4 = 71.78 \text{ g mole}^{-1}$$

$$1 + {}^{x}N_2O_4 = .56 \text{ or } {}^{x}N_2O_4 = 0.56$$

$$P_{N_2O_4} = (0.56)\,(0.3942) \text{ atm} = 0.0221 \text{ atm}$$

$$P_{NO_4} = (0.3942 - 0.0221) \text{ atm} = 0.3721 \text{ atm}$$

Specific Volume

The specific volume serves as a very important quantity in calculating the molecular weight of a gas.

$$PV = nRT = \frac{m}{M} RT$$

$$M = \frac{m}{V} \frac{RT}{P} = \rho \frac{RT}{P} \quad ...(7)$$

$$V_{ideal} = \frac{8.2 \times 10^{-2} \text{ 1 atm/mole} \times 273° K}{1 \text{ atm}}$$

$$= 22.4151/\text{g mole} \quad ...(8)$$

A plot between d/p and P at 0°C for argon

As P approaches zero d/p is 1.7826 g/1 atm

The molecular weight of Argon

$= 8.2057 \times 10^{-2}$ 1 atm/deg mole × 273 °K × 1.7826 g/1 atm

= 39.955 g/mole.

KINETIC MOLECULAR GAS MODEL

In the preceding sections of this chapter, we examined the physical behaviour of gases from an empirical point of view and paid no attention to questions like why gases occupy the entire space available to them, why they are highly compressible, how gases exert pressure, why different gases diffuse at different rates, etc. Since it is not possible to deduce the nature of gases from the measured properties alone, it becomes necessary to look for a model to account for the observed facts. In order that the model is simple, certain assumptions will have to be made about the nature of gases. We shall now examine such a model first by discussing the validity of assumptions and later by comparing its predictions with observed facts.

ASSUMPTIONS OF KINETIC THEORY

The following assumptions are made to describe an ideal gas :

(i) A gas consists of a large number of particles (atoms or molecules) that are so small and so far apart (on the average) that the volume of the molecules is negligible compared to the empty space between them. This assumption seems to be justified in view of the great compressibility of the gas. Calculations also show that about 99.9% of the total volume in nitrogen gas at STP is empty space. The average spacing between molecules of nitrogen at STP is worked out to be about ten times the molecular diameter.

When the gas is compressed, the average spacing between molecules is reduced.

(ii) There are no attractive forces between molecules and the molecules are completely independent of one other. The support for this assumption comes from the observation that even highly compressed gases expand spontaneously to occupy all the space available. This suggests that there is no appreciable interaction between molecules.

(iii) The molecules of a gas move randomly in straight lines colliding with one another and also with the walls of the container. The pressure exerted by a gas is the result of collisions of the molecules with the walls of the container. That particles of a gas are not at rest but in a state of continuous motion (called thermal agitation) cannot be directly observed because molecules are too small to be seen individually even through the best microscope. However, movement of particles of size in the neighbourhood of 1 p. diameter (*e.g.*, particles of smoke floating in air) can be observed microscopically. This phenomenon is called Brownian motion (after the British botanist who first observed the irregular motion of pollen particles suspended in water). It is found that the smaller the particles, the more violent is the irregular motion; also, the higher the temperature the more vigorous is the movement of the particles. In the absence of attractive forces between molecules, it is reasonable to assume straight line random motion for the molecules which are much smaller in size. Molecules, however, change the path as a result of collision with other molecules or the walls of the container.

(iv) In these collisions, there is no net loss of kinetic energy, although, there may be a transfer of energy between the partners in the collisions (elastic collisions). If there were loss of kinetic energy in collisions (inelastic collisions), the motion of molecules would eventually stop and the pressure and temperature would drop to zero. Since the pressure of a confined gas does not change with time (unless we heat or cool it), we conclude that a gas exerts pressure through elastic collisions with the wall.

(v) At any particular instant, different molecules in a sample of gas have different speeds and hence, different kinetic energies, but the average kinetic energy of the molecules is directly proportional to the absolute temperature.

Root Mean Square Velocity

The molecules of a gas in an enclosed space suffer continuous changes of momentum through collision with other molecules. As a result of such collision any individual molecule will acquire velocities ranging from zero to infinity. It is quite probable that any molecule in a given time may exhibit an average velocity zero as there are equal probability of its movement in any direction (both positive and negative) but the average of its square velocities shall always be positive.

$$\sqrt{u^2} = \sqrt{\frac{u_1^2 + u_2^2 + u_3^2 + u_4^2 \ldots u_n^2}{n}} \qquad \ldots(1)$$

Random Walk Model

Let the molecules be rigid spheres of diameter d moving at random at an average velocity $\overline{C}$ like Brownian motion. The two spheres will collide only whenever the distance between their centres will become d. Let the centre of a sphere be surrounded by an imaginary sphere of radius d. As the sphere is travelling with an average speed $\overline{C}$, in a unit time it will pass through a volume $\pi N d^2 \overline{C}$. Let there be N molecules per c.c. Then the sphere will experience $\pi N d^2 \overline{C}$ collisions per second. The quantity ndl is known as collision cross-section for the rigid sphere. It can be proved that the relative velocity after collision becomes $\sqrt{2\overline{C}}$.

The distance travelled by a sphere between collisions can be given by $\frac{1}{\sqrt{2}\, N \pi d^2}$ with velocity U_x, then this velocity U_x receives increments of velocity from random collision with other molecules. Feller has shown that the probability that a molecule will attain a velocity U_x can be given as:

$$p(U_x) = (2\pi\alpha^2)^{-1/2} \exp\left(U_x^2/2\sigma^2\right) \qquad \ldots(2)$$

Where a is known as the standard deviation from the mean. We also know that in one dimension

$$\frac{1}{2} Nm\, \overline{V}^2 = \frac{1}{2} RT \qquad \ldots(3)$$

or

$$\overline{V}^2 = \frac{R}{N}\frac{T}{m} \qquad \ldots(4)$$

Since velocity can be both positive or negative

$$\sigma^2 = V^2 \qquad \ldots(5)$$

Therefore, the probability that a molecule has a velocity between U_x and $U_x + dU_x$

$$P(U_x)\, dU_x = \left(\frac{m}{2\pi KT}\right)^{1/2} \exp\left(-mU_x^2/2\ KT\right) dU_x \qquad ...(6)$$

If there are large number of molecules N, then the fraction

$$\frac{dN(x)}{N} = \left(\frac{m}{2\pi KT}\right)^{1/2} \exp\left(-mCUx^2/2\,KT\right) dU_x \qquad ...(7)$$

Jeans has applied similar arguments to two dimensions where the simultaneous velocity components are U_x and U_y, then the velocity C regardless of the direction is

$$C^2 = U_x^2 + U_y^2 \qquad ...(8)$$

$$\frac{DN(c)}{N} = \frac{m}{2\pi KT} \exp\left(-C^2/2KT\right) Cdc \qquad ...(9)$$

Extending the arguments to three dimensions

$$c^2 = U_x^2 + U_y^2 + U_z^2 \; \frac{dN(C)}{dN} \qquad ...(10)$$

$$= 4\pi\left(\frac{m}{2\pi KT}\right)^{3/2} \exp\left(-mC^2/2KT\right)c^2 dc \qquad ...(11)$$

gives the plot of equation 48 at different temperatures.

Most Probable Velocity

The value of C, for which $\frac{dN(C)}{dN}$ is maximum, is called the most probable velocity C_p. It is given by

$$\frac{-dC_p}{dC} = \frac{-m}{2\pi KT}\ [\exp\ (-mC^2p/2KT)\ 2C_p\ (C^2{}_p)$$

$$+ [\exp\ (-mC^2{}_p/2\ KT)]\ (2C_p) \qquad ...(12)$$

Since $\qquad \frac{-dC_p}{dC} = 0$

$$c_p^2 = \frac{2KT}{m} \qquad ...(13)$$

Average Velocities

In this case of three-dimensional motion there are two types of average velocities viz. the average velocity C and average root mean square velocity $\left(\overline{C}\right)^{1/2}$

$$\overline{C} = \frac{1}{N}\int_0^{\alpha} CNd(C) \qquad ...(14)$$

From,

$$\overline{C} = 4\pi\left(\frac{m}{2\pi KT}\right)^{3/2}\int_0^{\alpha} \exp\left(-mC^2/2\ KT\right)C^3\,dC \qquad ...(15)$$

Let $\dfrac{mC^2}{2KT} = x^2$

$$Cdc = 2KT\,\frac{x}{m}dx \qquad ...(16)$$

or $C^2dc = \left(\dfrac{2KT}{m}\right)^2 x^3\,dx$

$$\overline{C} = 4\pi\left(\frac{m}{2\pi KT}\right)^{3/2}\left(\frac{2KT}{m}\right)^2\int_0^{\alpha} x^3\exp\left(-x^2\right)dx$$

since $x^3\exp\left(-x^2\right)dx = \dfrac{1}{2}$

$$\overline{C} = 2\pi\left(\frac{m}{2\pi KT}.\frac{2KT}{m}\right)^{3/2}\left(\frac{2KT}{m}\right)^{1/2} = \left(\frac{8KT}{\pi m}\right)^{1/2} \qquad ...(17)$$

$$\overline{C}^2 = 4\pi\left(\frac{m}{2\pi KT}\right)^{3/2}\int_0^{\alpha} \exp\left(-mC^2/2KT\right)C^4\,dc$$

$$\text{Let} = \frac{mC^2}{2KT}X^2$$

$$Cdc = \frac{mKT}{m}Xd_x$$

From equation (17)

$$C^3 = \left(\frac{2KT}{m}\right)^{3/2}X^3$$

$$C^4\,dC = \left(\frac{2KT}{m}\right)\left(\frac{2KT}{m}\right)^{3/2}X^4 \qquad ...(18)$$

$$\overline{C}^2\ 4\pi = \left(\frac{m}{2\pi KT}\right)^{3/2}\left(\frac{2KT}{m}\right)^{5/2}\int_0^{\alpha} X^4\exp\left(-x^2\right)d_x \qquad ...(19)$$

But $$X^4 \exp\left(-X^2\right)d_x = \frac{3}{8}\sqrt{\pi}$$

$$\overline{C}^4 = 4\pi \times \frac{3}{8}\sqrt{\pi}\left(\frac{m}{2\pi KT}\right)^{3/2}\left(\frac{2KT}{m}\right)^{5/2} = \frac{3KT}{m} \quad ...(20)$$

$$= \frac{3PV}{Nm} = \frac{3P}{\rho} \quad ...(21)$$

where ρ is the density of the gas and is given by $\frac{Nm}{V}$

The three characteristic velocities of a gas in three dimensions are:

Most probable velocity $C_P = \left(\frac{2\ KT}{m}\right)^{1/2}$

Average velocity $\overline{C} = \left(\frac{8\ KT}{\pi m}\right)^{1/2} = 1.13\ C_p$ ...(22)

Root mean square velocity $\left(\overline{C}\right)^{1/2} = \left(\frac{3\ KT}{m}\right)^{1/2} = 1.22\ C_p$

i.e., $\left(\overline{C}\right)^{1/2} : \overline{C} : C_p = 1 : 0.92 : 0.815$

The distribution of the three velocities and the effect of temperature on C where $\frac{\Delta N}{N}$ is plotted against molecular speed.

Average Kinetic Energy

$$\overline{K}.\overline{E}. = \frac{1}{2}\overline{m}\overline{C}^2 = \frac{m}{2N_0}\int_0^{\alpha} C^2 dm$$

$$= \frac{3}{2}KT = \frac{3}{2}\left(\frac{R}{N_o}\right)T \quad ...(23)$$

The average kinetic energy per molecule of an ideal gas is proportional to the absolute temperature and is independent of the kind of molecule.

It can be seen from the above that the fraction of the molecules depends on two factors.

(1) The square of the velocity

(2) $\exp\left(-\frac{1}{2}mC^2/KT\right)$

We know that $\frac{1}{2}mC^2 = \frac{1}{2}mX^2 + \frac{1}{2}m^2 + \frac{1}{2}mZ^2$

In the random walk model one velocity component is independent of the value of the other. Therefore, the kinetic energy of a molecule is the sum of its kinetic energy components and the magnitude of one velocity component is independent of the value of other velocity components.

Similarly,

$$\exp\left(-\frac{1}{2}\,mC^2\ KT\right)=\left(\exp-\frac{1}{2}\frac{mX^2}{KT}\right)$$

$$\left(\exp-\frac{1}{2}\frac{mY^2}{KT}\right)\left(\exp-\frac{1}{2}\frac{mZ^2}{KT}\right) \quad ...(24)$$

The kinetic energies associated with the velocity components must be additive but the probabilities of their individual values must be multiplicative.

DERIVATION OF THE EQUATION OF STATE OF AN IDEAL GAS

Let there be a spherical vessel of radius r. The molecules of gas are colliding with the surface of the vessel with velocity $\overline{U}$. Any force exerted on the surface of the wall by a colliding particle is balanced by an external opposite force. The net force being then zero, no motion of wall occurs. Let us isolate a gas particle and follow it. The particle that elastically collides with the wall of the vessel rebounding at an angle ϕ, equal to its angle of approach. Since the sphere is symmetrical the particle will always be in the same plane passing through the centre of the sphere. The distance between collisions is 2r cos ϕ. If δt is the time between the collisions, then

$$\delta t=\frac{c\cos\phi}{\overline{U}} \quad ...(1)$$

The number of collision per unit time is

$$\frac{1}{\delta t}=\frac{\overline{U}}{2n\cos\phi} \quad ...(2)$$

The component of the velocity along the direction perpendicular in the spherical surface is U cos ϕ.

The component of the momentum of the particle of mass m before collision is –mU cos ϕ and + mU cos ϕ after the collision. The change in the momentum = 2mU cos ϕ

The force exerted by the particle on the wall

$$= \frac{2, U \cos \phi \times \overline{U}}{2r \cos \phi} = \frac{m\overline{U}^2}{r} \qquad ...(3)$$

The outward force due to collision of A molecules

$$= \frac{A\, m\overline{U}^2}{r} \qquad ...(4)$$

The area of the sphere = $4\ \pi\ r^2$

$$\text{Pressure} = \frac{Am\overline{U}^2}{4\pi r^2 \times r} = \frac{am\overline{U}^2}{4\pi r^3} \qquad ...(5)$$

The volume V of the sphere

$$= \frac{4}{3}\pi r^3$$

$$\text{Pressure } P = \frac{{}^{\cdot}Am\ \overline{U}^2}{3V} \qquad ...(7)$$

or

$$PV = \frac{1}{3} A\ m\overline{U}^2 = \frac{2}{3} A \left(\frac{1}{2} m\overline{U}^2\right)$$

$$= \frac{2}{3}\ \text{A.K.E. (Kinetic energy)} \qquad ...(8)$$

$$= RT \text{ for 1 mole of gas} \qquad ...(9)$$

Average kinetic energy $= \frac{3}{2}$ RT

or T is a measure of KE (Kinetic energy).

Since in a gas at a fixed temperature A, w and $\overline{U}$ are fixed PV is constant which verifies the Boyles law.

At constant pressure and temperature. V is proportional to T which verifies the Charles law.

At constant pressure and temperature, V is proportional to A, which verifies the Avogadro's law.

Collision Frequency

Let us consider the molecules hitting a plane surface.

As already explained that all directions in space are equivalent, the results will be expected independent of direction. According to Boltzmann distribution law the molecules with a volecity of w,

$$\frac{dN(u)}{N} = \left(\frac{m}{2\pi KT}\right)^{1/2}\left[\exp\left(\frac{-mu^2}{2KT}\right)\right] \quad ...(10)$$

The number dn of molecules per unit volume reaching the wall with velocities between u and u + du

$$dz(u) = N\left(\frac{m}{2\pi KT}\right)^{1/2}\left[\exp\left(\frac{-mu^2}{2KT}\right)\right]udu \quad ...(11)$$

z will give the total surface collision frequency per unit area, which can be obtained by integrating the above from zero to infinity.

$$z = N\left(\frac{KT}{2\pi m}\right)^{1/2} z = N\left(\frac{KT}{2\pi m}\right)^{1/2} \quad ...(12)$$

$$= \frac{1}{4}N\overline{V} = N\left(\frac{RT}{2\pi M}\right)^{1/2} \quad ...(13)$$

$$= P\left(\frac{1}{2\pi m\,KT}\right)^{1/2} \quad ...(14)$$

The mass collision frequency mz is given by

$$mz = \rho\left(\frac{KT}{2\pi m}\right)^{1/2} = \rho\left(\frac{RT}{2\pi M}\right)^{1/2} \quad ...(15)$$

where ρ is the density of the gas.

Bimolecular Collision Frequency

Molecules undergo collision as they move through the gas. This will result in the change of direction and speed of the colloiding molecules. They will behave as hard spheres of radius r and undergo elastic collision so that the initial and final kinetic energies do not change.

Collision Diameter

The molecules are treated as rigid noninteracting spheres. A moving molecule will collide with other molecule whose centres come within a distance d_{12}. Thus, the molecular collision diameter d_{12} is the minimum distance of approach of two molecules colliding with zero initial kinetic energy. If r_1 and r_2 are the radii of the two molecules, then d_{12} is equal to the sum of r_1 and r_2.

The quantity πd_{12} is called scattering cross-section of the molecules and is represented by σ_{12}.

$$mz = \rho\left(\frac{KT}{2\pi m}\right)^{1/2} = \rho\left(\frac{RT}{2\pi M}\right)^{1/2} \qquad ...(16)$$

where μ is the viscosity coefficient.

BOYLE'S TEMPERATURE

There should be a temperature T_B for every gas, at which the gas behaves as an ideal gas. This temperature is known as Boyle's temperature, but it is not critical temperature.

$$T_B = \frac{a}{bR} \qquad ...(1)$$

In the case of the plot of β vs P, the three conditions are

(1) $T > T_B$, $b > \frac{a}{RT}$ — Positive slope

(2) $T < T_B$, $b < \frac{a}{RT}$ — Negative slope

(3) $T = T_B$, $b = \frac{a}{RT}$ — No change

LIQUEFACTION OF GASES

By the middle of the 19th century a number of gases like NH_3, CO_2, H_2S and N_2O were liquefied by the combined effect of compression and cooling. Liquefaction is easy to understand from the microscopic view point since the kinetic energy of the molecules decreases on cooling. As a result of this, some of the molecules get closer to one another till they change into the liquid state. High pressure also favours liquefaction because molecules are brought sufficiently close together to permit the attractive forces to be large enough to cause condensation. It is possible to liquefy most of the gases at atmospheric pressure, but it is quite impossible to liquefy many gases at ordinary temperatures. Gases like H_2, O_2 and N_2 cannot be liquefied at laboratory temperatures even under pressure of several thousand atmospheres. At sufficiently large pressure, the gas condenses to liquid provided the temperature is maintained at a value which is characteristic of that gas. In this process the homogeneous gas forms a heterogeneous gas. The essential condition for liquefaction was discovered by Andrews (1863). He demonstrated that there exists for each gas a certain temperature, called the critical temperature, above which it cannot be liquefied, no matter how great the pressure. Young studied the pressure-volume relation of isopentene at different temperatures and his results. The isotherm at 220°C represents the data obtained at a relatively high temperature. Its form is very close to that of a rectangular hyperbola and shows that at this temperature the gas follows Boyle's law rather closely.

Now consider the isotherm at 170°C which is below the critical point. As the vapour is compressed, the P-V curve follows AB which is roughly in accordance with Boyle's law. When point B is reached liquid isopentene is formed and this can be observed by the appearance of a meniscus between vapour phase and liquid phase. As the volume decreases further, more of the gas is transferred to the liquid phase while the pressure remains constant (note the flat portion of the curve). Finally, at C all of the gas is in the liquid phase and the curve CD is the isotherm of liquid isopentene. Since the liquids are relatively incompressible, the compression of the liquid from C to D results in little volume change and hence, the curve CD is very steep.

The flat part of the isotherm reveals an important fact that as long as both liquid and gas phases are present, the pressure of the gas in contact with the liquid must be the same, quite independent of whether a small or a large fraction of the volume is occupied by liquid. This equilibrium pressure is known as the vapour pressure of the liquid at that particular temperature. This corresponds to the heterogeneous mixture.

The specific volume V is given by

$$V_{mixture} = f_g V_g + f_l Y_l \qquad ...(2)$$

where f is a function of the gas and the liquid respectively. As isotherms are taken at successively higher temperature we find that the flat part, *i.e.*, the volume range in which the phases coexist, becomes shorter and shorter until for a particular isotherm at 187°C its length is reduced to a zero at the point P. This is the critical point a J the isotherm on which it lies is the critical isotherm. This temperature is known as the critical temperature.

The pressure at the critical point is known as the critical pressure (P_c), and the corresponding volume per mole is called the critical molar volume. It is to be noted that at the critical temperature, liquid and vapour of a substance become so similar that they can no longer be distinguished as individuals. The indexes of refraction, densities and molar volumes of the two phases become identical in the critical state. Andrews stated of this state: "If anyone asked whether it (the substance) is now in the gaseous or in the liquid state, the question does not arise I believe, admit of a positive reply."

LAW OF CORRESPONDING STATES

If the variables are converted to reduced variables such that $\frac{P}{P_c} = R_R$; $\frac{V}{V_c} = V_R$ and $\frac{T}{T_c} = T_R$, the Van der Waals equation becomes

$$\left(P_R + \frac{3}{V_r}\right)(3V_r - 1) = 8T_r \qquad ...(1)$$

This contains no constant specific to the particular substance.

The law of corresponding states relates that for a group of similar gases there is one functional form of the equation of state

$$P_r = f(V_r, T_r) \qquad ...(2)$$

By plotting β as a function of reduced pressure at various reduced temperatures, it is observed that many gases behave alike in terms ul three variables.

OTHER EQUATIONS OF STATE

The Van der Waals equation provides only a reasonable correct information in the range of moderate deviations from ideality at pressures below 1 atm. It contains three arbitrary constants a, b and R.

Viral Equation

The general form of equation is

$$PV = RT + \frac{B(T)}{V} + \frac{C(T)}{V^2} + \frac{D(T)}{V^3} \qquad ...(1)$$

Equation 127 is known as viral equation, and RT, B(T), C(T)... are known as first, second, third viral coefficients.

Clausius Equation

Clausius derived an equation containing four constants

$$\left(P + \frac{a}{(V+C)^2\ T}\right)(V - b) = RT \qquad ...(2)$$

$$\frac{RTC}{P_cV_c} = \frac{8(b+C)}{3(b+2c)} \qquad ...(3)$$

$\frac{RT_c}{P_cV_c}$ will be constant only when C = bK

$$\frac{RT_c}{P_cV_c} = \frac{8(1+K)}{3(1+2K)} \qquad ...(4)$$

The experimental value of K is found to be 1.63, which gives the value of $\frac{RT_c}{P_cV_c}$ = 3.36 a value lower than the experimental value,

D. Berthelots Equation

Berthelot gave an equation which gives better results at pressures not much above 1 atm. It also involves three constants.

$$\left(P + \frac{A}{TV^2}\right)(V - B) = RT \qquad ...(5)$$

Sometimes it is used in the reduced form

$$P_rV_r = R^1\ T_r\left[1 + \frac{9}{128}\frac{R_r}{T_r}\left(1 - \frac{6}{T_r^2}\right)\right] \qquad ...(6)$$

where $R^1 = \frac{RT_c}{P_cV_c}$

SOLVED EXAMPLES

Example 1:

A sample of oxygen is collected by displacing water from an inverted tube. The temperature is 25°C, the pressure is 750 mm Hg (barometric pressure) and the volume occupied is 280 cm³. What is the true volume of oxygen at STP?

Solution:

As the gas is collected over water barometer pressure (P_{bar}) does not correspond to the actual pressure of the gas

P_{bar} = + PH_2O

$$P_{O_2} = P_{bar} - P_{H_2O}$$

$$= (750.0 - 23.8) \text{ mm Hg} = 726.2 \text{ mm Hg}.$$

Volume of gas at STP can be calculated as:

Initial state: $P_1 = 726.2$ mm Hg; $V_1 = 280$ cm^3;

$n_1 = n$; $T_1 = 273 + 25 = 298°K$

$P_1V_1 = nRT_1$

Final state: $P_1 = 760$ mm Hg, $V_2 = ?$ $n_2 = n$; $T_2 = 273$ $P_2V_2 = nRT_2$

As nR is the same in the two states, therefore,

$$\frac{P_1V_1}{T_1} = \frac{P_2V_2}{T_2}$$

$$V_2 = \frac{P_1}{P_2} \times \frac{P_2V_2}{T_2}$$

$$= \frac{726.2 \text{ mm Hg}}{760 \text{ mm Hg}} \times \frac{273°K}{298°K} \times 280 \text{ cm}^3$$

$$= 245 \text{ cm}^3.$$

Example 2:

You are given three vessels A, B and C, all at the same temperature, with volumes respectively of 1.20 1,2.63 1 and 3.05 1. Vessel A contains 0.695 g of nitrogen gas at a pressure of 742 mm Hg, vessel B contains 1.10 g argon gas at a pressure of 383 mm Hg, vessel C is completely empty at the start of the experiment.

What will the pressure become in vessel C if the contents of A and B are completely transferred to C?

Solution:

Nitrogen was originally present in vessel A (V = 1.2 1) and has a pressure of 742 mm Hg. It is now transferred to vessel C, the volume of which is 3.05 1.

Argon was originally in vessel B (V = 2.63 1) and has a pressure of 383 mm Hg. This is transferred to vessel C(C = 3.05 1).

Now we can treat the two gases separately and find their pressures when present in vessel C. The total pressure in vessel C is equal to the sum of the pressures of nitrogen and oxygen in vessel C.

Gas	*Vessel*	*Initial conditions*	*Final conditions*	*Partial pressures (P) in vessel C*
				$P_1V_1 = P_2V_2$
Nitrogen	AC	$V_1 = 1.2$ 1	$V_2 = 3.05$ 1	$P_{N_2} = \frac{1.21 \times 742 \text{ mm Hg}}{3.05l}$
		$P_1 = 742$ mm Hg	$P_2 = ?$	$= 294$ mm Hg
Argon	BC	$V_1 = 2.631$	$V_2 = 3.05$ 1	$P_{Ar} = \frac{2.631 \times 383 \text{ mm}}{3.051}$
		$P_1 = 383$ Hg	$P_2 = ??$	

Pressure of gaseous mixture in C

$$= PN_2 + P_{Ar}$$
$$= 294 \text{ mm Hg} + 330 \text{ mm Hg}$$
$$= 624 \text{ mm Hg.}$$

Example 3:

(a) What is the average volume available to a molecule in a sample of nitrogen gas at STP?

(b) Assuming that N_2 molecule is approximately spherical with an effective diameter of 3.6Å, calculate

(i) The actual volume occupied by a N_2 molecule;

(ii) the percentage of the molar volume that is empty space;

(iii) The average distance between neighbouring gas molecules at STP.

Solution:

We know that one mole of an ideal gas at STP (*i.e.*, at 0°C and 1 atm pressure) occupies 22.4 litres.

Therefore, assuming ideal behaviour for nitrogen, one mole of this gas will occupy 22.4 litres or 22,400 cm^3 at 0°C and 1 atm, pressure.

There are 6.02×10^{23} molecules in one mole. Therefore, the number of molecules in 1 cubic centimeter is

$$\frac{6.02 \times 10^{23}}{22,400} = 2.69 \times 10^{19} \text{ molecules per cm}^3$$

and the volume available to a gas molecule is

$$\frac{1}{2.69\times10^{19}} = 3.72 \times 10^{-20} \text{ cm}^3 \text{ molecule}^{-1}$$

(b) Volume of a sphere $= \frac{4}{3}\pi r^3$

As diameter for nitrogen is 3.6 Å = 36 × 10^{-8} cm

$$r = \text{radius of nitrogen molecule} = \frac{3.6\times10^{-8}}{2}\text{cm}$$

$$= 1.8 \times 10^{-8} \text{ cm.}$$

$$\text{Volume of a nitrogen molecule} = \frac{4}{3}(1.8 \times 10^{-8})^3$$

$$= 2.44 \times 10^{-23} \text{ cm}^3 \text{ molecule}^{-1}$$

Volume of 1 mole of nitrogen molecules

$$= 2.44 \times 10^{-23} \times 6.02 \times 1023 = 14.7 \text{ cm}^3 \text{ mole}^{-1}$$

We find that the actual volume occupied by 1 mole of molecules is 14.7 cm^3 while the gas molar volume is 22,400 cm^3. Thus, the difference 22,400 – 14.7 = 223,85.3 cm^3 is apparently the empty space.

$$\text{per cent empty space} = \frac{\text{Volume empty}}{\text{Volume available}} \times 100$$

$$= \frac{223,85.3}{224.00} \times 100 = 99.9\%$$

The average distance between any molecule and its nearest neighbour can be obtained from the volume available to a gas molecule: Let us assume that the volume available to a molecule has the shape of sphere of radius R. Hence, the volume available to a molecule is 4/3 R^3 and this is equal to 3.72 × 10^{-20} cm^3 as found in (a)

$$4/3\ R^3 = 3.72 \times 10^{-20} \text{ cm}^3$$

$$R^3 = \frac{3 \times 3.72 \times 10^{-20}}{4x}$$

$$R = 20.7 \times 10^{-8} \text{ cm}$$

Thus, on an average, the volume available to a gas molecule can be assumed to be the volume of a sphere with R = 20.7 × 10^{-8} cm^3, and accordingly the average distance between the centres of any two neighbouring molecules in the gas sample will be 2R.

Therefore, the average distance between a gas molecule and its nearest neighbour is 2R = 20.7 × 10^{-8} = 41.4 × 10^{-8}cm = 41.4Å.

Another way of computing the distance between a gas molecule and its nearest neighbour in a sample of gas is by assuming that the gas volume is divided into small cubes of edge-length and the volume of this small cube is the volume, on an average, available to a gas molecule. The length of this small cube (1) will then roughly be equal to the average distance between the gas molecules assuming that the molecules are for most part evenly spaced at the centres of identical cubes. If 1^3 is the volume of each small cube, then the value of 1 can be computed from the relation

$$1^3 = 3.72 \times 10^{-20}\text{cm}^3$$

where 3.72×10^{-20} is the volume available to each molecule

or $1 = 33.4 \times 10^{-8}$ cm = 33.4Å

From these calculations, we find that the average distance between two neighbouring molecules at STP is about 33.4×10^{-8} cm or 33.4Å (students should find out why this value is different from that obtained previously). The diameter of the gas molecule is only 3.6Å and the distance between molecules in the gas is, therefore, about ten times their diameter.

Similar calculations for solid (density 1.03 gm/cm^3) and liquid (density 0.8 gm/cm^3) nitrogen show that in these states, the average distance between molecules is about 3.56Å and 3.8Å respectively (which are nearly equal to the molecule diameter of nitrogen).

Hence, in solid or liquid states, individual molecules touch each other and it is difficult to compress the substances in these states unlike in the gaseous state.

Example 4(a):

Naphthalene on oxidation gives phthalic anhydride, a very important raw material for glyptal type of resins. When it is heated to 30°C, it exerts a vapour pressure of 0.177 mm. Calculate the effusion rate when M = 128.16 g/g.mole.

Solution:

$$0.177 \text{ mm} = 2.36 \times 10^5 \text{ g-cm/sec}^2$$

$$Z_m = \frac{\alpha}{4} P \left(\frac{3M}{RT}\right)^{1/2}$$

$$= \frac{0.92}{4} \times 2.36 \times 10^5 \left(\frac{3 \times 128.16}{8.314 \times 303 \times 10^7} \right)^{1/2}$$

$$= 0.23 \times 2.36 \times 10 \left(\frac{128.16}{8.314 \times 10^7} \right)^{1/2}$$

$$= 6.73 \text{ g/cm}^2 \text{ sec.}$$

Example 4(b):

Compute the relative rates of diffusion of hydrogen and carbon dioxide. Molecular mass of hydrogen is 2 and that of carbondioxide is 44.

Solution:

$$\frac{r_{H_2}}{r_{CO_2}} = \frac{(M_{CO_2})^{1/2}}{(M_{H_2})^{1/2}} = \left(\frac{44}{4} \right)^{1/2}$$

$$= \frac{6.63}{1.41} = 4.7$$

Hydrogen diffuses 3.7 times faster than carbon dioxide.

Example 4(c):

The speeds of diffusion of CO_3 and O_3 were found to be as 0.29 and 0.271. What is the molecular mass of O_3 if the molecular mass of CO_2 is 44?

Solution:

$$\frac{r_{O_3}}{r_{CO_2}} = \left(\frac{M_{CO_2}}{M_{O_3}} \right)^{1/2}$$

$$\frac{0.271}{0.29} = \left(\frac{44}{M_{O_3}} \right)^{1/2}$$

Squaring both sides

$$\frac{(0.271)^2}{(0.29)^2} = \frac{44}{M_{O_3}}$$

or $$M_{O_3} = 44 \times \frac{(0.29)^2}{(0.271)^2} = \frac{44 \times 0.29 \times 0.29}{0.271 \times 0.271}$$

$= 50.4$ (Mol. mass of ozone is 50.4).

Example 5:

A given sample of ideal gas occupies a volume of 11.21 at 0.863 atm. If the temperature is kept constant, to what pressure will you have to go to change the volume to 15.01?

Solution:

For the gas in the two states we have

Initial state: $P_1 = 0.863$ atm; $V_1 = 11.21$; $n_1 = n$; $T_1 = T$; $P_1V_1 = nRT$

Final state: $P_2 = ?$; $V_2 = 15.01$; $n_2 = n$; $T_2 = T$; $P_2V_2 = nRT$

In this problem nRT is the same in the two states and so

$$P_1V_1 = P_2V_2$$

$$0.863 \text{ atm} \times 11.21 = P_2 \times 55.01$$

$$P_2 = \frac{0.863 \text{ atm} \times 11.21}{15.01}$$

$$= 0.644 \text{ atm.}$$

Example 6:

250 cm³ of a gas is collected over acetone at –10°C and 85 cm Hg. If the gas weighs 1.34 g and the vapour pressure of acetone at –10° is 39 mm Hg, what is the molecular mass of the gas?

Solution:

By Dalton's law

$$P = P_g + P_{acetone}$$

or $$P_g = P - \rho_{acetone}$$

$$= (850 - 39) \text{ mm Hg} = 811 \text{ mm Hg}$$

$$PV = nRT = \frac{g}{M} RT$$

or $$M = g.\frac{RT}{PV}$$

$$R = 0.0821 \frac{1\text{atm}}{\text{mole}\,°\text{K}}$$

$$t = 273 - 10 = 263°\text{K}$$

$$V = 250\ cm^3 = \frac{250}{1000} - 1 = 0.25\ 1$$

$$P = 811\ mm\ Hg = \frac{811 mm\,Hg}{760\ mm\ Hg} \times 1\ atm = \frac{811}{760}\ atm$$

$$g = 1.34\ g$$

Substituting,

$$M = \frac{1.34\ g \times 0.0832 \dfrac{1}{mole\,°K} \times 263°K}{\dfrac{811}{760}\ atm \times 0.251}$$

Example 7:

A lighter-than-air balloon is designed to rise to a height of 50 kilometres at which point it will be fully inflated. At that altitude the atmosphere pressure is 1.52 mm Hg and the temperature is –5°C. If the full volume of the balloon is one hundred thousand litres, how many kilograms of helium will be needed to inflate the balloon?

Solution

To calculate the number of kilograms of helium required we shall first calculate the number of moles and then convert these to kilograms.

The balloon has flexible walls, hence, the volume of the balloon varies with the applied pressure and at any height the pressure inside the balloon will be substantially equal to the atmospheric pressure. As the atmospheric pressure at a height of 50 km when the balloon is fully inflated is 1.52 mm Hg, the pressure inside the balloon will also be 1.52 mm Hg.

Hence, for the helium at a height of 50 km, we have,

$$P = 1.52 mm\ Hg$$

$$V = \text{Volume of balloon when fully inflated}$$

$$= 100{,}0001$$

$$= 1.0 \times 10^{51}$$

$$n = ?$$

$$x = 0.821 \frac{1\ atm}{mol°K}$$

$$T = 273–5 = 268°K$$

and $n = \dfrac{PV}{RT}$

Before substituting the given numerical values we have to see that the variables in the equation are expressed in .he proper units. If we choose to use the value of R expressed in 1 – atm/mole°K, then the units of the terms in the equation are of necessity to those that appear in R and any quantity or quantities which do not have those units must be converted before substituting in the gas law.

As R is in –1 –atm/mole –°K, the pressure must be in atm, V in L and Tin kelvin.

Therefore, P = 1.52 mm Hg × 1 – atom/760 mm

Hg = 2.0×10^{-3} atm.

V = 1.0×10^{5} 1 and r = 268°K

Substituting the numerical values

$$n = \frac{PV}{RT} = \frac{20 \times 10^{-3} \text{ atm} \times 1.0 \times 10^{5} \text{ 1}}{0.0821 - \dfrac{1-\text{atm}}{\text{mole}^\circ\text{K}} \times 269^\circ\text{K}}$$

= 9.090 moles.

Since one mole of helium weighs 4 gm the mass of helium is

$$9.09 \text{ moles} \times \frac{4.00}{1\text{mole}} = 36.36 \text{ gm}$$

= 0.03636 kilograms.

Example 8:

The density of oxygen at 25°C is 1.43 g/1 at 1 atm pressure. At what pressure will oxygen have a density twice this value?

Solution:

To solve this problem we use the relation

$$P = \rho\frac{RT}{M}, \frac{P}{\rho} = \frac{RT}{M}$$

For the gas in the two states, we have

Initial state: P_1 = 1 atm; 1 = 1.43 g/1; T_1 = T; $\dfrac{P_1}{\rho_1} = \dfrac{RT}{M}$

Final state: P_2 = ?; $\rho_2 = 2 \times 1.43$ g/1; T_2 = T; $\dfrac{P_2}{2} = \dfrac{RT}{M}$

Here $\frac{RT}{M}$ is the same in the two states and so

$$\frac{P_1}{\rho_1} = \frac{P_2}{\rho_2}$$

or
$$P_2 = \frac{P_2}{\rho_1} \times P_1$$

Substituting the numerical values

$$P_2 = \frac{2 \times 1.43 \text{ g/l} \times 1 \text{ atm}}{1 \times 1.43 \text{ g/l}} = 2 \text{ atm.}$$

Example 9:

A sample of gas weighing 5.75g occupies a volume of 3.41 at 50° and 715 mm Hg pressure. What is the molecular mass of the gas?

Solution:

To solve this problem, we use the equation

$$P = \frac{gRT}{VM}$$

or
$$M = \frac{gRT}{PV}$$

In this problem, we have

$g = 5.75\text{gm}$

$R = 0.0821 \frac{1 \text{ atm}}{\text{mole}°\text{K}}$

T = 273 + 50 = 323 °K the pressure used must be in atm and V in 1

P = 7.15mm Hg

As R is $\frac{1- \text{atm}}{\text{mole}-°\text{K}}$

P = 715mm Hg

$= 715\text{mm Hg} \times \frac{1 \text{ atm}}{760 \text{ mm Hg}}$

$= \frac{715}{760} \text{ atm}$

V = 3.41

Substituting the values,

$$M = \frac{5.5g \times 0.0821 \dfrac{1\,atm}{mole°K} \times 323°K}{\dfrac{715}{760}\,atm \times 3.41}$$

$$= 47.7 \text{ g/mole.}$$

Example 10(a):

One method for the commercial production of chlorine uses the electrolysis of molten sodium chloride. The chemical reaction that occurs is

$$2NaCl\ (1) \rightarrow 2Na(1) + Cl_2(g)$$

Solution:

How many litres of chlorine will be produced from one kg of molten sodium chloride if the gas is measured at 27°C and 1 atm pressure?

From the chemical equation, it is apparent that for each 2 mole (2 × 58.5 = 117.0 g) of NaCl electrolysed, there will be formed 2 moles of Na(I) and one mole (35.5g) of chlorine. This information can be written as

$$2 \text{ moles aCl} \rightarrow 1 \text{ mole } Cl_2$$

$$117 \text{ g NaCl} \rightarrow \text{mole } Cl_2$$

Therefore,

$$1000 \text{ g NaCl} \rightarrow \frac{1\,mole}{117g} \times 1000 \text{ g } Cl_2$$

$$n = 8.55 \text{ moles } Cl_2$$

Assuming that chlorine obeys ideal gas law,

PV = nRT, we have

$$V = \frac{nRT}{P}$$

$$R = 0.0821 \frac{1\ atm}{mole\ °K}$$

P = 1 atm

T = 273 + 27 = 300°K.

Hence,

$$V = \frac{8.55\ mole \times 0.082 \dfrac{1\,atm}{Mole°K} \times 3000°K}{1\ atm}$$

$$= 2.106 \times 10^21.$$

Example 10(b):

The density of hydrogen at 0°C and 760 mm Hg pressure is 0.00009 g/cm³. What is the root mean square speed of hydrogen molecules?

Solution:

$$PV = \frac{1}{3} m N \bar{u}^2$$

$$\text{Therefore, } \bar{u}^2 = 3 \frac{PV}{mN}$$

As m is the mass of a molecule and N is the number of molecules, therefore, mN, the total mass; so = Mass/Volume = mN/V

$$\text{and} \qquad \bar{U}^2 = \frac{3P}{\rho}$$

$$\text{or} \qquad U_{rms} = \left(\bar{U}^2\right)^{1/2} = \left(\frac{3P}{\rho}\right)^{1/2}$$

$$\text{But} \qquad P = 76 \times 13.59 \times 981 \text{ dynes/cm}^2$$

$$= \left(\frac{3 \times 76 \times 13.50 \times 981}{0.0009 \text{ g/cm}^3} \text{ dynes/cm}^2\right)^{1/2}$$

$$= 183.8 \times 10^8 \text{ cm/sec} = 1.838 \text{ km/sec.}$$

Example 11:

A tube 1 metre long and 10.0 cm² cross section is filled with nitrogen at a pressure of .54 atm at 25°. The ends are then sealed with hydrogen permeable material. One end is put in contact with a source of hydrogen at STP while the other end is in contact with vacuum. A steady state is obtained. Estimate the flow rate of hydrogen.

Solution:

$$\frac{1}{A} m_k = -D_k \frac{C_k(2) - C_k(1)}{D}$$

$$\frac{1}{10\text{cm}^2} m_k = \frac{.67\text{cm}^2/\text{sec} \times .54 \text{ atm}}{1 \text{ metre}}$$

$$m_k = 0.0217 \text{ y/hour.}$$

Example 12:

Calculate the kinetic energy of an Avogadro's number of molecules of gas molecules (that is one mole of gas) at 0°C.

Solution:

Kinetic energy of Avogadro's number of molecules = $N_o \overline{KK}$

$$= N_o \cdot \frac{3}{2} RT \left(R = kN_o\right)$$

The value of R is 1.987 cal/mole °K, and T 273°K, thus, kinetic energy of Avogadro's number of molecules

$$= \frac{3}{2} 1.987 \frac{\text{cal}}{\text{mole.K}} \times 273 \text{ K}° = 814 \text{ cal/mole.}$$

Note that in this calculation, the nature of gas molecules is unimportant. All ideal gases at 0°C will have this amount of kinetic energy.

The average kinetic energy of gas molecule $\left(\overline{KE}\right)$ is generally expressed in ergs and not in calories and at 0°C, it has the value.

$$\overline{KE} = \frac{3}{2} KT$$

$$= \frac{3}{2} \times 1.381 \times 10^{-16} \text{ erg/molecule K} \times 273°\text{K}$$

$$= 5.66 \times 10^{-14} \text{ erg/molecule.}$$

Example 13(a):

Calculate the root mean square speed of oxygen molecules in the lungs at normal body temperature 37°C.

$$u_{rms} = \left(\frac{3RT}{M}\right)^{1/2}$$

Solution:

In order to give U_{rms} the units of cm/sec, the value of R must be expressed in mechanical energy units, *i.e.*, in ergs (1 erg equal 1 g.cm^2/ sec^2).

$$R = 8.31 \times 10^7 \text{ ergs/mole}$$

$$°K = 8.31 \times 10^7 \text{ g.cm}^2 \text{ sec}^{-2} \text{ mole}^{-1} \text{ °K}^{-1}$$

$$T = 273 + 37 = 310°K$$

$$M = 32 \text{ g/mole}$$

Substituting these values in the equation

$$u_{rms} = \left(\frac{3 \times 8.31 \times 10^7 \text{ g.cm}^2 \sec^{-2} \text{mole}^{-1} \,^\circ\text{K} \times 310^\circ \text{K}}{32.0 \text{ g.mole}^{-1}} \right)^{1/2}$$

$$= 4.92 \times 10^4 \text{ cm sec}^{-1}.$$

Example 13(b):

Calculate the pressure exerted by 5.0 moles of carboy dioxide in a 1-litre flask at 47°C, using: (a) the Van der Waals equations, and (b) the ideal gas kw. Compare these results with the experimentally observed pressure of 82 atm.

Solution:

(a) The Van der Waals equation is,

$$\left(P + \frac{n^2 a}{V^2}\right)(V - nb) = nRT$$

$$P = ?$$

$$n = 5.0 \text{ moles}$$

$$V = 1 \text{ litre}$$

$$r = 47 + 273 = 320^\circ\text{K}$$

$$R = 0.0821 \text{ 1.atm mole}^{-1} \,^\circ\text{K}^{-1}$$

From the table

$$a = 3.592 \text{ litre}^2 \text{ atm. mole}^{-2};$$

$$b = 0.0427 \text{ litre mole}^{-2}$$

Substituting these values

$$\left(P + \frac{25 \times 3.592}{2}\right)(1.0 - 5.0 \times 0.0427) = 5.0 \times 0.0821 \times 320$$

$$(P + 25 \times 3.592)(0.7865) = 5.0 \times 0.0821 \times 320$$

$$P + \frac{5.0 \times 0.0821 \times 320}{0.7865} - 25 \times 3.92 = 77.2 \text{ atm.}$$

(b) For an ideal gas, PV = nRT

$$P \times 1.0 = 5.0 \times 0.0821 \times 320$$

$$P = 5.0 \times 0.0821 \times 320 = 131.3 \text{ atm.}$$

The pressure calculated from the Van der Waals equation is closer to the experimental value (85 atm). This is to be expected because the total pressure of the gas is high enough to cause significant deviations from the ideal behaviour.

Example 13(c):

Two parallel plates are each of 100.0 cm^2 area and are spaced 1.0 cm apart. One plate is kept at 10.0°C by some means, the other plate at 0°C by being in thermal contact with a mixture of ice and water. The region between the plates is filled with air at atmospheric pressure (K is 0.586 × 10^{-4} calcm sec. °C). The latent heat of fusion of ice is 79.7 cal/g. Calculate the rate of melting of ice in the freezing mixture.

$$\frac{dQ}{dT} = AK\left(\frac{dT}{dz}\right) = 100\ \text{cm}^2 \times 0.586 \times 10^{-4}\ \text{cal/cm sec°C} \times 10°\text{C} \times 60\ \text{sec}$$

$$= \frac{3516 \times 10^{-2}}{797}\ \text{g/min} = 0.0441\ \text{g/min.}$$

Example 14:

At 20°C, water formed 29 drops when flowing through the capillary of a stalagmometer, while an equal volume of ether formed 86 drops. If the densities of water and ether are 0.997 and 0.70 gm per cc respectively, find the surface tension of ether if that of water is 72.8 dynes/cm.

Solution:

We have $$\gamma_2 = \frac{n_1}{n_2} \cdot \frac{\rho_2}{\rho_1} \cdot \gamma_1$$

$$= \frac{29}{86} \times \frac{0.07}{0.997} \times 718 = 17.24\ \text{dynes/cm}$$

Example 15:

At 293 K, $10^{-2}dm^3$ of water formed 29 drop sand the same volume of an organic liquid formed 86 drops in the same stalagmometer. Density of the organic liquid is 0.7 g cm^{-3}. If the surface tension of water is 7.2 × 10^{-2} N m^{-1}, what would be the surface tension of the organic liquid ?

Solution:

If n_1, ρ, and γ_1 are the number of drops, density and the surface tension of the organic liquid, and n_2, d_2 and γ_2 are the number of drops, density and the surface tension of water respectively, then we have

$$\frac{\gamma_1}{\gamma_2} = \frac{n_1\rho_1}{n_1\rho_2}$$

or $$\gamma_1 = \frac{\gamma_2 n_2 \rho_1}{n_1 \rho_2}$$

Since $n_2 = 29$ $\quad n_1 = 86$

$\rho_2 = 1 \text{ g cm}^{-3}$ $\quad \rho_1 = 0.70 \text{ g cm}^{-3}$

$= 1 \times 10^3 \text{gm}^{-3}$ $\quad = 7.0 \times 10^2 \text{ gm}^{-3}$

$\gamma_2 = 7.2 \times 100^{-2}\text{Nm}^{-1}$ $\quad \gamma_1 = ?$

Therefore

$$\gamma_1 = \frac{29(7.0\times10^2 \text{ kg m}^{-3})(7.2\times10^{-2} \text{ N m}^{-1})}{86\,(10^3 \text{ kg m}^{-3})}$$

$$= 1.726 \times 10^{-2} \text{ m}^{-1}$$

or $= 17.26 \text{ dynes cm}^{-1}$

Example 16(a):

Equal volumes of an organic liquid and water gave 55 drops and 35 drops respectively. The densities of water and the organic liquid are 0.996 and 0.80g cm^{-3} *and the surface tension of water is* 7.2×10^{-2} *Nm*$^{-1}$*. Calculate the surface tension of the organic liquid. How many times a water drop is heavier than a drop of the organic liquid ?*

Solution:

$$\frac{\gamma_1}{\gamma_2} = \frac{n_2 d_1}{n_1 d_2}$$

or $$\gamma_1 = \frac{\gamma_1 n_2 d_1}{n_1 d_1}$$

Since

$n_1 = 55$ $\quad n_2 = 35$

$d_1 = 0.80\text{g cm}^{-3}$ $\quad d_2 = 0.996\text{g cm}^{-3}$

$= 8.0 \times 10^2 \text{ kg m}^{-3}$ $\quad = 9.96 \times 10^2 \text{ kg m}^{-3}$

$\gamma_1 = ?$ $\quad \gamma_2 = 7.2 \times 10^{-2} \text{ Nm}^{-1}$

Therefore

$$\gamma_1 = \frac{(7.2 \times 10^{-2}\ Nm^{-1})(35)\ (8.0 \times 10^2\ kg\,m^{-3})}{(55)\ (9.96 \times kg\,m^{-3})}$$

$$= 2.63 \times 10^{-2}\ Nm^{-1}$$

Now, $$\frac{m_{water}}{m_{organic\ liquid}} = \frac{\gamma_{water}}{\gamma_{organic\ liquid}}$$

$$= \frac{7.2 \times 10^2\ Nm^{-1}}{2.63 \times 10^{-2}\ Nm^{-1}} = 2.74.$$

Example 16(b):

Water requires 120.5 seconds to flow through a viscometer and the same volume of acetone requires 49.5 seconds. If the densities of water and acetone at 293 K are 9.982 × 10^2 kg m^{-3} and 7.92 × 10^2 kgm^{-3} respectively and the viscosity of water at 293 K is 10.05 pascal second, calculate the viscosity of acetone at 293 K.

Solution:

Since, $$\frac{\eta_{water}}{\eta_{Acetone}} = \frac{t_1\rho_1}{t_2\rho_2}$$

$$\eta_{Acetone} = \frac{\eta_{water}\ t_2\rho_2}{t_1\rho_2}$$

$$= \frac{(1.005 \times 10^1\ Pas)\ (49.5s)\ (7.92 \times 10^2\ kg\,m^{-3})}{(12.5s)\ (9.982 \times 10^2\ kg\ m^{-3})}$$

Example 16(c):

The coefficients of viscosity of two liquids at 298 K are 1.408 × 10^{-3} kg m^{-1} s^{-1} and 1.594 × 10^{-3} kg m^{-1} s^{-1} and their densities at the same temperature are 8.07 × 10^{-3} kg m^{-3} and 10.17 × 10^2kg m^{-3} respectively. If the time of flow in an Ostwald viscometer for the first liquid is 100 seconds, calculate the time of flow for the second liquid.

Solution:

$$\frac{\eta_1}{\eta_2} = \frac{t_1\rho_1}{t_2\rho_2}$$

or $$t_2\ \frac{\eta_2}{\eta_1} = \frac{t_1\rho_1}{\rho_2}$$

$$= \frac{(1.594 \times 10^{-3}\ kg\,m^{-1}s^{-1})\ (100s)\ (8.07 \times 10^2\ kg\,m^{-3})}{(1.408 \times 10^{-3}\ kg\,m^{-1}\ s^{-1})\ (10.7 \times 10^2\ kg\ m^{-3})}$$

Example 17:

A steel ball of density 7.90 gm per cc and 4 mm diameter requires 55 sees to fall a distance of 1 meter through a liquid of density 1.10 gm per cc. Calculate the viscosity of the liquid in poises.

Solution :

According to equation (8), we have

$$\eta = \frac{2r^2\,(\rho - \rho')g}{9v}$$

where, r = Radius = 0.2 cm

v = Velocity = 100/55 cm sec^{-1}

$$\eta = \frac{2 \times (0.2)^2\ (7.90 - 1.10) \times 980}{9 \times \dfrac{100}{55}} = 32.58 \text{ poises}$$

Example 18(a):

A sphere of 5 × 10^{-2} cm and density 1.10 gm per cc falls at constant velocity through a liquid of density 1.00 gm per cc and viscosity 1.00 poise. What is the velocity of the falling sphere?

Solution:

Rearranging equation (8) we have

$$v = \frac{2r^2\,(\rho - \rho')g}{9\eta}$$

On substituting

$$v = \frac{2 \times (5 \times 10^{-2})^2 \times (1.10 - 1.00) \times 980}{9 \times 1.00}$$

$$= 5.434 \times 10^{-3} \text{ cm/sec.}$$

Example 18(b):

Two immiscible liquids, water and chlorobenzene are boiled at 97.89 kNm^{-2} pressure and the mixture boils at 363K. The ratio of the weight of chlorobenzene to water collected in the distillate is 2.47. Calculate the molecular weight of chlorobenzene if vapour pressure of water at this temperature is 70.11 kNm^{-2}.

Solution:

Vapour pressure of chlorobenzene

$$= 97.89 \text{ kNm}^{-2} - 70.11 \text{ kNm}^{-2} = 27.78 \text{ kNm}^{-2}$$

$$\frac{\text{Weight of chlorobenzene}}{\text{Weight of water}} = \frac{P^\circ_A \, M_A}{P^\circ_B \, M_B}$$

i.e.
$$2.47 = \frac{27.78 \text{ kNm}^{-2} \times M_A}{70.11 \text{ kNm}^{-2} \times 0.018 \text{ kgmol}^{-1}}$$

$\therefore$ M_A *i.e.*, molecular weight of chlorobenzene

$$= \frac{2.47 \times 70.11 \text{ kNm}^{-2} \times 0.018 \text{ kgmol}^{-1}}{27.78 \text{ kNm}^{-2}} = 0.1122 \text{ kgmol}^{-1}$$

$$\text{Relative molar mass} = \frac{0.1122 \text{ kgmol}^{-1}}{0.001 \text{ kgmol}^{-1}} = 112.2.$$

Example 18(c):

The radius of a given capillary is 1.05×10^{-4} m. A liquid whose density is $0.80 g/cm^3$ rises in the capillary up to a height of 6.25×10^{-2}m. Calculate the surface tension of the liquid assuming the contact angle θ to be zero.

Solution:

Here $r = 1.05 \times 10^{-4}$m $\quad d = 0.80 g/cm^3$

$h = 6.25 \times 10^{-2}$m $\quad = 8.0 \times 10^2 kg/m^3$

$g = 9.8 ms^{-2}$

Since, $\gamma = 1/2 \text{ rhdg}$

$$\therefore \gamma = \frac{1}{2}(1.05\times10^{-4}\text{m})\,(6.25\times10^{-2}\text{m})\,(8.0\times10^{2}\text{kg m}^{-3})(9.8\text{ms}^{-2})$$

$$= 2.5725 \times 10^{-2} \text{kg s}^{-2}$$

$$= 2.5725 \times 10^{-2} \text{Nm}^{-1} \qquad (1 \text{ N} = 1 \text{kg ms}^{-2})$$

Example 18(d):

The surface tension of toluene at 293K is 0.0284 Nm^{-1} and its density at this temperature is 0.866g cm^{-3}. What is the largest radius of the capillary that will permit the liquid to rise 2×10^{-2} m ?

Solution:

Since $\gamma = \frac{1}{2} \text{ rhdg}$

Therefore $r = \dfrac{2r}{hdg}$

Here $\gamma = 2 \times 10^{-2}$ m $\qquad$ h = 0.866 g cm^{-3}

$\gamma = ?$ $\qquad$ $= 8.66 \times 10^2$ kg m^{-3}

$\gamma = 0.0284$ N m^{-1} $\qquad$ g = 9.8 m s^{-2}

Substituting these values in the above equation, we get

$$r = \frac{2(0.084 \text{ N m}^{-1})}{(2 \times 10^{-2}\text{ m})\,(8.66 \times 1^2 \text{ kg m}^{-3})\,(9.8 \text{ ms}^{-2})}$$

$= 3.347 \times 10^{-4}$ kg^{-1} s^{2}m^2

$= 3.347 \times 10^{-4}$m $\qquad$ (1N = 1 kg ms^{-2})

$= 0.03347$ cm

Example 19:

At a pressure of 101.3kNm^{-2} a mixture of water and nitrobenzene distils over at 372K. The vapour pressure of water at 372K is 97.70 kN m^{-2}. Estimate the proportion by weight of nitrobenzene in the distillate.

Solution:

$$\frac{\text{Weight of nitrobenzene } W_A}{\text{Weight of water } W_B} = \frac{P^\circ_A\, M_A}{P^\circ_B\, M_B}$$

Molecular weight of nitrobenzene = 0.123 kgmol^{-1}

Molecular weight of water = 0.018kg mol^{-1}

$P^\circ_B = P - P^\circ_A = 101.3$ k Nm^{-2} – 97.70 kNm^{-2} = 3.60 kNm^{-2}

$$\therefore \qquad \frac{W_A}{W_B} = \frac{3.60 \text{ kNm}^{-2}}{97.70 \text{ kNm}^{-2}} \times \frac{0.123 \text{ gmol}^{-1}}{0.018 \text{ kgmol}^{-1}}$$

$\therefore$ WA : WB = 0.25 : 1.00

$\therefore$ Proportion of nitrobenzene in the distillate

$$= \frac{0.25}{1.25} \times 100 = 20\%.$$

Example 20:

In a steam distillation of an insoluble oil, the mixture boils at 368K when the external pressure is 100 kNm^{-2}. If the vapour pressure of water at this temperature is 84.0 kNm^{-2}. If the molecular weights of water and *oil are 0.018 kg mol^{-1} and 0.380 kg mol^{-1} respectively, calculate the amount of steam required to distil 0.1 kg of oil.*

Solution:

$$\frac{\text{Weight of steam } W_A}{\text{Weight of oil } W_B} = \frac{P^\circ_A \ M_A}{P^\circ_B \ M_B}$$

$$P^\circ_A = 84.0 \text{ kNm}^{-2}$$

$$P^\circ_B = 100 - 84.0 = 16 \text{ kNm}^{-2}$$

$$\frac{W_A}{W_B} = \frac{84.0 \text{ kNm}^{-2}}{16.0 \text{ kNm}^{-2}} \times \frac{0.018 \text{ kg mol}^{-1}}{0.380 \text{ kgmol}^{-1}}$$

Weight of oil in the distillate = 0.1 kg.

$$\therefore \quad W_A = \frac{84.0 \text{ kNm}^{-2}}{16.0 \text{ kNm}^{-2}} \times \frac{0.018 \text{ kg mol}^{-1}}{0.380 \text{ kgmol}^{-1}} \times 0.1 \text{ kg}$$

$$= 0.0248 \text{ kg}$$

i.e., weight of steam necessary to distil 0.1 kg of oil is 0.0248 kg.

4

Steady and Stable of the Motion

INTRODUCTION

The motion of a top is said to be steady if it spins about is vertex O in such a way that its axis OC makes the same angle with the vertical OZ throughout the motion.

For steady motion $\theta = \alpha$ (constant), *i.e.* the angular velocity about the vertical OZ is constant.

In this case the axis of the top describes a cone about the vertical with uniform angular velocity. Thus precessional velocity ϕ is constant.

In case the motion steady.

Put $\theta = \alpha, \ddot{\theta} = 0$ and $\dot{\phi} = \omega$ in equation

$$A\ddot{\theta} - A\dot{\phi}^2 \sin\theta \cos\theta + Cn\dot{\phi}\sin\theta - mgh\sin\theta = 0$$

Thus, we have ω^2 A cos α – Cnω + mgh= 0 ...(1)

$$\text{or} \quad \omega \frac{Cn \pm \sqrt{(C^2n^2 - 4A\, mgh\cos\theta}}{2A\cos\theta}$$

i.e. there are in general two values of the precessional velocity ω; the two precessional velocities, call them ω_1, ω_2 ω_3; are real and distinct if

$$C^2n^2 > 4A\, mgh\cos\alpha$$

This is the necessary condition of steady motion.

Cor. If $\alpha = \frac{1}{2}\pi$, then from (1), $\omega = \left(\frac{mgh}{Cn}\right)$

Thus if the top is given an angular velocity n about OC when OC is horizontal and OC be then given an angular velocity (mgh/Cn) about OZ, then OC will continue to revolve uniformly in a horizontal plane round the vertical OZ.

If the top is set in motion in the usual manner. Then n is very great, thus

$$\omega = \frac{1}{2A\cos\alpha}\left[Cn \pm Cn\left\{1 - \frac{4Amgh\cos\alpha}{C^2n^2}\right\}^{1/2}\right]$$

$$= \frac{1}{2A\cos\alpha}\left[Cn \pm Cn\left\{1 - \frac{1}{2}.\frac{4Amgh\cos\alpha}{C^2n^2}\right\}\right]$$

$$= \frac{Cn}{2A\cos\alpha}\left[1 \pm \left\{1 - \frac{2Amgh\cos\alpha}{C^2n^2}\right\}\right]$$

$$= \frac{Cn}{A\cos\alpha} \text{ or } \frac{mgh}{Cn}$$

(by taking positive and negative signs respectively)

Thus $\omega = \dfrac{Cn}{A\cos\alpha}$ or $\dfrac{mgh}{Cn}$ approx. When n is very great. Obviously the first value of ω is large and second is very small, when n is very great.

STABLE MOTION

When axis is vertical : A top is executing steady motion with angular velocity n about its axis which is vertical; to show that the motion is stable.

If the body be disturbed in such a way that θ is very small whilst ψ is unaltered initially, so the general equations of motion of the top are

$$A\ddot{\theta} - A\dot{\phi}^2\sin\theta\cos\theta + Cn\dot{\phi}\sin\theta = mgh\sin\theta$$

and $A\dot{\phi}^2\sin^2\theta\cos\theta + Cn\cos\theta = D$

Since motion is steady with axis vertical; thus we have

$\theta = 0, \ddot{\theta} = 0, \dot{\phi} = n, \therefore D = Cn$

Also $C^2n^2 > 4A$ mgh is the condition for steady motion when $\alpha = 0$. Let the motion be slightly disturbed. The disturbed motion is a general motion of the top; hence its equations are

$$A\ddot{\theta} - A\dot{\phi}\sin\theta\cos\theta + Cn\dot{\phi}\sin\theta = mgh\sin\theta \quad ...(1)$$

and $A\dot{\phi}\sin^2\theta + Cn\cos\theta = Cn$ ...(2)

Eliminating ϕ between (1) and (2), we have

$$A^2\ddot{\theta} - \frac{C^2n^2}{(1+\cos\theta)^2}\sin\theta\cos\theta + \frac{C^2n^2}{(1+\cos\theta)}\sin\theta = Amgh\sin\theta$$

Since θ is small, therefore, writing θ for $\sin\theta$ and 1 for $\cos\theta$ the above equation reduces to

$$A^2\ddot{\theta} - \frac{C^2n^2\theta}{4} + \frac{C^2n^2\theta}{2} = \text{Amgh}\,\theta \text{ or } \ddot{\theta} = -\left(\frac{C^2n^2 - 4\text{Amgh}}{4A^2}\right)\theta,$$

(the coefficient of q is negative, $\because C^2n^2 > 4\text{Amgh}$)

The above equation is simple harmonic motion.

i.e. the motion of the axis of the figure is simple Harmonic about the vertical OZ from which θ is measured. In other words, axis of the top, if disturbed slightly from its vertical position in steady motion, will tend to come back to its original position, *i.e.* the steady motion, in which axis of the figure is vertical, is stable.

The period of oscillation

$$= 2\pi\sqrt{\left(\frac{4A^2}{B^2n^2 - 4\text{Amgh}}\right)} = \frac{4\pi A}{\sqrt{(C^2n^2 - 4\text{Amgh})}} \text{ (where axis is not vertical)}$$

STABLE MOTION

A top is executing steady motion, with its axis inclined at a constant angle α to the vertical, and precessional velocity ω; to show that motion is stable. Here, we shall show that, the axis of the top, when disturbed slightly, from its position in steady motion, will tend to come back to the original position.

Equations of motion of the top are

$$A\ddot{\theta} - A\dot{\phi}^2\sin\theta\cos\theta + Cn\dot{\phi}\sin\theta = mgh\sin\theta$$

and $A\dot{\phi}\sin^2\theta + Cn\cos\theta = D$

Since the motion, originally, is steady with $\theta = \alpha, \dot{\phi} = \omega$.

Thus the above equation reduces to

$Cn\omega - A\omega^2\cos\alpha = mgh$ and $D = A\omega\sin^2\alpha + Cn\cos\alpha$.

Let the steady motion be slightly disturbed, so that motion is then general one and so its equations are

$$A\ddot{\theta} - A\dot{\phi}^2\sin\theta\cos\theta + Cn\dot{\phi}\sin\theta = mgh\sin\theta \quad ...(1)$$

$$\text{and } A\dot{\phi}\sin^2\theta + Cn\cos\theta = A\omega\sin^2\alpha + Cn\cos\alpha \quad ...(2)$$

Eliminating $\dot{\phi}$ between (1) and (2), we get

$$A^2\ddot{\theta}\sin^3\theta - \{A\omega\sin^2\alpha + Cn(\cos\alpha - \cos\theta)\}^2\cos\theta + Cn\sin^2\theta \times$$

$$\{A\omega\sin^2\alpha + Cn(\cos\alpha - \cos\theta)\} = \text{Amgh}\sin^4\theta$$

Putting $\theta = \alpha + X$ Where X is small, the above equation reduces to

$$A^2\ddot{X}\sin^3(\alpha+X)-[A\omega\sin^2\alpha+Cn\{\cos\alpha-\cos(\alpha+X)\}]^2\cos(\alpha+X)$$

$$+Cn\sin^2(\alpha+X)[A\omega\sin^2\alpha+Cn\{\cos\alpha-\cos(\alpha+X)\}]$$

$$=Amgh\sin^4(\alpha+X)$$

Since X is small, neglecting quantities of second and higher orders of smallness, we have

$$A^2\ddot{X}(\sin\alpha+X\cos\alpha)^3-(A\omega\sin^2\alpha+Cn\{\cos\alpha-(\cos\alpha-$$

$$X\sin\alpha)\}]^2(\cos\alpha-X\sin\alpha)+Cn(\sin\alpha+X\cos\alpha)^2[A\omega\sin^2\alpha$$

$$+Cn\{\cos\alpha-(\cos\alpha-X\sin\alpha)\}]=Amgh(\sin\alpha+X\cos\alpha)^4$$

or $A^2\ddot{X}\sin^3\alpha-(A\omega\sin^2\alpha+CnX\sin\alpha)$

$$=A\,mgh(\sin^4\alpha+4X\cos\alpha\sin^3\alpha)$$

or $A^2\ddot{X}\sin^3\alpha-(A^2\omega^2\sin^4\alpha+2XA\,Cn\omega\sin^3\alpha)(\cos\alpha-X\sin\alpha)$

$$+Cn(\sin^2+2X\sin\alpha\cos\alpha)(A\omega\sin^2\alpha+CnX\sin\alpha)$$

$$=A\,mgh(\sin^4\alpha+4X\cos\alpha\sin^3\alpha)$$

or $A^2\ddot{X}\sin^3\alpha-(A^2\omega^2\sin^4\alpha+2X\,ACn\omega\sin^3\alpha+XA^2\omega^2\sin^5\alpha)$

$$+Cn(\sin^2\alpha+2X\sin\alpha\cos\alpha)A\omega\sin^2\alpha+XC^2n^2\sin^3\alpha$$

$$=A\,mgh(\sin^4\alpha+4X\cos\alpha\sin^3\alpha)$$

or $A^2\ddot{X}-A^2\omega^2\sin\alpha\cos\alpha+Cn\omega A\sin\alpha+XA^2\omega^2\sin^2\alpha+XC^2n^2$

$$=Amgh(\sin\alpha+4X\cos\alpha)$$

or $A^2X-A^2\omega^2\sin\alpha\cos\alpha+(A\omega^2\cos\alpha+mgh)A\sin\alpha+XA^2\omega^2\sin^2\alpha$

$$+X(A\omega^2\cos\alpha+mgh)^2(1/\omega^2)=Amgh(\sin\alpha+4X\cos\alpha)$$

or $A^2\ddot{X}+X[A^2\omega^4(\sin^2\alpha+\cos^2\alpha)+2A\,mgh\omega^2\cos\alpha+m^2g^2h^2]$

$$\times[1/\omega^2]\,4XA\,mgh\cos\alpha$$

or $A^2\ddot{X}+X[A^2\omega^4-2A\,mgh\,\omega^2\cos\alpha+m^2g^2h^2](1/\omega^2)=0$

or $\ddot{X}=-\dfrac{A^2\omega^4-2Amgh\omega^2\cos\alpha+m^2g^2h^2}{A^2\omega^2}X$ [S.H.M]

(the coefficient of X is obviously negative)

This shows that the axis of the top, if disturbed slightly from its position in steady motion will tend to come back to the original (predisturbed), position, *i.e.* the steady motion, with axis inclined at α to the vertical, is stable.

The period of oscillation $= 2\pi \dfrac{A\omega}{(A^2\omega^4 - 2Amgh\,\omega^2\cos\alpha + m^2g^2)^{(1/2)}}$

In case the top is set in motion in usual manner, then n is very large and two values of the precessional velocity are approximately given by

$$\omega = \frac{Cn}{A\cos\alpha.}\frac{mgh}{Cn}$$

When n is large, the first value is large and the second small. For first value of ω, the period of oscillation is given by

$$\frac{2\pi A\omega}{A\omega^2}\ \text{(approx.)} = \frac{2\pi\cos\omega}{Cn}$$

And for second value of ω, the period of oscillation is given by

$$\frac{2\pi A\omega}{mgh}\ \text{(approx.)} = \frac{2\pi A}{Cn}$$

LIMITS OF θ

A top is set spinning with angular velocity n about its axis. Initially he axis being instantaneously at rest and inclined at an angle i to the vertical. To show that in the subsequent motion, the angle θ between the axis and the vertical oscillates between

$$i \le \theta \le \cos^{-1}\{p - \sqrt{[(p^2 - 2p\cos i + 1)]}\} \quad \text{where} \quad p = \left(\frac{C^2n^2}{4Amgh}\right),$$

and the letters have their usual meanings.

By equation (3) and (4), the two equations of motion of top are

$$A\dot{\phi}\sin^2\theta + Cn\cos\theta = D \quad \text{and} \quad A(\dot{\theta}^2 + \dot{\phi}^2\sin^2\theta) + 2mgh\cos\theta = E$$

where D and E are constants.

Initially $\theta = i, \theta = \dot{0}, \psi = 0,$

$\therefore$ D = Cn cos i and E = 2 mgh cos i.

Thus the above equations reduce to

$$A\dot{\phi}\sin^2\theta + Cn\cos\theta = Cn\cos i \qquad ...(1)$$

$$\text{and} \quad A(\dot{\theta}^2 + \dot{\phi}^2\sin^2\theta) + 2mgh\cos\theta = 2mgh\cos i \qquad ...(2)$$

Now eliminating ϕ. between (1) and (2), we get

$$A\dot{\theta}^2\sin^2\theta + C^2n^2(\cos i - \cos\theta)^2 + 2Amgh(\cos\theta - \cos i)\sin^2\theta = 0$$

$$\text{or} \quad A^2\dot{\theta}^2\sin^2\theta = -2Amgh(\cos\theta - \cos i)\sin^2\theta - C^2n^2(\cos\theta - \cos i)^2$$

$$= -2Amgh\,(\cos\theta - \cos i)\sin\theta - 4Amgh\,p\,(\cos\theta - \cos i)^2$$

$$= 2Amgh\,(\cos\theta - \cos i)\,[-\sin^2\theta - 2p\,(\cos\theta - \cos i)]$$

$$= 2Amgh\,(\cos i)\ [\cos^2\theta - 2p\cos\theta - 1 + 2p\cos i]$$

$$= 2Amgh\,(\cos\theta - \cos i)\ [(\cos\theta - p)^2 - (p^2 - 2p\cos i + 1])$$

$$= 2Amgh\,(\cos\theta - \cos i)\ \{\cos\theta - (p - \sqrt{[p^2 - 2p\cos i + 1]})\}$$

$$\times\{\cos\theta - (p + \sqrt{[p^2 - 2p\cos i + 1]})\} \qquad ...(3)$$

Now $\dot{\theta}$ will vanish when $\theta = i$ or θ_1 or θ_2

where $\cos\ \theta_1 = p - \sqrt{[(p^2 - 2p\cos i + 1)]}$

and $\cos\theta_2 = p + \sqrt{[(p^2 - 2p\cos i + 1)]}$

The second expression is greater than 1, *hence* θ_2 *is imaginary.*

But $\cos\theta_1 < \cos i$. $\quad \therefore\ \theta_1 > i=$

Again from (2) θ^2 is negative if cos θ> cos *i.e.* θ<i.

Hence possible values of θ are i and θ_1 where

$\cos\ \theta_1 = p\ - \sqrt{(p^2 - 2p\cos i + 1)}$

Again from (3) θ^2 is negative if

$\cos\ \theta > p\ - \sqrt{(p^2 - 2p\cos i + 1)}$ *i.e.* $\theta > \theta_1$

Thus, we conclude that top is never at a less inclination than i or at a greater inclination than θ_1, *i.e.* the motion is included between these two limits.

SPINNING TOP OF THE MOTION

Definition : An interesting example of rigid body motion occurs when a symmetrical rigid body having one point on the axis of symmetry fixed in space is set spinning in a gravitational field. A top is a *rigid body which is a solid of revolution about the axis of symmetry.* The adjoining figure shows such a top rotating about a fixed point O on its axis of symmetry., Thus we idealise the ordinary top by supposing it to terminate in a sharp point (or vertex) and to spin on a floor rough enough to prevent slipping.

EQUATION OF MOTION OF A TOP

(Derived from Euler's Equation) : Let us assume that a top spins with its vertex O fixed in contact with a floor rough enough to prevent slipping,

Let OC be the axis of the top and G its centre of gravity such that OC = h.

Let OX, OY, OZ be the fixed axes and OA, OB, OC the principal axes at O.

Let C, A be the moments of inertia about OC and OA, since the top is symmetrical about OC, so B = A.

Let at zero time OC be in the plane ZOX and at time t it makes an angle θ to OZ and the plane ZOC makes the angle ϕ with the fixed plane ZOX. Let ω_1, ω_2, ω_3 be the angular velocities of the top about OA, OB and OC respectively and L, M, N be the moments of the external forces about OA, OB and OC respectively. Again the external forces acting on the top are its weight mg acting vertically downwards through G, the C.G. of the top in a direction parallel to OZ and the reactions at point O (0, 0, 0).

The direction cosines of OZ with respect to OA, OB and OC are. $-\sin\theta\cos\psi$, $\sin\theta\sin\psi$, $\cos\theta$

or $\quad -\sin\theta, 0, \cos\theta \qquad (\because$ In case of top $\psi = 0)$

Thus the weight – mg acting at G (0, 0, h) has the resolved parts mg $\sin\theta$; 0, $-mg\cos\theta$, parallel to OA, OB and OC respectively.

By $\quad L = \Sigma (y_1 Z_1 - z_1 Y_1)$ etc.,

we have $\quad L = 0$, $M = hmg\sin\theta$, $N = 0$

Therefore Euler's equations give

$$A\dot{\omega}_1 - (B - C)\omega_2\omega_3 = L = 0;$$

$$B\dot{\omega}_2 - (C - A)\omega_3\omega_1 = M = hmg\sin\theta \qquad ...(1)$$

$$C\dot{\omega}_3 - (A - B)\omega_1\omega_2 = N = 0$$

Again Euler's geometrical relations are

$$\omega_1 = \dot{\theta}\sin\psi - \dot{\phi}\sin\theta\cos\psi, \omega_2 = \dot{\theta}\cos\psi + \dot{\phi}\sin\theta\sin\psi,$$

$$\omega_3 = \dot{\psi} + \dot{\phi}\cos\theta$$

which reduce to $\omega_1 = -\dot{\phi}\sin\theta. \omega_3 = \dot{\phi}\cos\theta, \omega_2 = \dot{\theta}$ since for top $\dot{\psi} = 0$.

Again B = A, thus the Euler's third equation gives

$\dot{\omega}_3 = 0$ or $\omega_3 = n$ (constant). or $\phi\cos\theta = n$

Putting the values of ω_1, ω_2, ω_3.

Euler's first equation $A\omega_1 - (B - C)\omega_2\,\omega_3 = 0$

deduces to $A\dfrac{d}{dt}(-\dot{\phi}\sin\theta) - (A - C)\dot{\theta}\dot{\phi}\cos\theta = 0$ as B = A or $-\ A\phi^2\sin\theta - 2A\dot{\theta}\dot{\phi}\cos\theta + C\dot{\theta}\dot{\phi}\cos\theta = 0$

or $-\ A\phi\sin\theta - 2A\dot{\theta}\dot{\phi}\cos\theta + Cn\dot{\theta} = 0$ ...(2)

And Euler's second equation $B\dot{\omega}_2 - (C - A)\omega_3\omega_1 = mgh\ \sin\theta$ reduces to $A\frac{d}{dt}(\dot{\theta}) + (C - A)\dot{\phi}^2\ \sin\theta\cos\theta = mgh\sin\theta$ as $B = A$

or $A\ddot{\theta} - A\dot{\phi}^2\sin\theta + C\dot{\phi}^2\sin\theta\cos\theta = mgh\sin\theta$

$A\ddot{\theta} - A\dot{\phi}^2\sin\theta + Cn\dot{\phi}\sin\theta = mgh\sin\theta$

Equations (1) and (2) are the *equations of motion of the top* and are of second order.

Multiplying (1) by sin θ and integrating, we have

$A\dot{\phi}\sin^2\theta + Cn\cos\theta = D\ (\text{cosnt.})$ ...(3)

Again multiply (1) by $2\dot{\phi}\sin\theta$ and (2) by $2\dot{\theta}$ and subtracting, we get

$2A\ddot{\theta}\dot{\theta} + 2A\dot{\theta}\dot{\phi}^2\sin\theta\cos\theta + 2A\dot{\phi}\ddot{\phi}\sin^2\theta = 2mgh\sin\theta$

Integrating, we get

$A(\dot{\theta}^2 + \dot{\phi}^2\sin^2\theta) + 2mgh\cos\theta = E(\text{const.})$ ...(4)

Equations (3) and (4) are also the equations of motion of the top but are of first order. Equation (4) is called equation of Energy. The motion due to change in θ is called *mutation* and due to change in ϕ is called precision. The general motion of the top about its fixed vertex O is a combination of these two.

EQUATIONS OF MOTION OF A TOP

(Deduced from the Principle of Energy and Momentum) : Since for motion of top B = A and $\psi = 0$

$\therefore \omega_1 = -\dot{\phi}\sin\theta,\ \omega_2 = \dot{\theta},\ \omega_3 = n = \dot{\phi}\cos\theta$

$\therefore \text{K.E.} = \frac{1}{2}[A(\omega_1^2 + \omega_2^2) + C\omega_3^2] = \frac{1}{2}[A\ \dot{\phi}^2\sin^2\theta + \dot{\theta}^2) + Cn^2]\quad (\because B = A)$

Again potential energy = mgh cos θ where *h* is the distance of C.G. from the vertex O and θ inclination of OC to the vertical. By the principle of energy, the sum of the kinetic and the potential energy remains constant throughout motion. Thus by the principle of energy we have

$$\frac{1}{2}[A(\dot{\phi}^2\sin^2\theta + \dot{\theta}^2) + Cn^2] + mgh\cos\theta = \text{const.}$$

or $A(\dot{\theta}^2 + \dot{\phi}^2\sin^2\theta) + mgh\cos\theta = E$ ...(1)

Since the external force (weight) is parallel to OZ and the reactions at O cut OZ; so the sum of their moments about OZ is zero. Thus by the principle of momentum, the moment of momentum about OZ is constant throughout the motion.

Again A ω_1, Aω_2, Cω_3 are the moments of momentum of the top about OA, OB, OC respectively; and the direction cosines of OZ about these axes are given by – sin θ, 0 cos θ.

Resolving these momentums about OZ, we have moment of momentum about OZ –A ω_1 sinθ + Cω_3 cos θ.

Therefore by the principle of momentum, we have

$-A\,\omega_1 \sin\theta + C\omega_3 \cos\theta = \text{const.}$

or $A \sin\theta\ (-\dot{\phi}\sin\theta) + Cn\cos\theta = D$ (constant). ...(2)

Equations (1) and (2) ae same equations of motion of a top.

EQUATIONS OF MOTION OF A TOP

(*Deduced from Lagrange's Equations*) : For motion of a top, we have

$B = A, \dot{\psi} = 0$ and $\omega_1 = -\dot{\phi}\sin\theta, \omega_2 = \dot{\theta}, \omega_3 = n = \dot{\phi}\cos\theta$

∴ The kinetic energy is

$$\therefore T = \frac{1}{2}m[A(\omega_2^2 + \omega_2^2) + C\omega_3^2] \qquad (\because B = A)$$

$$= \frac{1}{2}m[A(\dot{\phi}^2\sin^2\theta + \dot{\theta}^2) + C\dot{\phi}^2\cos^2\theta]$$

And the work function W is W = – mgh cos θ where h is distance of C.G. form the vertex of O. and θ is inclination of OC to the vertical OZ.

Here the generalised co-ordinates are θ and ϕ.

Lagrange's θ- equation is $\frac{d}{dt}\left(\frac{\partial T}{\partial\dot{\theta}}\right) - \frac{\partial T}{\partial\theta} = \frac{\partial W}{\partial\theta}$ which gives

$$\frac{d}{dt}(mA\dot{\theta}) - \{mA\dot{\phi}^2\sin\theta\cos\theta - C\dot{\phi}^2\cos\theta\sin\theta\} = mgh\sin\theta$$

or $\quad A\ddot{\theta} - A\dot{\phi}^2\sin\theta\cos\theta + Cn\dot{\phi}\sin\theta = mgh\sin\theta$ ...(1)

And Lagrange's ϕ-equation is $\frac{d}{dt}\left(\frac{\partial T}{\partial\dot{\phi}}\right) - \frac{\partial T}{\partial\phi} = \frac{\partial W}{\partial\phi}$

which gives $\frac{d}{dt}(mA\dot{\phi}\sin^2\theta + C\dot{\phi}\cos^2\theta) = 0$

or $\quad A\dot{\phi}\sin^2\theta + Cn\cos\theta = D$ (const.) ...(2)

Equation (1) and (2) are same equations of the motion of a top.

STEADY MOTION

The motion of a top is said to be steady if it spins about is vertex O in such a way that its axis OC makes the same angle with the vertical OZ throughout the motion.

For steady motion $\theta = \alpha$ (constant), *i.e.* the angular velocity about the vertical OZ is constant.

In this case the axis of the top describes a cone about the vertical with uniform angular velocity. Thus precessional velocity ϕ is constant.

In case the motion steady.

Put $\theta = \alpha, \ddot{\theta} = 0$ and $\dot{\phi} = \omega$ in equation

$$A\ddot{\theta} - A\dot{\phi}^2 \sin\theta \cos\theta + Cn\dot{\phi}\sin\theta - mgh\sin\theta = 0$$

Thus, we have ω^2 A cos α – Cnω + mgh= 0 ...(1)

or $$\omega \frac{Cn \pm \sqrt{(C^2n^2 - 4A\, mgh\cos\theta}}{2A\cos\theta}$$

i.e. there are in general two values of the precessional velocity ω; the two precessional velocities, call them ω_1, ω_2 ω_3; are real and distinct if

$$C^2n^2 > 4A\, mgh\cos\alpha$$

This is the necessary condition of steady motion.

Cor. If $\alpha = \frac{1}{2}\pi$, then from (1), $\omega = \left(\frac{mgh}{Cn}\right)$

Thus if the top is given an angular velocity n about OC when OC is horizontal and OC be then given an angular velocity (mgh/Cn) about OZ, then OC will continue to revolve uniformly in a horizontal plane round the vertical OZ.

If the top is set in motion in the usual manner. Then n is very great, thus

$$\omega = \frac{1}{2A\cos\alpha}\left[Cn \pm Cn\left\{1 - \frac{4Amgh\cos\alpha}{C^2n^2}\right\}^{1/2}\right]$$

$$= \frac{1}{2A\cos\alpha}\left[Cn \pm Cn\left\{1 - \frac{1}{2}.\frac{4Amgh\cos\alpha}{C^2n^2}\right\}\right]$$

$$= \frac{Cn}{2A\cos\alpha}\left[1 \pm \left\{1 - \frac{2Amgh\cos\alpha}{C^2n^2}\right\}\right]$$

$$= \frac{Cn}{A\cos\alpha} \text{ or } \frac{mgh}{Cn}$$

(by taking positive and negative signs respectively)

Thus $\omega = \dfrac{Cn}{A\cos\alpha}$ or $\dfrac{mgh}{Cn}$ approx. When n is very great. Obviously the first value of ω is large and second is very small, when n is very great.

MOTION IN TWO DIMENSIONS

Introduction : *To obtain the equation of motion of a rigid body under impulsive forces.*

By 3.01, when forces are finite, the equations of motion are given by

$$M\frac{d^2x_G}{dt^2} = \Sigma X \ \ i.e. \ \frac{d}{dt}\left(M\frac{dx_G}{dt}\right) = \Sigma X \quad \text{...(i)}$$

$$M\frac{d^2y_G}{dt^2} = \Sigma Y \text{ i.e} \frac{d}{dt}\left(M\frac{dy_G}{dt}\right) = \Sigma Y \quad \text{...(ii)}$$

and $Mk^2\,\ddot{\theta} = \Sigma(x'Y - y'X)$ *i.e.* $\dfrac{d}{dt}(Mk^2\dot{\theta}) = \Sigma(x'Y - y'X)$ ...(iii)

where (x_G, y_G) are the co-ordinates of the centre of gravity.

When large forces act for a very short time then we say that the forces are impulsive.

Again when the forces are impulsive, let τ be the short time during which the impulsive forces act on the body.

Then integrating (i), we have $\left[M\dfrac{dx_G}{dt}\right]_0^{\tau} = \Sigma\left\{\displaystyle\int_0^{\tau} X dt\right\}$

$$i.e. \left[M\frac{dx_G}{dt}\right]_0^{\tau} = \Sigma X', \text{ where } X' = \int_0^{\tau} X dt \quad \text{...(iv)}$$

is the impulse acting at the point (x, y) parallel to x-axis.

Now let u_G and u'_G be the velocities, of the C.G., parallel to x-axis, just before and just after the application of the impulsive forces, then evaluating (iv), we get

$$M(u'_G - V_G) = \Sigma X' \quad \text{...(v)}$$

is the impulse acting at the point (x, y) parallel to x-axis.

Now let u_G and u'_G be the velocities, of the C.G., parallel to x-axis, just before and just after the application of the impulsive forces, then evaluating (iv), we get

$$M(u'_G - v_G) = \Sigma X' \quad \text{...(v)}$$

Integrating (ii), we have

$$\left[M\frac{dy}{dt}\right]_0^{\tau} = \Sigma\left\{\int_0^{\tau} Y\,dt\right\} = \Sigma Y',$$

where $Y' = \int_0^{\tau} Y\,dt$ ...(vi)

i.e. $M(v' - u) = SY'$

where v_G, V'_G are the velocities of the centre of gravity just, before and after, the application of the impulsive forces.

Hence we conclude that change in the momentum of the mass M, supposed collected at the centre of gravity, in the direction is equal to the sum of the impulses in that direction.

Also $$\left[Mk^2\dot{\theta}\right]_0^{\tau} = \Sigma\left[x'\int_0^{\tau} Ydt - y'\int_0^{\tau} Xdt\right]$$

i.e. $$\left[Mk^2\dot{\theta}\right]_0^{\tau} = \Sigma\left(x'Y' - y'X'\right)$$ ...(viii)

Now let ω and ω' be the angular velocities of the body just before and after the application of the impulsive forces, then ...(vii)

$\Rightarrow$ $M\,k^2\,(\omega' - \omega) = \Sigma\,(x'Y' - y'X')$ ...(viii)

This equation (viii) **states** that change in the moment of momentum about the centre of gravity is equal to the sum of the moments of the impulses of the forces about the centre of gravity.

WORK DONE BY AN IMPULSE

A body whose mass is M, is acted upon at a given point P by a blow of impulse X. If V and V_1 be the velocities of P in the direction of X just before and just after the action of X, show that the change in the kinetic energy of the body, that is the work done on it by the impulse, is

$$\frac{1}{2}(V + V_1)X.$$

Proof:

The section of the body through the direction of the impulse and through the centre of gravity G, of the body. Now take a line through G parallel to the direction of X as x-axis and another line through G, perpendicular Gx as y-axis.

Just before the impulse, let u and v be the velocities of G parallel to x-axis and y-axis respectively and ω the angular velocity about G. Similarly let u_1, v_1, ω_1 be the corresponding quantities, just after the action of the impulse.

Now consider a point $P \equiv (\lambda, y)$, where the impulse X is imparted, and let M be the mass of the body.

Hence equations of motion of the body are :

$$M(u_1 - u) = X \quad ...(1)$$

$$M(v_1 - v) = 0 \quad ...(2)$$

and $$Mk^2 (\omega_1 - \omega) = - y X. \quad ...(3)$$

Thus change in K.E. $= \frac{1}{2}M\left[u_1^2 + v_1^2 + k^2\omega_1^2\right] - \frac{1}{2}M\left(\omega_1^2 - \omega^2\right)$

$$= \frac{1}{2}M\left(u_1^2 - u^2\right) + \frac{1}{2}M\left(v_1^2 - v^2\right) + \frac{1}{2}M\left(\omega_1^2 - \omega^2\right)$$

$$= \frac{1}{2}M\left(u_1^2 - u^2\right) + \frac{1}{2}(u_1 + u)X - \frac{1}{2}yX(\omega_1 + \omega)$$

[using (1), (2) and (3)]

$$= \frac{1}{2}M\left(u_1^2 - u^2\right) + \frac{1}{2}X[(u_1 - y\omega_1) + (u - y\omega)] = \frac{1}{2}X(V_1 + V)$$

$\because$ V = velocity of the point P parallel to GX, before

$$\begin{bmatrix} \text{the impulse} = \text{velocity of G parallel GX} + \\ \text{velocity of P relative G} \Rightarrow V = u - y\omega \\ \text{Similarly, } V_1 = u_1 - y\omega_1 \end{bmatrix}$$

Theorem:

A rod, of length 2a, is held in a position inclined at an angle α to the vertical, and is that let fall on a smooth inelastic horizontal plane.

Show that the end which hits the plane will leave it immediately after the impact, if the height through which the rods falls is greater than,

$$\frac{1}{18} \text{ a sec } \alpha \text{ cose c}^2 \alpha \ (1 + 3 \sin^2 \alpha)^2.$$

Proof:

Let v be the vertical velocity of the rod AB just before the impact when the rod is inclined at an angle α to the vertical. Let R be the impulsive reaction of the plane and mg the weight of the rod.

Just after the impact of the end A, let V be the velocity of G and ω be its angular velocity. Just after the impact the inclination of the rod also remains α to the vertical.

Now change in momentum = impulse

$\Rightarrow$ m (V – v) = – R ...(i)

and change in moment of momentum about the centre of gravity G = moment of impulse R about G.

$$\Rightarrow \frac{m^2 a\omega}{3} = Ra \sin \alpha \quad ...(ii)$$

Also, since the end A does not rebound after impact (since the plane is inelastic), we have vertical velocity of the end A = 0.

$$\Rightarrow V - a\omega \sin \alpha = 0 \quad ...(iii)$$

Whence eliminating V and R between (i), (ii) and (iii), we get

$$\omega = \frac{3V \sin \alpha}{a\left(1 + 3\sin^2 \alpha\right)} \quad ...(iv)$$

Now assume that after impact, the end A remains in contact with the plane and AB moving about A. When the rod is inclined at θ to the vertical, let S be the normal reaction. Now vertical height of G above the plane is a cos θ motion is under finite forces S and mg.

$$\therefore S - mg = \frac{d^2}{dt^2}(a \cos \theta) \quad ...(v)$$

and $$m = \frac{a^2}{3}\ddot{\theta} \quad ...(vi)$$

$$\therefore m = \frac{a^2}{3}\ddot{\theta} - mg\, a \sin \theta = -ma^2 \sin \theta \left[\sin \theta \ddot{\theta} + \cos \theta \ddot{\theta}^2\right]$$

[eliminating S between (v) and (vi)]

i.e. $$m\frac{a^2}{3}\ddot{\theta}\,(1 + 3 \sin^2 \theta) = mg\, a \sin \theta - ma^2 \sin \theta \cos \theta \dot{\theta}^2 \quad ...(vii)$$

Now eliminating $\ddot{\theta}$ between (vii) and (vi), we get

$Sa \sin \theta\,(1 + 3 \sin^2 \theta) = mg\, a \sin \theta \cos \theta\; \dot{\theta}^2$.

Initially, when $\theta = \alpha, \dot{\theta} = \omega$ and S = R

$\therefore Ra \sin \alpha\,(1 + 3 \sin^2 \alpha) = mg\alpha \sin \alpha - ma^2 \sin \alpha \cos \alpha\, \omega^2$

At this time, the end A will leave the plane, if R is negative, *i.e.* if $ma^2 \sin \alpha \cos a\omega^2 > mga \sin \alpha$

$$\Rightarrow \omega^2 > \frac{g}{a \cos \alpha} \Rightarrow \frac{9v^2 \sin^2 \alpha}{a^2\left(a + 3\sin^2 \alpha\right)^2} > \frac{g}{a \cos \alpha'}$$

$$\therefore \frac{v^2}{2g} > \frac{1}{18} a \sec\alpha \operatorname{cosec}^2\alpha (1 + 3\sin^2 ga)^2,$$

⇒ height through which the rod falls is greater than

$\frac{1}{18}$ a sec α cosec^2 α (1 + 3 sin^2 α)^2.

AN IMPERFECTLY ELASTIC SPHERE IMPINGING ON A FIXED PLANE

An imperfectly elastic uniform sphere, rotating with an angular velocity ω about an axis perpendicular to the plane of motion of its centre, impinges on a horizontal plane, find the resulting change in its motion.

First of all Assume that the Plane is Rough Enough to Prevent Any Sliding

Let u and v be the components of the linear velocities in the two assigned directions and ω the angular velocity just before impact while u′,v′ be the components and ω′ the angular velocity after the impact.

Let F be the impulsive friction and R the normal impulsive reaction, then the equation of motion are :

$$m\,(u' - u) = -F \qquad ...(1)$$

$$m\,(v' + v) = R \qquad ...(2)$$

and $$mk^2\,(\omega' - \omega) = Fa \qquad ...(3)$$

Since P, the point of contact is instantaneously reduced to rest, there being no sliding hence

$$u' - aw' = 0 \qquad ...(4)$$

Also by Newton's law

$$v' = ev \qquad ...(5)$$

where e is the coefficient of restitution. Thus, we obtain

$$u' = a\omega' = \frac{1}{7}(5u + 2a\omega) \qquad ...(6)$$

and $$F = \frac{2}{7}m(u - a\omega) \qquad ...(7)$$

Case (i) If u = a ω then F = 0, *i.e.* there is no frictional force,

Hence u = u′ and ω = ω′

which shows that u and w are unchanged. $\left(\because k^2 = \frac{2a^2}{5}\right)$

Case (ii) If u<aω, F is – ive, *i.e.* F, acts $(\rightarrow)$. Hence ω′<ω and u′>u.

Since u>a ω, the velocity of the point of contact P (= u – aω) is negative *i.e.* the direction is $(\leftarrow)$ and the angular velocity is decreased after impact.

Case (iii) If u> aω, F is + ive, *i.e.* F, acts $(\leftarrow)$. Hence w′>ω and u′ <u. Hence when the point of contact before impact is moving $\rightarrow$, the angular velocity is increased, the horizontal velocity is diminished.

i.e. in this case the sign of ω is changed, so we have

$$u' = a\omega' = \frac{1}{7}(5u - 2a\omega), \qquad \text{...(8)}$$

$$\text{and } F = \frac{2}{7}m(u + a\omega) \qquad \text{...(9)}$$

Now there are further two cases.

(i) If $u = \frac{2}{5}a\omega$, then u′ = 0 = w′ *i.e.* after the impact the sphere rebounds from the plane vertically with no spin.

(ii) If $u < \frac{2}{5}a\omega$, t then u¢ is + ive and the sphere rebounds towards the direction from which it came. In both the above cases, vertical velocity after impact is v′ = ev and R = m (1+ e)u.

Case of Pure Rolling

In the case of pure rolling, we must have F<μR in order that the point of contact may be instantaneous brought to rest

$$i.e.\ \frac{2}{7}(u - a\omega) < \mu(1 + e)v. \qquad \text{...(A)}$$

But, if $\frac{2}{7}(u - a\omega) > \mu(1 + e)v$, the friction is not sufficient to bring the point of contact P to instantaneous rest, and hence in this case equation (4) will not hold good. Also instead of F, *we shall write the maximum frictional force* ⇒ , (u′ – u) = μR...(A)

$$m\,(v' + v) = R \qquad \text{...(B)}$$

$$\text{and} \quad mk^2\,(\omega' - \omega) = mRa \qquad \text{...(C)}$$

$$\therefore u' = u - \mu v\,(1 + e),\ v' = ev \ \text{ and } \omega' = \omega + \frac{5\mu}{2a}v\,(1 + e) \qquad \text{...(D)}$$

In case (iv) F<μR so that the friction may be sufficient to bring the point of contact to rest.

$$\Rightarrow \frac{2}{7}(u + a\omega) < \mu v(1 + e).$$

But if $\frac{2}{7}(u + a\omega) > \mu v(1 + e).$ ...(E)

The friction will be insufficient and then we have similar type of equations (A), (B), (C) with sign of ω changed. Solving these we get

$m' = u - \mu v\ (1 + e)$, $v' = ev$ and

$$\omega' = \frac{5\mu}{2a} v(1 + e) - \omega.$$

Hence by virtue of (E), it will be possible for u to be less than μv (1 + e), if ω be large enough. Hence if the ball is sufficiently large enough, under-cut u′ can be negative *i.e.* the ball rebounds.

SOLVED EXAMPLES

Example 1:

If a top is made by running a thin pin through the centre of a circular disc, of radius 3 inches, so that length of the pin below the disc is 2 inches, prove that, for steady motion in which rim does not touch the ground, the number of revolutions per second about the axis must exceed $\frac{80}{3\pi(13)^{1/4}}$.

Solution:

In this case we shall discuss the steady motion when the rim of the disc just goes on touching the ground. Let axis OC of the top make an angle α with the vertical OZ, then it makes the angle $\frac{1}{2}\pi - \alpha$ with the horizontal OX.

Again $\cos\left(\frac{\pi}{2} - \alpha\right) = \cos NOC = \frac{OC}{ON} = \frac{2}{\sqrt{(13)}}$

i.e. $\Rightarrow \sin\alpha = \frac{2}{\sqrt{(13)}}, \cos\alpha = \frac{3}{\sqrt{(13)}}$

The condition for steady motion is $C^2n^2 > 4Amgh \cos\alpha$ ...(1)

where A and C have their usual meanings.

Here A = moment of inertia about OA

= moment of inertia about a line through O

⊥ to OC

$$= m\left\{\frac{1}{2}\left(\frac{1}{4}\right)^2 + \left(\frac{1}{6}\right)^2\right\} \quad [\because \text{ radius of the disc}]$$

$$= 3'' = \frac{1}{4}\text{ft. and } CO = 2'' = \frac{1}{6}\text{ft.} = m\left(\frac{1}{64}+\frac{1}{36}\right) = \frac{25m}{576}$$

$$C = \text{moment of inertia about OC} = \frac{1}{2}m\left(\frac{1}{4}\right)^2 = \frac{m}{32}$$

and h = distance of C.G. of the top from the vertex

$$= \text{distance of C.G. of the disc from } O = 2'' = \frac{1}{6}\text{ft.}$$

Putting these values in (1) the required condition is

$$\left(\frac{m}{32}\right)^2 n^2 > 4\left(\frac{25m}{576}\right)mg\left(\frac{1}{6}\right)\frac{3}{\sqrt{(13)}} \text{ or } n^2 > \frac{4.25.1.3}{576.0}, \frac{32.32}{\sqrt{(13)}}$$

$$\text{or } n^2 > \frac{25.32.32}{2(13)^{1/2}} \quad n > \frac{160}{2(13)^{1/4}} \text{ or } \frac{n}{2\pi} > \frac{160}{3\pi(13)^{1/4}}$$

$$\text{or number of revolutions per second} > \frac{80}{3\pi(13)^{1/4}}$$

i.e. number of revolutions per second must exceed

$$= \frac{80}{3\pi(13)^{1/4}} = 4.5 \text{ approx.}$$

Example 2:

Show that the vertical pressure of the top on the point of support is equal to its weight when the inclination of its axis to the vertical is given by the least root of the equation

$$3\,Amgh\cos^2\theta - \cos\theta\,(C^2n^2 + 2\,Amgbh) + (C^2n^2a - Amgh) = 0$$

where a and b are constants depending on the initial circumstances of the motion

Solution:

We have equations of motion of the top as

$$-A\ddot{\phi}\sin\theta - 2A\dot{\phi}\,\dot{\theta}\cos\theta + Cn\dot{\theta} = 0 \quad ...(1)$$

$$A\ddot{\theta} + Cn\,\dot{\phi}\sin\theta - A\dot{\phi}^2\sin\theta\cos\theta = Mgh\sin\theta \quad ...(2)$$

$$A\dot{\phi}\sin^2\theta + Cn\cos\theta = D \quad ...(3)$$

$$\text{and } A(\dot{\theta}^2 + \dot{\phi}^2\sin^2\theta) + 2Mgh\cos\theta = E \quad ...(4)$$

Let the vertical component of the reaction at the vertex O be Y, thus for vertical motion of the top, we have

$$M\frac{d^2}{dt^2}(h\cos\theta) = Y - Mg$$

In case Y = M g, then $\frac{d^2}{dt^2}(\cos\theta) = 0$ ($\because$ h is constant)

or $\sin\theta\ddot{\theta} + \cos\theta\dot{\theta}^2 = 0.$

Now, putting for $\ddot{\theta}$ and $\dot{\theta}^2$ from (2) and (4)
and writing E = 2 Mgh b, we get

$$\sin\theta\,(A\dot{\phi}^2 \sin\theta\cos\theta - Cn\dot{\phi}\sin\theta + Mgh\sin\theta)$$

$$+\cos\theta\,(-A\dot{\phi}^2\sin^2\theta + 2M\,ghb + 2Mgh\,b\cos\theta) = 0$$

or $\sin\theta\,(-Cn\dot{\phi}\sin\theta + Mgh\sin\theta)$

$+\cos\theta\,(2M\,ghb - 2M\,gh\,b\cos\theta) = 0$

or $2M\,gh\,(\sin^2\theta + 2b\cos\theta - 2\cos^2\theta) = Cn\dot{\phi}\sin^2\theta.$

In this put for ϕ from (3) and putting D = Cna
We obtain M gh ($\sin^2\theta + 2b\cos\theta - 2\cos^2\theta$)

$$= \frac{Cn}{A} Cn\,(a - \cos\theta)$$

or AM gh ($1 + 2b\cos\theta - 3\cos^2\theta$) = C^2n^2 ($a - \cos\theta$)

or 3AM gh $\cos^2\theta$ –cos θ C^2 n^2 + 2AM gh b) + (C^2n^2 a –AM gh) = 0

Example 3:

Show that in order that it may be possible for a uniform cone, of height h and vertical angle 2α, to spin a top on a rough horizontal plane it is necessary that

$$n^2 > \frac{5g(4 + \tan^2\alpha)\sin\alpha}{h\tan^4\alpha}$$

Solution:

The spinning is possible so long as the slant side just touches the horizontal plane. The motion then will be steady with the axis OC of the top making angle $\frac{1}{2}\pi - \alpha$ with the vertical OZ.

The condition for steady motion is

$C^2n^2 > 4A$ mgh cos α ...(1)

where A is M.I. about OA and C is M.I about OC (the axis on the top). In this case A = Moment of inertia about OA

= moment of inertia about a line through O ⊥ to OC

$$= \frac{3m}{20}(a^2 + 4h^2) = \frac{3m}{20}(h^2\tan^2\alpha + 4h^2) \qquad \left(\because \tan\alpha = \frac{a}{h}\right)$$

$$=\frac{3mh^2}{10}(4+\tan^2\alpha)$$

C = moment of inertia about OC (the axis of the top)

= moment of inertia about the axis of the cone

$$=\frac{3ma^2}{10}=\frac{3mh^2}{10}\tan^2\alpha$$

The distance of the C.G. of the top from the vertex is $\frac{3h}{4}$ and the axis of the top is making an angle $\frac{1}{2}\pi-\alpha$ with the vertical.

Putting these values in the condition for steady motion, we have

$$\left(\frac{3mh^2\tan^2\alpha}{10}\right)^2 n^2 > 4\frac{3mh^2}{20}(4+\tan^2\alpha)mg\frac{3h}{4}\cos\left(\frac{1}{2}\pi-\alpha\right)$$

or $n^2 > \dfrac{5g(4+\tan^2\alpha)\sin\alpha}{h\tan^4\alpha}$

Example 4(a):

If initially the axis of the top is horizontal and it is set spinning with angular velocity ω *in* a horizontal plane, prove that the axis will start to rise if Cn ω> *mgh, and that when Cn ω = 2m gh, the axis will rise to an angular distance.* $\cos^{-1}$ (A ω / Cn), *provided that Aω < Cn, and will be there at an instantaneous rest.*

Solution:

We have equations of motion of the top as

$$A\ddot{\theta}+Cn\,\dot{\phi}\sin\theta-A\dot{\phi}^2\sin\theta\cos\theta=mgh\sin\theta \quad ...(1)$$

$$A\,\dot{\phi}\sin^2\theta+Cn\cos\theta=D=A\omega \quad ...(2)$$

Initially $\dot{\theta}=\frac{1}{2}\pi$ and $\dot{\phi}=\omega$,

∴ D = Aω

Thus initially from (1) $A\dot{\theta}=-(cn\,\omega-mgh)$

which is negative if Cn ω> mgh. In other words, the axis will start to rise if Cn ω> mgh.

Now, eliminating ϕ between (1) and (2) we get

$$A\ddot{\theta} + Cn\frac{A\omega - Cn\cos\theta}{A\sin^3\theta}\sin\theta - A\left(\frac{A\omega - Cn\cos\theta}{A\sin^2\theta}\right)^2\sin\theta\cos\theta$$

$= \text{mgh} \sin\theta$

or $$A\ddot{\theta} + Cn\frac{A\omega - Cn\cos\theta}{A\sin^3\theta} - \frac{B\omega - Cn\cos\theta}{A\sin^2\theta}\cos\theta = \frac{Cn\omega\sin\theta}{2}$$

(putting $\text{mgh} = \frac{1}{2}\, Cn\,\omega$)

$$A\ddot{\theta} = \frac{A^2\omega^2 + C^2n^2}{A}\frac{\cos\theta}{\sin^3\theta} + Cn\,\omega\left\{\frac{1}{2}\sin\theta - \left(\frac{1}{\sin\theta} + \frac{2\cos\theta}{\sin^3\theta}\right)\right\}$$

Integrating it, $$A\dot{\theta}^2 = -\frac{A^2\omega^2 + C^2n^2}{A\sin^2\theta} + Cn\,\omega\left(-\cos\theta + \frac{2\cos\theta}{\sin^2\theta}\right) + C'$$

where C' is constant.

But initially $\theta = \frac{1}{2}\pi, \dot{\theta} = 0,$

$$\therefore C' = \frac{A^2\omega^2 + C^2n^2}{A}$$

$$\therefore A\dot{\theta}^2 = -\frac{A^2\omega^2 + C^2n^2}{A\sin^2\theta} + Cn\omega\frac{(-\sin^2\theta\cos\theta + 2\cos\theta)}{\sin^2\theta} + \frac{A^2\omega^2 + C^2n^2}{A}$$

or $$A^2\dot{\theta}^2\sin^2\theta = -(A^2\omega^2 + C^2n^2)\cos^2\theta + Cn\,\omega A(-\sin^2\theta + 2)\cos\theta$$

$$= -(A^2\omega^2 + C^2n^2)\cos^2\theta + Cn\,\omega A(1 + \cos^2\theta)\cos\theta$$

$$= \cos\theta\,(A\omega - Cn\cos\theta)\,(Cn - A\omega\cos\theta)$$

For instantaneous res $\dot{\theta} = 0$

in which cos cosq $(A\,\omega - Cn\cos\theta)\,(Cn - A\,\omega\cos\theta) = 0$

i.e. when $\cos\theta = 0$ or $\theta = \frac{1}{2}\pi$ which is the initial positions.

and when $A\,\omega - Cn\cos\theta = 0$ or $\cos\theta = \frac{A\omega}{Cn}$

i.e. $\theta = \cos^{-1}\frac{A\omega}{Cn}$ provided $A\omega < Cn$

and when $Cn - A\omega\cos\theta$ or $\theta = \cos^{-1}\left(\frac{Cn}{A\omega}\right)$, which is impossible when $A\omega < Cn$.

Example 4(b):

AB, BC are two equal similar rods freely hinged at B and lie in a straight line on a smooth table. The end A is struck by a blow perpendicular to AB; show that resulting velocity of A is $3\frac{1}{2}$ times of B.

Solution:

Let P be the impulsive force applied at A. Just after the blow let μ_1, ω_1 and μ_2, ω_2 be the liner and angular velocities of the rods AB and BC respectively. When the blow is struck, there will be an impulsive action between the two rods at B. Let Q be the impulsive action at the joint B in opposite directions on two rods. Also let 2a be the length of each rod and m be the mass, then equations of motion for the rods AB and BC are

$$\left.\begin{aligned} &m(u_1 - 0) &&...(1)\\ &m\cdot\frac{1}{3}a^2\omega_1 = (P-Q)a, &&...(2)\end{aligned}\right\} \textit{for the rod AB}$$

$$\left.\begin{aligned} &\text{And } m\mu_2 = Q &&...(3)\\ &\text{and } m\frac{1}{3}a^2\omega_2 = Qa &&...(4)\end{aligned}\right\} \textit{for the rod BC.}$$

But the rods are connected at B so the velocity of B, as deduced from each rod must be equal,

$$\textit{i.e. } u_1 - a\omega_1 = -(u_2 + a\omega_2) \Rightarrow u_1 = \frac{5P}{4m}; a\omega_1 = \frac{9P}{4m} \qquad ...(5)$$

$$\therefore \frac{\text{velocity of A}}{\text{velocity of B}} = \frac{u_1 + a\omega_2}{u_1 - a\omega_1} = \frac{\frac{5P}{4m} + \frac{9P}{4m}}{\frac{5P}{4m} - \frac{9P}{4m}} = -\frac{7}{2} = -3\frac{1}{2}$$

$$\Rightarrow \text{velocity of A} = 3\frac{1}{2} \text{ times the velocity of B.}$$

Example 4(c):

Three equal uniform rods AB, BC, CD are freely jointed and placed in a straight line on a smooth table. The rod AB is struck at its end A by a by a blow which is perpendicular to its length, find the resulting motion and show that the velocity of the centre of AB is 19 times that of CD and its angular velocity 11 times that of CD.

Solution:

Let each rod be of length 2a and mass m and P be the impulse of the blow at A. Hence there will be impulsive action at B and C in opposite directions, which also taken as X and Y respectively.

Just, before the blow, the three rods are at rest and after the blow let u_1, ω_1; u_2, ω_2; u_3, ω_3 be the velocities of C.G.'s and the angular velocities of the rods AB, BC and CA respectively. Then we have

For, AB, $mu_1 = P - X,\ m\frac{a^2}{3}\omega_1 = pa + Xa\cdot$ ...(A)

For, BC, $mu_2 = X - Y,\ m\frac{a^2}{3}\omega 3 = Xa + Ya\cdot$ ...(B)

For, CD, $mu_3 = Y,\ m\ \frac{a^2}{3}\omega_3\ = Y\ a.$ ...(C)

Now, velocity of B should be the same as deduced from motion of AB and BC and similarly velocity of C should be the same as deduced from the motion of BC and CD respectively.

For B $u_1 - a\omega_1 = u_2 + a\omega_2$ ...(1)

For C $u_2 - a\omega_2\ = u_3 + a\omega_3$ ...(2)

$$\Rightarrow u_1 - a\omega_1 = \frac{P-X}{m} - \frac{3(P+X}{m} = \frac{-2P-4X}{m}$$

From (B) , we have $u_2 + a\omega_2 = \frac{X-Y}{m} + \frac{3(X+Y)}{m} = \frac{4X+2Y}{m}$

$\therefore\quad (1) \Rightarrow 2P - 4X = +\,2\ Y$

or $\qquad 4X + Y = -P.$ ...(3)

Again from (B), we obtain $u_2 - a\omega_2 = \frac{X-Y}{m} + \frac{3(X+Y)}{m} = \frac{-2X-4Y}{m}$

and $u_3 + a\omega_3 = \frac{Y}{m} + \frac{3Y}{m} = \frac{4Y}{m}.$ [using C]

$\Rightarrow -2X - 4Y = 4Y$ or $X + 4Y = 0.$ [using (2)] ...(4)

Solving (3) and (4), we get $X = -\ \frac{4P}{15}$ and $Y = \frac{P}{15}$

$$\Rightarrow u_1 = \frac{1}{m}\left(P + \frac{4P}{15}\right) = \frac{19P}{15m}, \omega_1 = \frac{3}{ma}\left(P - \frac{4P}{15}\right) = \frac{11P}{5ma}$$

$$u_2 = \frac{1}{m}\left(-\frac{4P}{15} - \frac{P}{15}\right) = -\ \frac{P}{3m}, \omega_2 = \frac{3}{ma}\left(-\frac{4P}{15} - \frac{P}{15}\right) = \frac{-5P}{5ma}$$

$$u_3 = \frac{1}{15m}, \omega_3 = \frac{3}{ma}\left(\frac{P}{15}\right) = \frac{P}{5ma}$$

$$\Rightarrow u_1 : u_3 = \frac{19P}{15m} : \frac{P}{15m} = 19 : 1 \textit{ i.e. } u_1 = 19u_3$$

i.e. velocity of centre of AB is 19 times that of CD.

$$\text{Further, } \omega_1 : \omega_2 = \frac{11P}{5ma} : \frac{P}{5ma} = 11 : 1 \textit{ i.e. } \omega_1 = 11\omega_3$$

i.e. the angular velocity of AB is 11 times that of CD.

Example 4(d):

Two uniform rods, AB and BC, of the same material are smoothly joined to B and placed in a horizontal line; the rod BC is struck at G by a blow at right angles to it, find the position of G that the angular velocities of AB and BC may be equal in magnitude.

Solution:

Let AB = 2a and BC = 2b and let m be mass of each rod per unit length.

Also let G_1 be the centre of gravity of the rod BC and G_2 that of the rod AB. Let P be impulse applied at a point G of the rod BC such that $GG_1 = x$.

After the application of the blow let u_1 be the velocity of G_1 the centre of gravity of BC, and ω_1 its angular velocity, u_2 and ω_2 similar quantities for the rod AB, in the direction as shown in the above figure.

When the blow is struck, let the impulsive action at B between the two rods be Q, in opposite directions on the two rods.

Initially the rods BC and AB were at rest

$$\therefore\ m_2bu_1 = P - Q(\text{motion of } G_1) \qquad ...(1)$$

$$m\ 2b\frac{b^2}{3}\omega_1 = Px + Qb, \text{ (motion about } G_{2)} \qquad ...(2)$$

$$m\ .\ 2au_2 = Q; \qquad (\text{motion about } G_2) \qquad ...(3)$$

$$\text{and } m\ 2a\frac{a^2}{3}\omega_2 + Qa; \quad (\text{motion about } G_2) \qquad ...(4)$$

Further the rods are connected at B, so the velocity of B, form each rod, must be the same *i.e.*

$$u_1 - b\omega_1 = u_2 + a\omega_2 \qquad ...(5)$$

Now we have two cases:

Case (i) If $\omega_1 = \omega_2$ *(equal in magnitude and same in direction)*

$$\text{Equations (3) and (4)} \Rightarrow u_2 = \frac{a\omega_2}{3} = \frac{a\omega_1}{3} \qquad ...(6)$$

and equations (5) $\Rightarrow u_1 - b\omega_1 = \frac{a\omega_1}{3} + a\omega_1$, *i.e.* $u_1 = \left(\frac{4a+3b}{3}\right)\omega_1$

Also (2), Þ $xP = 2m\ \frac{b^2}{3}\omega_1 - Qb = = \frac{2mb^3}{3}\omega_1 - 2mabu_2$ [using (3)].

$= 2mb\ \frac{4a+3b}{3}\omega_1 + \frac{2ma^2\omega_1}{3}$ {after substituting for u_1 and u_2}

$= \frac{2m}{3}(a^2 + 4ab + 3b^2)\,\omega_1 = \frac{2m}{3}(a+b)\,(a+3b)\omega_1$

$\Rightarrow x = \frac{2m(b^2 - a^2}{3} \cdot \frac{3}{2m(a+b)(a+3b)} = \frac{b-a}{a+3b}$.

Case (ii)

If $\omega_1 = -\omega_2$ (magnitude equal but opposite in direction)

In this case (3) (4) $\Rightarrow u_2 = \frac{a\omega_1}{3} = -\frac{a\omega_2}{3}$

and (5), $\Rightarrow u_1 - b\omega_1 = u_2 + aw_2 = -\ \frac{aw_1}{3} - 0w_1, \Rightarrow u_1 = \frac{3b-4a}{2}\omega_1$ ·

$\therefore xP = \frac{2mb^3}{3}\omega_1 - Qb = \frac{2mb^3}{3}\omega_1 - 2mabu_2$ [using (2)]

$= \frac{2mb^3}{3}\omega_1 + \frac{2ma^2b}{3}\omega_1\left(\therefore u_2 = \frac{a\omega_1}{3}\right) = \frac{2mb}{3}(b^2 - a^2)\omega_1,$

Now $P = 2mbu_1 + Q$ {from (1) }

$= \frac{2mb(3b-4a)}{3}\omega_1 - \frac{2ma^2}{3}\omega_1$ [substituting for u_1 and u_2]

$= \frac{2m}{3}(3b^2 - 4ab - a^2)\omega_1$ ·

$\Rightarrow x = \frac{2mb(b^2+a^2)}{3}\frac{3}{2m(3b^2-4ab-a^2)} = \frac{b(b^2+a^2)}{3b^2-4ab-a^2}$.

Example 5:

Of two inelastic circular discs with milled edged. Each of mass m and radius a, one is rotating with angular velocity ω round its centre O, which is fixed on a smooth plane, and the other is moving with spin in the plane with velocity v directed towards O. Find the motion immediately afterwards, and show that the energy lost by the impact is

$\frac{1}{2}m\left(v^2 + \frac{a^2\omega^2}{5}\right)$.

Solution:

The disc P is rotating about its centre O with angular velocity ω (say) while disc Q is moving with linear velocity v towards O as shown in the figure. Further let F be the impulsive friction at the point of contact, K. But the discs are inelastic, so they will not rebound, and hence after the impact, the velocity of the disc Q will be along O_1O.

Further let the velocity of Q be u, along the common tangent. Also let ω_1 and ω_2 be angular velocities of P and Q after the impact, then equations of motion are:

$$\frac{ma^2}{2}(\omega_1 - \omega) = -Fa \qquad \text{(for the disc P)} \quad ...(1)$$

$$mu = F \quad ...(2)$$

and $$\frac{ma^2}{2}\omega^2 = Fa \qquad \text{[for the disc P]} \quad ...(3)$$

But the discs P and Q touch each other at K, so the velocity of the point of contact K as deduced from discs must be equal,

$$i.e.\ a\omega_1 = u + a\omega_2 \quad ...(4)$$

Now putting the value of u, ω_1 and ω_2 from (2), (1) and (3) in (4), we easily obtain

$$-\frac{2F}{m} + a\omega = \frac{F}{m} + \frac{2F}{m} \quad i.e.\ F = \frac{ma\omega}{5}.$$

$$\therefore a\omega_1 = -\frac{2a\omega}{5} + a\omega = \frac{3a\omega}{5}:$$

and $$u = \frac{a\omega}{5}: \text{ Also } a\omega_2 = \frac{2a\omega}{5}.$$

$\therefore$ Total K.E. after the impact = K.E. of the disc P + K.E. of the disc Q

$$= \frac{1}{2}m\frac{a^2}{2}\omega_1^2 + \frac{1}{2}m\left(u^2 + \frac{a^2}{2}w_2^2\right)$$

$$= 2m\left[\frac{9a^2\omega^2}{50} + \frac{a^2\omega^2}{50} + \frac{4a^2\omega^2}{50}\right] = \frac{3a^2m\omega^2}{20} = E_2 \text{ (say)}$$

But K.E. before the impact

$$= \frac{1}{2}m\frac{a^2}{2}\omega^2 + \frac{1}{2}mv^2 = E_1 \text{ (say)}$$

$$= \frac{1}{2}m\frac{a^2}{2}\omega^2 + \frac{1}{2}mv^2 - \frac{3ma^2\omega^2}{20} = \frac{1}{2}m\left(v^2 + \frac{a^2\omega^2}{5}\right)$$

Example 6:

If a hollow lawn tennis ball of elasticity e has on striking the ground supposed perfectly rough, a vertical velocity v and angular velocity ω about a horizontal axis, find its angular velocity after impact and that the range of the rebound will be $\frac{4a\omega}{5g}ev$.

Solution:

Just before the impact, v is vertical velocity and ω the angular velocity while, just after impact let u′, v′ be the horizontal and vertical velocities and ω' the angular velocity ( ), then equations of motion are

$$mu' = -F \qquad ...(1)$$

and $$m(v' + v) = R \qquad ...(2)$$

and $$m\frac{2}{3}a^2(\omega' - \omega) = Fa \qquad ...(3)$$

The point of contact P is reduced to rest instantaneously, there being no sliding,

$\Rightarrow u' - a\omega' = 0$ also $v' = ev$.

Now eliminating from (1), (3) we get

$$\frac{2}{3}a(\omega' - \omega) = -u'$$

or $$\frac{2}{3}(u' - a\omega) = -u' \Rightarrow \frac{5}{3}u' = \frac{2}{3}a\omega$$

$$\text{or } u' = \frac{2}{5}a\omega \qquad ...(4)$$

$$\therefore \text{ range after rebound} = \frac{2(\text{horizontal velo.})(\text{vert.velo.})}{g}$$

$$= \frac{2u'v'}{g} = \frac{2\frac{2}{5}a.\omega.ev}{g} = \frac{4a\omega}{5g}ev.$$

Example 7:

An imperfectly elastic sphere descending vertically comes in contact with a fixed rough point, the impact taking place at a point distant α from the lowest point, and the coefficient of elasticity being e. Find the motion, and show that the sphere will start moving horizontally after the impact if

$$\alpha = \tan^{-1}\sqrt{\left(\frac{7e}{5}\right)}$$

Solution:

Before the impact the sphere is descending vertically say with velocity V; implies that it has then no horizontal velocity and no angular velocity. Just after the impact let u_1 and v_1 be the velocities of the sphere along and perpendicular to PG and ω its angular velocity as marked in the above figure.

Then the equations of motion are :

$$m(v_1 - V\sin\alpha) = -F \quad \text{...(1)}$$

$$m(v_1 + V\cos\alpha) = R \quad \text{...(2)}$$

$$\text{and } m\frac{2a^2}{5}\omega = Fa. \quad \text{...(3)}$$

After impact there is no sliding velocity *i.e.* velocity of the point P along tangent at P is zero.

$\therefore v_1 - a\omega = 0 \Rightarrow v_1 = a\omega$

Also $u_1 = eV\cos\alpha$ (by Newton's Law)

Form (1) and (3), we get $(v_1 - V\sin a) = -\frac{2a}{5}\omega$

$\Rightarrow v_1 - V\sin\alpha = -\frac{2}{5}v_1 \Rightarrow v_1 = \frac{5}{7}V\sin\alpha.$

$\therefore v_1 = a\omega = \frac{5}{7}V\sin\alpha.$

The sphere will start moving horizontally after impact if the vertical velocity of sphere = 0 *i.e.* if $u_1\cos\alpha - v_1\sin\alpha = 0$.

or $\tan\alpha\frac{u_1}{v_1}$ or $\tan\alpha\frac{eV\cos\alpha}{\frac{5V}{7}\sin\alpha}$ or $\tan^2\alpha = \frac{7}{5}e \Rightarrow \alpha = \tan 1\sqrt{\left(\frac{7e}{5}\right)}$

Example 8:

A rough imperfectly elastic ball is dropped vertically and when its velocity is V, a man suddenly moves his racket forward in its own plane with velocity U, and thus subjects the ball to pure cut in a downward direction making an angle α with the horizon. Show that, on striking the rough found, the ball will not proceed beyond the point of impact, provide

$$(U - v\sin\alpha)(1-\cos\alpha) > (1+e)\left(1+\frac{a^2}{k^2}\right)V\sin\alpha\cos\alpha.$$

Solution:

Let P be a point on the ball such that it is hit by the man. Obviously the plane of the racket is tangential to the ball. Since the man moves the racket in its own plane with velocity U, the velocity of the point of contact P will also be U along the tangent at P.

Just before the impact the velocity of the ball is given to be V, there being no horizontal velocity and no angular velocity.

Just after the impact, let u,v be the horizontal and vertical velocities and ω the angular velocity as marked in the figure. *The we have for the motion of ball.*

Just before the impact, the moment of momentum about P

$= mV\ a \sin\alpha.$ $\left[\because \text{moment of momentum} = mk^2\dot{\theta} + mv\,p\right]$

After impact, moment of momentum about P

$= mk^2\omega - mua\cos\alpha + mv\,a\sin\alpha$

$\therefore$ Change in moment of momentum

$= (mk^2\omega - mua\cos\alpha + mv\,a\sin\alpha) - mV\,a\sin\alpha$

But the impulse is applied at P, hence moment of the impulse at P is zero Now

$mk^2\omega - mua\cos\alpha + mv\,a\sin\alpha - mV\,a\sin\alpha = 0$

($\therefore$ Change in moment of momentum = moment of the impulse).

But velocity of the point P of the racket = velocity of the point of the ball

$\Rightarrow U = u\cos\alpha - v\sin\alpha + a\omega$ ...(2)

Again multiplying (1) by cos α and (3) by sin α and adding, we get

Also by Newton's experimental law, we have

$(mk^2\omega_1 - mu_1a) - (mk^2\omega - mua) = 0$ ...(4)

($\because$ change in moment of momentum = moment of the impulse) Also, the point of contact has no horizontal velocity so we get $u_1 + a\omega_1 = 0$ ...(5)

(4) and (5) $\Rightarrow u_1\left(1+\frac{k^2}{a^2}\right) = u - \frac{k^2}{a}\omega$ ...(6)

From (6), it is clear that u_1 will be negative if $u < \frac{k^2}{a}\omega$.

Thus we can say that the ball will not proceed beyond the point of contact, if

$$u < \frac{k^2}{a}\omega, \text{ or } \frac{Uk^2\cos\alpha + V\sin\alpha\cos\alpha\left\{a^2 + e\left(k^2 + a^2\right)\right\}}{a^2 + k^2}$$

$$< \frac{k^2 a\left(U - V\sin\alpha\right)}{a\left(a^2 + k^2\right)}$$

if $k^2 (U - V \sin \alpha) > Uk^2 \cos \alpha + V \sin \alpha \cos \alpha \{a^2 + e (k^2 + a^2)\}$

$= Uk^2 \cos \alpha - Vk^2 \sin \alpha \cos \alpha$

$\qquad + Vk^2 \sin \alpha \cos \alpha + V \sin \alpha \cos \alpha \{a^2 + e (k^2 + a^2)\}$

$= k^2 (U - V \sin \alpha) \cos \alpha + V \sin \alpha \cos \alpha \{(a^2 + k^2) + e (k^2 + a^2)\}$

$= k^2 (U - V \sin \alpha) \cos \alpha + (1 + e) (k^2 + a^2) V \sin \alpha \cos \alpha.$

Example 9:

Three equal rods, AB, BC, CD are freely jointed and placed in a straight line on a smooth table. The rod AB is struck at its end A by a blow which is perpendicular to the length; find the resulting motion and show that the velocity of the centre of AB is 19 times that of CD, and its angular velocity 11 times that of CD.

Solution:

Let P be the impulse of the blow applied at A and let $(u_1, \omega_1,)$ $(u_2, \omega_2,)$ $(u_3, \omega_3,)$ be the velocities angular velocities of the rods AB, B C, CD respectively just after the blow.

Again let Q and R be the impulsive reactions at B and C respectively, then we have

$$mu_1 = P + Q \qquad ...(1)$$

$$m\frac{a^2}{3}\omega_1 = (P - Q)a \qquad ...(2) \qquad \text{[for the rod AB]}$$

$$mu_1 = - R + Q \qquad ...(3)$$

$$m\frac{a^2}{3}\omega_2 = (Q + R)a \qquad ...(4) \qquad \text{[for the rod BC]}$$

$$mu_3 = R \qquad ...(5)$$

$$- m\frac{a^2}{3}\omega_3 = - Ra \qquad \text{[for the rod CD]} ...(6)$$

As the rods are connected at B and C, the velocity of B as considered from AB = the velocity of A as considered from BC.

$\Rightarrow$ velocity of G_1 + velocity of B relative to G_1

= velocity of G_2 + velocity of B relative to G_2.

$\Rightarrow - u_1 + a\omega_1 = u_2 + a\omega_2$ ...(7)

Similarly for C, we have $a\omega_2 = u_2 = a\omega_3 + u_3$ ...(8)

Now from (7) and {(1) to (4)}, we easily obtain

3 (P – Q) (– (P + Q) = 3 (R + Q) + Q – R)

$\Rightarrow$ 2P = 8Q + 2R $\Rightarrow$ P = 4Q + R ...(9)

Also from (8) and {(3) to (6)},

we get 3 (Q + R) – (– R + Q) = 3R + R $\Rightarrow$ Q = – 4R

from (9) and (10), we get easily P = – 15R and Q = – 4R.

$$\therefore \frac{\text{velocity of } G_1}{\text{velocity of } G_3} = \frac{u_1}{u_3} = \frac{15R + 4R}{R} = 19$$

$$\text{and } \frac{\text{angular velocity of AB}}{\text{angular velocity of CD}} = \frac{\omega_1}{\omega_3} = \frac{3(15R + 4R)}{3R} = 11$$

Example 10(a):

Two equal uniform rods, AB and BC, are freely jointed at B and turn about a smooth joint at A. When the rods are in a straight line, ω being the angular velocity of AB and u the velocity of the centre of mass of BC, BC impinges on a fixed inelastic obstacle at point D, show that the rods are instantaneously brought to rest if $BC = 2a\dfrac{2u - a\omega}{3u + 2a\omega'}$ where 2a is the length of either rod.

Solution:

when the rods AB and BC are in a straight, line, ω is the angular velocity of AB and u the velocity of G_2 the centre of gravity of BC, and let ω_1 be the angular velocity of BC about G_2 before impinging on an inelastic obstacle at D, such that BC = x.

There will be an impulsive reaction between the two rods at B denoted by Q acting on opposite directions as marked. Bow the velocity of B as deduced from AB = the velocity of B as deduced from BC.

$\Rightarrow$ 2aω = velocity of G_2 + velocity of B relative to G_2 = $(u - a\omega_1)$

Take moments about A to remove the unknown reaction at A, we get

$$m \cdot \frac{4}{3}a^2\omega = -Q \cdot 2a \Rightarrow m\frac{2a\omega}{3} = -Q \qquad ...(1)$$

For the rod BC, we get mu = – (P – Q) ...(2)

$$\text{and} \qquad m\frac{a^2}{3}\omega_1 = -\{P(x - a) + Qa\} \qquad ...(3)$$

$$\therefore (2)\text{and}(3) \Rightarrow m\left(u + \frac{a}{3}\omega_1\right) = -P - P\frac{(x - a)}{x} = -\frac{Px}{a} \qquad ...(4)$$

Also (2) and (1), imply $-P = \left(mu + \frac{2ma\omega}{3}\right)$

With this substitution in equation (4), we readily obtain

$$\frac{x}{a}\left(u+\frac{2a\omega}{3}\right)=\left(u+\frac{a}{3}\omega_1\right)=\left[u+\frac{(u-2a\omega)}{3}\right]$$

$$\Rightarrow \frac{x}{a}\left(\frac{3u+2a\omega}{3}\right)=\left(\frac{4u-2a\omega}{3}\right)=\frac{2(2u-a\omega)}{3}\Rightarrow x=\frac{2a(2u-a\omega)}{(3u+2aw)}$$

Example 10(b):

A disc of any form moving in its plane without rotation with velocity V at right angles to a fixed plane strikes the plane so that the distances of its centre of gravity from the point of impact and from the plane are r and p. Assuming the plane to be elastic and sufficiently rough to prevent sliding, show that the impulsive pressure and friction are respectively

$$\frac{m(1+e)V(p^2+k^2)}{r^2+k^2} \quad \textit{and} \quad \frac{m(1+e)Vp(r^2-p^2)^{1/2}}{r^2+k^2};$$

where k is the radius of gyration. Also show that loss of kinetic energy is

$$\frac{1}{2}\frac{m(1+e^2)(K^2+p^2)V^2}{(r^2+k^2)};$$

Solution:

Let the disc strike the horizontal plane at K. Before the impact, V is the vertical velocity of G. After the impact let (u, v, w) be the velocities and angular velocity of the disc. Angular velocity of K relative to G is r ω perp. to GK where GK = r.

Now equations of motion are,

$m\ [v - (-V)\ R$ ...(1)

$mu = F$ ...(2)

$mk^1\,\omega = -Fp + R\,r\sin\theta$

or $mk^2\,\omega = -Fp + R\sqrt{(r^2-p^2)}$ ...(3)

p being the perp. from G on the tangent at K.

Also since there is no sliding.

∴ tangential velocity of the point K = velocity of G + Velocity of K relative to G.

$$= u - \frac{p}{r}(\omega r) = 0 \Rightarrow u - p\omega = 0 \qquad ...(4)$$

Also Newton's rule (V = ev) $\Rightarrow v + \sqrt{(r^2 - p^2)}\omega = eV$ ...(5)

Now eliminating u, ω from (4) with the help of (2) and (3), we readily obtain

$$\frac{F}{m} = p\left(-\frac{Fp}{mk^2} + \frac{R\sqrt{r^2 - p^2}}{mk^2}\right) \Rightarrow \frac{F}{m}\left(1 + \frac{p^2}{k^2}\right) = \frac{Rp\sqrt{r^2 - p^2}}{mk^2}$$

$$\Rightarrow F(k^2 + p^2) = Rp\sqrt{r^2 - P^2} \qquad ...(6)$$

Now $\frac{R}{m} - V + \sqrt{(r^2 - p^2)}\left\{-\frac{Fp}{mk^2} + \frac{R(r^2 - p^2)}{mk^2}\right\} = eV$

Now $\frac{R}{m} - V + \sqrt{(r^2 - p^2)}\left\{-\frac{Fp}{mk^2} + \frac{R\sqrt{r^2 - p^2}}{mk^2}\right\} = eV$

[putting the values of v, ω in (5) from (1)].

This gives, $\frac{R}{m}\left\{1 + \frac{r^2 - p^2}{k^2}\right\} - \frac{Rp\sqrt{(r^2 - p^2)}}{mk^2} = V(1 + e)$

$$\Rightarrow \frac{R}{mk^2}\left[k^2 + r^2 - p^2\right] - \frac{Fp^2(r^2 - p^2)}{(k^2 + p^2)mk^2} = V(1 + e)$$

$$\Rightarrow R = \frac{mV(1 + e)\cdot(k^2 + p^2)}{k^2 + r^2}.$$

Second part

$$\therefore F = \frac{Rp\sqrt{(r^2 - p^2)}}{k^2 + p^2} = \frac{mV(1 + e)\sqrt{(r^2 - p^2)}}{k^2 + r^2}p \qquad \text{[using (6)]}$$

$\therefore$ loss of K.E. $= \frac{1}{2}$R (velocity of the point K in the direction of impulse before impact + the velocity after impact)

$$= \frac{1}{2}R[V - tV] = \frac{1}{2}(1 - e)RV = E \text{ say}$$

[$\because$ velocity in the direction of F is zero, hence loss of K.E. in that direction is zero].

$$\Rightarrow E = \frac{1}{2}mv^2\left(1-e^2\right)\frac{k^2+p^2}{k^2+r^2} \qquad \text{[putting the value of R]}$$

Example 10(c):

Four freely jointed rods, of the same material and thickness, form a reactable of sides 2a and 2b and of mass M′. When lying in this form on a horizontal plane, an inelastic particle of mass M moving with velocity V in a direction perp. to the rod of length 2a impinges on it at a distance c from its centre. Show that the K.E. lost in the impact is

$$\frac{1}{2}V^2 \div \left[\frac{1}{M}+\frac{1}{M'}\left(1+\frac{3a+3b}{a+3b}\frac{c^2}{a^2}\right)\right]$$

Solution:

Let AB be a rod of length 2b and mass m_2 and BC of length 2a and mass m_1. Let u be the linear velocity of G_1, the C.G. of BC, and w the angular velocity just after the action of the impulse (*i.e.* when the particle of mass M strikes at E such that ($G_1E = c$). Let I be the impulse applied. Obviously the rods AB and CD will not rotate and remain parallel.

Now velocity of AB = velocity of B = velocity of G_1 + velocity of B rel. to $G_1 = u - a\omega$ Velocity of CD = velocity of C = velocity of G_1 + velocity of C rel. to $G_1 = u + a\omega$ and the velocity of the point E just after the blow

= Velocity of G_2 + velocity of E rel. to $G_1 = u + c\omega$.

(Since the a.v of BC is in opposite direction to the AD)

we also have

$$\frac{m_1}{2a} = \frac{m_2}{2b} = \frac{m_1+m_2}{2(a+b)} = \frac{M'}{4(a-b)}\left(\because m_1+m_2 = \frac{M'}{2}\right).$$

Now equation of motion for the system (all the four rods) is

$$M'u = I \qquad \text{...(1)}$$

and for the particle, of mass M, we have M (V – u – c ω) = I ...(2)

Taking moment of momentum about G, the centre for all the rods, we have

$$b + m_1\frac{a^2}{3}\omega - m_1ub + m_1\frac{a^2}{3}\omega - m_2(u-a\omega)a + m_2a\left(u+c\omega\right) = Ic$$

$$\Rightarrow 2\left(m_1\frac{a^2}{3}\omega + 2m_2a^2\omega\right) = Ic \Rightarrow \frac{2a^2\omega}{3}[m_1+3m_2] = Ic$$

$$\Rightarrow \frac{a^2\omega}{3}\left[\frac{M'a+3M'b}{(a+b)}\right] = cM\,(V - c\omega - u),$$

[putting the value of I and m_1, m_2]

$$\Rightarrow \frac{M'a^2\omega}{3}\cdot\frac{a+3b}{a+b} = cM(V - c\omega - u)$$

using (1) and (2), we get $M'u = M\,(V - c\omega - u)$ ...(3)

$$\Rightarrow (M + M')u = MV - Mc\omega$$

Now putting the value of in (3) from (4), we obtain

$$M'\frac{a^2}{3}\omega\cdot\frac{a+3b}{a+b} = cM\ (V - c\ \omega) - \frac{cM^2(V - c\omega -)}{M+M'}$$

$$= c\ M\ (V\ \text{-}\ c\ w)\ M + M' - M)\ /\ (M + M') = \frac{cMM'(V - c\omega)}{(M+M')}$$

$$\Rightarrow c\omega\left[\frac{M}{M+M'} + \frac{a^2}{3c^2}\frac{a+3b}{a+b}\right] = \frac{VM}{M+M'}$$

$$\Rightarrow c\omega = \frac{MVK}{M(1+K)+M'} \quad \text{where } K = \frac{3c^2}{a^2}\frac{a+b}{(a+3b)}$$

$$\therefore\ (M + M')\ u = M\left[v - \frac{MVK}{M(1+K)+M'}\right] = \frac{(M+M')MV}{M\,(1+K)+M'}$$

$$\Rightarrow u = \frac{MV}{M\,(1+K)+M'}$$

Now loss of K.E. = loss of the energy of the four rods + loss of energy of the particle.

$$\frac{1}{2}I\cdot\left[V + (u + c\omega)\right] - \frac{1}{2}I.(u + c\omega) = E \text{ say}.$$

[$\therefore$ *Impulse on the particle and rod is equal and opposite*]

$$\therefore E = \frac{1}{2}I.\,V = \frac{1}{2}V.\,M[V - (u + c\omega)]$$

$$= \frac{1}{2}\frac{MM'V^2}{M(1+K)+M'} = \frac{v^2}{2\left[\frac{1}{M} + \frac{1}{M'}\left\{1 + \frac{3c^2}{a^2}\frac{(a+b)}{(a+3b)}\right\}\right]}$$

Example 11:

AB, BC, CD three equal uniform rods hinged freely at B and C are lying on a smooth horizontal table, so that ABC and BCD are at right angles on opposite sides of BC. A blow is given to A in the direction AC, prove that D begins to move in a direction $\tan^{-1}\left(\frac{7}{4}\right)$ with CD.

Solution:

Let P be the impulse applied at A in the direction of AC. *Before impulse* the system is at rest. *After impulse,* let the velocity of G_1, the centre of gravity of AB, be (u_1, v_1) and the angular velocity of this rod AB be ω_1. Similarly the angular velocities of the other rods BC and CD are ω_2 and w_3 respectively.

Hence, we obtain

velocity of B along AB = u_1 ...(1)

velocity if B perp. to AB = velocity of G_1 + velocity of B rela. to $G_1 = v_1 + a\omega_1$ ...(2)

velocity of G_2 along AB = velocity of B + velocity of G_2 relative to $B = u_1 - a\omega_2$ ...(3)

velocity of G_2 perp. to AB = the same as that of $G_1 = v_1 + a\omega_1$ ...(4)

velocity of C along AB = velocity of B + velocity of C relative to $B = u_1 - 2a\omega_2$ ...(5)

velocity of C perp. to AB = the same as that of G_2

or $G_1 = v_1 + a\omega_1$ ... (6)

velocity of G_3 along AB = the same as that of C = $u_1 - 2a\omega_2$ ...(7)

velocity of G_3 perp. to AB = velocity of G_2 + velocity of G_3 relative to $G_2 = v_1 + a\omega_1 + a\omega_3$ ...(8)

velocity of D along AB = the same as C = $u_1 - 2a\omega_2$ ...(9)

velocity of D perp, to AB = velocity of C + velocity of D relative to $C = v_1 + a\omega_1 + 2a\omega_3 + a\omega_3$ ...(10)

Taking moments about C for the rod CD, we have

$$\frac{1}{3}a^2\omega_2 + a(v_1 + a\omega_1 + a\omega_3) = 0 \Rightarrow 3v_1 + 3a\omega_3 = 0 \quad ...(11)$$

Again taking moments about B for the rods BC and CD, we get

$$\frac{[illegible]}{3}[illegible] + a^2\omega_2 - 2a\left(u_1 - a\omega_2\right) = 0 \quad ...(12)$$

$$\Rightarrow \frac{a^2}{3}\omega_2 - 2a\left(u_1 - 2a\omega_2\right) - a(u_1 - a\omega_2) = 0 \quad \{\text{using (11)}\}$$

$$\Rightarrow 9u_1 - 16a\,\omega_2 = 0 \quad ...(13)$$

For all the rods, taking moments about A, we get

$$\frac{1}{3}a^2(\omega_1+\omega_2+\omega_3)-2a(u_1-2a\omega_2)+3a(v_1+a\omega_1+a\omega_3)$$

$$-(u_1-a\omega_2)+2a(v_1+a\omega_1)+av_1^{\cdot}=0 \qquad ...(14)$$

$$\Rightarrow 15v_1+13a\omega_1+6a\omega_3=0 \qquad ...(15)$$

Now resolving all the velocities perpendicular to AB, we have

$$(u_1\cos 45^0-v_1\cos 45^0)+(u_1-a\omega_2)\cos 45^0-(v_1+a\omega_1)\cos 45^0$$

$$+(u_1-2a\omega_2)\cos 45^0-(V_1+a\omega_1+a\omega_3)\cos 45^0=0$$

or $3(u_1-v_1)-2a\omega_1-3a\omega_2-aw_2=0$ ...(16)

If the direction of motion of D makes an angle θ with DC, then obtain

$$\tan\theta=\frac{v_1+a\omega_1+2a\omega_2}{u_1-3a\omega_2}=\frac{3}{-\frac{2a\omega_2}{9}}=-3\left(\frac{\omega_3}{\omega_2}\right) \text{[using (11) and (13)]}$$

From equation (11) and (15), we get

$$\frac{v_1}{-17}=\frac{a\omega_1}{21}=\frac{a\omega_2}{-3}\Rightarrow v_1=\frac{17a\omega_3}{3};a\omega_1=-7a\omega_3$$

Putting these values in (16 and making use of (13), we have

$$\frac{16a\omega_2}{3}-17a\omega_3+14a\omega_3-3a\omega_2-a\omega_3=0$$

$$\Rightarrow\frac{7a\omega_2}{3}=4aw_3\Rightarrow\frac{3\omega_3}{\omega_2}=\frac{7}{4}$$

$$\therefore \tan\theta=-\frac{3\omega_3}{\omega_2}=-\frac{7}{4}$$

Example 12:

AB and Cd are two equal similar rods connected by a string BC; AB, BC and CD form three sides fo the square. The point A of the rod AB is struck by a blow in a direction perp. to the rod, show that the initial velocity of A is seven times that of D.

Solution:

Let u_1 be the velocity of G_1, the C.G. of AB and ω_1 the angular velocity of AB and u_2, ω_2 those of CD.

As a matter of fact, the initial motion of B must be perp. to Ab, so that the tension in the string at B must be along BC.

Let P be the impulse applied at A perp. to AB, then we have

$$mu_1 = P + T \qquad ...(1)$$

$$m\frac{a^2}{3}\omega_1 = (P - T)a \qquad ...(2)$$

$$mu_2 = T; \qquad ...(3)$$

$$m\frac{a^2}{3}\omega_3 = T.a \qquad ...(4)$$

As AB and CD are connected by a string, the velocity of B = the velocity of C

$$\Rightarrow a_1\ \omega - u_1 = u_2 + a\omega_2 \qquad ...(5)$$

Substituting the values of u_1, ω_1, u_2, ω_2,.

[From (1) to (4) in (5), we have 3 (P – T) – (P + T) = T + 3T ⇒ T = (P/4)

$$\therefore \frac{\text{initial velocity of A}}{\text{initial velocity of D}} = \frac{u_1 + a\omega_1}{a\omega_2 - u_2} = \frac{\frac{5T}{m} + \frac{9T}{m}}{\frac{3T}{m} - \frac{T}{m}} = \frac{7}{1}.$$

Example 13:

A light rod ABC has three particles each of mass m attached to it at A, B, C. The rod is struck by a blow P at right angles to it is a point, distant from A equal to BC. Prove that the K.E. set up is

$$\frac{1}{2}\frac{P^2\left(a^2 - ab + b^2\right)}{m\left(a^2 + ab + b^2\right)} \text{ where } AB = a,\ BC = b.$$

Solution:

Let the three particles each of mass m be placed at A, B, C of a light rod ABC, and let the impulse P be applied at O such that AO = BC = b, where AB = a, BC = b.

Let $\dot{u}$ be the velocity of C and ω, the angular velocity of the rod just after the blow, then the velocity of B is u + bω, the velocity of A is u + (a + b) ω and the velocity of the point O at which the impulse is applied is u + aω. Since the system was at rest initially, the velocity of O just before the impact is zero.

Now equation the total momentum per. to rod, to the impulse, we have

$$mu + m\,(u + b\omega) + m\,\{u + (a + b)\,\omega\} = P \Rightarrow 3u + \omega\,(a + 2b) = \frac{P}{m}$$

Also taking moments about O, we easily obtain

$$\{u + (a + b)\}\ b - m\ (u + b\ \omega)\ (a - b) - mua = 0$$

$$\Rightarrow\ u\ (a - b) = b^2\omega.$$

$$\therefore \frac{u}{b^2} = \frac{\omega}{a-b} \Rightarrow \frac{3u + (a+2ab)\omega}{3b^2 + (a+2b)(a-b)} = \frac{(P/m)}{\left(a^2+ab+b^2\right)}$$

$$\Rightarrow u = \frac{b^2}{a^2+ab+b^2}\frac{P}{\dot{m}} \text{ and } \omega = \frac{(a-b)}{\left(a^2+ab+b^2\right)}\frac{P}{\dot{m}}$$

K.E. set up $= \frac{1}{2}P$ (velocity of the point O) $= \frac{1}{2}\ \frac{P^2\left(a^2+b^2-ab\right)}{m\left(a^2+b^2+ab\right)}$

Aliter. K.E. $= \frac{1}{2}$ m $\{u + (a + b)\ \omega\}^2 + (u + b\omega)^2 + u^2]$

$$= \frac{1}{2}\ \frac{P^2}{m}\left[\frac{a^2}{\left(a^2+ab+b^2\right)} + \frac{a^2b^2+b^4}{\left(a^2+b^2+ab\right)^2}\right]$$

but $u + (a + b)\ \omega = \frac{a^2}{a^2+ab+b^2}\frac{P}{m}$ and $u + b\omega = \frac{ab}{a^2+ab+b^2}\frac{P}{m}$

$$\therefore \text{ K.E. } = \frac{1}{2}\frac{P^2}{m}\left[\frac{(a^2+b^2)^2 - a^2b^2}{(a^2+b^2+ab)^2}\right] = \frac{1}{2}\ \frac{P^2}{m}\frac{(a^2+b^2-ab)}{(a^2+b^2+ab)}$$

Revision at a Glance

(i) If u and u¢ are the velocities of the C.G. parallel to x axis, just before and just after application of the impulsive forces, then we have

$M\ (u' - u) = \Sigma\ X'$, Similarly $M\ (v' - v) = \Sigma\ Y'$.

$\Rightarrow$ change in momentum of mass, supposed collected at the centre of gravity, in any direction is equal to the sum of the impulses in that direction.

(ii) If ω, ω' be the angular velocities of the body just before and just after the application of the impulsive forces, then we have

$Mk^2\ (\omega' - \omega) = S\ (x'\ Y' - y'\ X')$

$\Rightarrow$ change in moment of momentum about the centre gravity is equal to the sum of the moments of the impulses of the forces about the centre of gravity.

Example 14:

Two uniform rods AB and BC are freely joined at B and laid on a horizontal table. AB is struck by a horizontal blow of impulse P in a direction

perpendicular to AB at a distance c form its centre; the length of AB, BC being 2a and 2b and their masses M′ find the motion immediately after the blow.

Solution:

Let u_1, u_2 be the velocity of G_1 and G_2, the centres of gravity of the rods aB and BC respectively, and w_1, w_2 be the angular velocities of the rods just after blow. Let D be a point in AB where the impulse P is imparted, and $DG_1 = c$.

Before the impulse, the rods are at rest. There will be an impulsive action at B between the two rods AB and BC when the blow is struck. Let this action Q in opposite directions on the two rods, then for the motion of AB, we have

$$M(u_1 - 0) = P - Q \qquad ...(1)$$

and $$M\frac{a^2}{3}\omega_1 = Pc - Qa \qquad ...(2)$$

and for the motion of BC, we have

$$M'(u_2 - 0) = Q ...(3)$$

and $$M'\frac{a^2}{3}\omega_2 = -Qb \qquad ...(4)$$

Now as the rods are connected at B, the velocity of the point of the rods AB is the same as that of the some point B of rod BC.

$$\therefore u_1 + aw_1 = u_2 - bw_2 \qquad ...(5)$$

substituting the values of u_1, w_1; u_2, w_2 from (1), (2), (3), (4) in (5), we have

$$\frac{P-Q}{M} + 3\frac{(Pc - Qa)}{aM} = \frac{Q}{M'} - \frac{3Q}{M'}\left(1 + \frac{3c}{a}\right)$$

$$\therefore u_1 = \frac{P}{M}\left[1 - \frac{1}{4}\frac{M'}{M+M'}\left(1+\frac{3c}{a}\right)\right] \text{ and}$$

$$\omega_1 = \frac{3P}{Ma}\left[\frac{c}{a} - \frac{1}{4}\frac{M}{M+M'}\left(1+\frac{3c}{a}\right)\right]$$

$$u_1 = \frac{1}{4}\frac{P}{M'+M'}\left(1+\frac{3c}{a}\right) \text{ and}$$

$$\omega_1 = -\frac{3}{4}\frac{P}{b(M+M')}\left(1+\frac{3c}{a}\right)$$

Hence the four quantities which determine the motion have been obtained

Example 15:

Four equal uniform rods, AB BC, and DE are freely joined at B, C and D and lie on a smooth table in the form of a square. The rod AB is struck by a blow at A at right angles to AB from the inside of the square; show that the initial velocity of A is 79 times that of E.

Solution:

Let the impulse applied at A be P from inside, so that u_1, u_2, u_3, u_4 are the velocities of G_1, G_2, G_3 and G_4 in the direction of blow,. Further let w_1 and w_2 be the angular velocities of AB and CD respectively. The angular velocities of BC and DE are zero, because the impulsive reactions upon these two rods are along the rods themselves.

Then we have :

$$\left.\begin{array}{ll} mu_1 = P + X_1 & \text{...(1)} \\ \text{and}\, m\dfrac{a^2}{3}\omega_1 = Pa - X_1 a & \text{...(2)} \end{array}\right] \qquad \text{(for the rod AB)}$$

where AB = 2a and X_1 is the impulsive action at B.

$$mu_2 = X_2 - X_1 \qquad \text{(for the rod BC)}$$

Where X_2 is the impulsive action at C.

$$\left.\begin{array}{ll} mu_3 = (X_2 + X_3) & \text{...(4)} \\ \text{and}\, m\dfrac{a^2}{3}\omega_3 = X_1 a - X_3 a & \text{...(5)} \end{array}\right]$$

where X_3 is the impulsive action at D. $mu_4 = X_3$ (for the rod DE) ...(6)

Now the velocity of the point B as deduced from AB must be equal to the velocity as deduced from BC.

$$\therefore\ u_1 - aw_1 = u_2 \qquad \text{...(7)}$$

Similarly; for the point C, $u_2 = u_3 - a\omega_3$. ...(8)

and for the poind D, $u_3 + a\omega_3 = u_4$. ...(9)

whence substituting the values of u_1, ω_1 and u_2 in (7), we readily obtain

$P + X_1 - 3\ (P - X_1) = X_2 - X_1$ *i.e.* $5X_2 = 2P$ ...(10)

Also substituting the value of u_1, u_3 and ω_3 and u_4 in (9), we obtain

$X_2 - X_1 = -\ (X_2 + X_3) = X_3$ *i.e.* $2X_2 - 0$ ...(11)

Finally substituting the values of u_3, ω_3 and u_4 in (9), we obtain

$-\ (X_2 - X_3) + 3\ (X_2 - X_3) = X_3$ *i.e.* $2X_2 - 5X_3 = 0$ ...(12)

Solving (11) and (12), we get

$$X_2 = \frac{5}{21}X_1$$

and $$X_3 = \frac{2}{21}X_1 \quad \therefore (10) \Rightarrow 5x_1 - \frac{5}{21}x_1$$

$$= 2P \text{ i.e. } X_1 = \frac{21P}{50}$$

Hence $$X_2 = \frac{P}{10} \text{ and } X_3 = \frac{P}{25}$$

Putting these values in (1), (2) and (6), we get

$$mu_1 = P + \frac{21P}{50}$$

i.e. $$u_1 = \frac{71P}{50m} \text{ and } a\omega_1 = \frac{3}{m}\left(P - \frac{21P}{50}\right)$$

i.e. $$a\omega_1 = \frac{87P}{50m} \text{ and } mu_4 = \frac{P}{25} \textit{ i.e. } u_4 = \frac{P}{25m}$$

$$\Rightarrow \frac{\text{velocity of A}}{\text{velocity of E}} = \frac{u_1 + a\omega_1}{u_4} = \frac{\left(\frac{71P}{50m}\right) + \left(\frac{87P}{50m}\right)}{\left(\frac{P}{25m}\right)} = \frac{79}{1}.$$

Example 16:

Two rods AB and BC; of length 2a and 2b and of masses proportional to their lengths, are freely joined at B are lying in a straight line. A blow is communicated to the end A; show that the resulting kinetic energy when the system is free is to the energy when C is fixed as

$(4a + 3b)(3a + 4b) : 12(a + b)^2$.

Solution:

Case I. When the system is free. Let the mass of the unit length be m, then length of the rod AB = 2a ⇒ mass of AB = 2ma; length of the rod BC =2B ⇒ mass of BC = 2mb; Further let G_1 be the C.G. of the rod AB and G_2 that of BC.

Let P be the impulse applied at A at right angles to Ab, then there will be an impulsive action between the two rods at B. Let the impulse be Q, in opposite directions on the two rods AB and BC respectively. Just after the impulse, let u_1 be the velocity of the centre of gravity of AB and ω_1 its angular velocity; u_2 and ω_2 similar quantities for BC. Since the rods AB, BC are started from rest, the equations of motion of the rod AB are:

$$2mau_1 = P + Q \text{ (motion of } G_1) \quad ...(1)$$

$$\text{and } 2ma\frac{a^2}{3}\omega_1 = (P - Q)a \text{ [taking moment about } G_1] \quad ...(2)$$

Similarly for the motion of BC, $2mbu_2 = Q$ (motion of G_2) ...(3)

and $2mb\ \dfrac{b^2}{3}\omega_2 = Qb$ [taking a moment about G_2] ...(4)

The rods are connected at B, so the velocity of the point B of the rod AB = velocity of the point B referred to BC

i.e.
$$u_1 - a\omega_1 - u_2 - b\omega_2 \quad ...(5)$$

Substituting the values for ω_1, u_1; u_2, ω_2 in (5), we get

$$\frac{3(P-Q)}{2a} - \frac{P+Q}{2a} = \frac{Q}{2b} + \frac{3Q}{2b} \Rightarrow Q = \frac{pb}{2(a+b)}$$

$\therefore$ velocity of the point A at which the below has been given

$$= u_1 + a\ \omega_1 = \frac{P+Q}{2am} + \frac{3(P-Q)}{2am} + \frac{P(4a+3b)}{2am(a+b)}$$

Now substituting the value of Q in (1), (2), (3), and (4), we get

$$u_1 = \frac{1}{2ma}\left\{P + \frac{bP}{2(a+b)}\right\} - \frac{(2a+3b)P}{4ma(a+b)},$$

$$a\omega_1 = \frac{3}{2ma}\left\{P - \frac{bP}{(a+b)}\right\} = \frac{3(b+2a)P}{4ma(a+b)},$$

$$u_2 = \frac{1}{2mb}\left\{\frac{bP}{2(a+b)}\right\} = \frac{P}{4ma(a+b)},$$

$$\text{and } b\omega_2 = \frac{3}{2mb}\left\{\frac{bP}{2(a+b)}\right\} = \frac{3P}{4m(a+b)},$$

So in this case total kinetic energy,

= K.E. of the rod AB + K.E. of the rod BC.

$$= \frac{1}{2}\cdot 2ma\left[u_1^{\ 2} + \frac{a^2\omega_1^2}{3}\right] + \frac{1}{2}\cdot 2mb\left[u^2 + \frac{a^2\omega_1^2}{3}\right]$$

$$= ma\left[\left\{\frac{2a+3b^2P^2}{16m^2a^2(a+b)^2}\right\} + \left\{\frac{3(b+2a)^2P^2}{16m^2a^2(a+b)^2}\right\}\right]$$

$$+ mb\left[\left\{\frac{P^2}{16m^2a^2(a+b)^2}\right\} + \left\{\frac{3P^2}{16m^2(a+b)^2}\right\}\right]$$

$$= ma\left[\frac{(2a+3b)^2 P^2}{16m^2a^2(a+b)^2} + \frac{3(b+2a)^2 P^2}{16m^2a^2(a+b)^2}\right]$$

$$+ mb\left[\frac{P^2}{16m^2(a+b)^2} + \frac{3P^2}{16m^2(a+b)^2}\right]$$

$$= \frac{P^2(4a+3b)(a+b)}{4ma(a+b)^2} = \frac{(4a+3b)P^2}{4ma(a+b)} = E_1 \text{ say.}$$

Case II. When C is kept fixed.

In this case, for the motion of AB, we get

$$2mau_1 = P + Q. \qquad ...(1)$$

and $2ma\dfrac{a^2}{3}\omega_1 = (P - Q)a$

For the motion of BC, we also have

$$2mb\ \frac{4b^2}{3}\omega_2 = 2bQ$$

[taking moments about C]. ...(2)

Now velocity of B referred to AB = velocity of B referred to BC.

$$\Rightarrow \quad -u_1 + a\omega_1 = 2b\ \omega_2 \qquad ...(4)$$

From (1) and (2), we get $u_1 - a\omega_1 = (-2P + 4Q).\dfrac{1}{2ma}$

Substituting from this and from (3) in (4) we readily get

$$(-2P + 4Q)\ \frac{1}{2ma} = -3\ \frac{Q}{2mb} \text{ or } Q = \frac{(2bP)}{(3a+4b)}$$

Substituting this value of Q in (1), (2) and (3), we get

$$u_1 = \frac{1}{2ma}\left(P + \frac{2bP}{3a+4b}\right) = \frac{(3a+6b)P}{2ma(3a+4b)},$$

and $a\omega_1 = \dfrac{3}{2ma}\left(P - \dfrac{2bP}{3a+4b}\right) = \dfrac{3}{2ma}\cdot\dfrac{3a+2b\cdot P}{(3a+4b)}.$

Also $b\omega_2 = \dfrac{3}{2ma}\left(\dfrac{bP}{3a+4b}\right)$

∴ Total K. E. of AB + K.E. of BC.

$$= \frac{1}{2}\cdot 2ma\left(u_1^{\ 2} + \frac{1}{3}a^2\omega_1^{\ 2}\right) + \frac{1}{2}\cdot 2mb\left(\frac{4b_2\omega_2^2}{3}\right)$$

$$= ma\left[\frac{1}{4m^2a^2}\left(\frac{3a+6b}{3a+4b}\right)^2 P^2 + \frac{1}{3}\frac{9}{4m^2a^2}\left(\frac{3a+2b}{3a+4b}\right)P^2\right]$$

$$+ mb\left[\frac{4}{3}\cdot\frac{9}{4m^2b^2}\frac{b^2P^2}{(3a+4b)^2}\right]$$

$$= 12b\,\frac{\left(3a^2+7ab+4b^2\right)}{4mab\left(3a+4b^2\right)}P^2 = \frac{3(a+b)}{ma(3a+4b)}P^2 = E_2 \text{ Say}$$

$$\Rightarrow \frac{E_1}{E_2} = \left\{\frac{4a+3b)P^2}{4ma(a+b)}\right\} / \left\{\frac{3(a+b)P^2}{ma(3a+4b)}\right\}$$

$$\text{or } \frac{E_1}{E_2} = \{4a+3b\}/\{12(a+b)^2.$$

Example 17:

AB, BC are two equal similar rods freely hinged at B and lie in a straight line on a smooth table. The end A is struck by a blow perpendicular to AB; show that resulting velocity of A is $3\frac{1}{2}$ *times of B.*

Solution:

Let P be the impulsive force applied at A. Just after the blow let μ_1, ω_1 and μ_2, ω_2 be the liner and angular velocities of the rods AB and BC respectively. When the blow is struck, there will be an impulsive action between the two rods at B. Let Q be the impulsive action at the joint B in opposite directions on two rods. Also let 2a be the length of each rod and m be the mass, then equations of motion for the rods AB and BC are

$$\left.\begin{aligned} &m(u_1-0) && ...(1)\\ &m\cdot\frac{1}{3}a^2\omega_1 = (P-Q)a, && ...(2)\end{aligned}\right\}\textit{ for the rod AB}$$

$$\left.\begin{aligned} &\text{And } m\mu_2 = Q && ...(3)\\ &\text{and } m\frac{1}{3}a^2\omega_2 = Qa && ...(4)\end{aligned}\right\}\textit{ for the rod BC.}$$

But the rods are connected at B so the velocity of B, as deduced from each rod must be equal,

$$\textit{i.e. } u_1 - a\omega_1 = -(u_2 + a\omega_2) \Rightarrow u_1 = \frac{5P}{4m};\ a\omega_1 = \frac{9P}{4m} \qquad ...(5)$$

$$\therefore \frac{\text{velocity of A}}{\text{velocity of B}} = \frac{u_1 + a\omega_2}{u_1 - a\omega_1} = \frac{\frac{5P}{4m} + \frac{9P}{4m}}{\frac{5P}{4m} - \frac{9P}{4m}} = -\frac{7}{2} = -3\frac{1}{2}$$

$\Rightarrow$ velocity of A $= 3\frac{1}{2}$ times the velocity of B.

Example 18:

A rectangular lamina, whose sides are of lengths 2a and 2b, is at rest when one corner is ought and suddenly made to move with prescribed speed V in the plane of the lamina; show that the greatest angular velocity which can thus be imparted to the lamina is

$$\frac{3V}{4\sqrt{(a^2 + b^2)}}.$$

Solution:

The corner A is suddenly made to move with prescribed velocity V, in the plane of the lamina, such that V makes an angle θ (say) with BA. Just after the impulse, let μ, v, be the velocities of the centre of gravity, G, parallel and perpendicular to AB, and let ω, be the angular velocity of the lamina.

Now $DG = \sqrt{(a^2 + b^2)}$;

So $\cos\beta = \dfrac{b}{\sqrt{a^2 + b^2}}$,

$\therefore$ equations of motion are $mu = X$ ...(1)

$$mv = Y \qquad ...(2)$$

and $m\dfrac{a^2 + b^2}{3}\omega = Xb + Ya$ ...(3)

Now velocity of A parallel BA = velocity of G parallel to BA + velocity of A parallel to BA relative to G.

$$\Rightarrow V\cos\theta = u + \sqrt{(a^2 + b^2)}\,\omega\cos\beta = u + \sqrt{a^2 + b^2}\,\omega\frac{b}{\sqrt{(a^2 + b^2)}}$$

$\Rightarrow V\cos\theta = v + a\,\omega$ ...(4)

and velocity of A perpendicular to AB = velocity of G perpendicular to AB + velocity of A perpendicular to AB relative to G.

$$\Rightarrow V\sin\theta = v + \sqrt{a^2 + b^2}\,\omega\sin\beta = v + \sqrt{a^2 + b^2}\,\omega\frac{b}{\sqrt{a^2 + b^2}}$$

$\Rightarrow V \sin\theta = v + a\,\omega$

Substituting the values of X and Y [using (1) and (2)] in (3), we have

$$\frac{a^2+b^2}{3}\omega = ab + a = [(V\cos\theta - b\omega)\, b + V\sin\theta - a\omega)]$$

[using (4), (5)]

$$\Rightarrow \frac{4\left(a^2+b^2\right)}{3}\omega = V\,(b\cos\theta + a\sin\theta) \qquad \text{...(6)}$$

Differentiating (6), with respect to " θ", we get

$$\frac{4}{3}\left(a^2+b^2\right)\frac{d\omega}{d\theta} = V\,(-b\sin\theta + a\cos\theta)$$

For ω to be maximum, we must have $\frac{d\omega}{d\theta} = 0$,

$$i.e.\ -b\sin\theta + a\cos\theta = 0, \Rightarrow \frac{\cos\theta}{b} = \frac{\sin\theta}{a} = \frac{1}{\sqrt{\left(a^2+b^2\right)}} \qquad \text{...(7)}$$

$$\therefore \text{(6) and (7) give } \frac{4}{3}\left(a^2+b^2\right) = V\left[\frac{b^2}{\sqrt{\left(a^2+b^2\right)}} + \frac{a^2}{\sqrt{a^2+b^2}}\right]$$

$$\text{or} \quad \omega = \frac{3V}{4\sqrt{\left(a^2+b^2\right)}}$$

Example 19:

Two equal uniform rods. AB and BC are freely joined at B and turn about a smooth joint at A. When the rods, are in a straight line, ω being angular velocity of AB and u the velocity of the centre of mass BC; BC impinges on a fixed inelastic obstacle at a point D; show that the rods are instantaneously brought to rest if BD = 2 a $\frac{2u - a\omega}{2u + 2a\omega}$ *where 2a is the length of the either rod.*

Solution:

Let ω_1 be the angular velocity of BC before the impulse P is given at D. Here BD = x say so that $G_1D = x - a$.

Obviously there will be an impulsive action between the two rods at B and let it be X, in opposite directions on the two rods at B and let it be X, in opposite directions on the two rods. Now, further if the rods are instantaneously brought to rest, then the equation of motion of rod AB is

$$m\cdot\frac{4}{3}a^2\,w = X.\ 2a$$

(taking moments about B) ...(1)

Also equations of motion of BC are mu = P – X ...(2)

$$m\cdot\frac{1}{3}a^2\omega = P\ (x - a) + Xa.\ [\because G_1 C = x - a] \quad ...(3)$$

But the rods are connected at B so the motion of B as deduced from each rod must be the same.

$\Rightarrow 2a\omega$ = vel. of G_1+vel. rel. to $G_2 = u - a\omega_1$.

Putting the values of P and ω_1 from (2) and (4) in (3), we obtain

$$m\cdot\frac{1}{3}a(u-2a\omega) = (mu + X)\ (x - a) + Xa = mu\ (x - a) + Xx.$$

$$\Rightarrow m\frac{1}{3}a(u-2a\omega) = mu\ (x - a) + m\cdot\frac{2}{3}a\omega\cdot x \text{ [putting for X from (1)]}$$

$$\Rightarrow \frac{1}{3}a(u-2a\omega+3u) = \frac{1}{3}x(3u+2\omega a) \Rightarrow x = BD = 2a\ \frac{2u-a\omega}{3u+2a\omega}\cdot$$

Example 20:

Three particles of equal masses are attached to the ends A and C and the middle point B of a light rod ABC and the system is at rest on a smooth table. The particle C is struck a blow at right angles to the rod. Show that the energy communicated to the system when A is fixed to the energy communicated when the system is free as 24 : 25.

Solution:

Let P be the impulse of the blow imparted at B and mass of each particle at A, B, C be m where AC = 2a.

Ist Case. Let A be not fixed. i.e. system is free.

The C.G. of the three particles is at B and the system is at rest before the action of the impulse. Let u be the velocity of the point B. When total mass 3m is supposed to be placed at B, let ω be the angular velocity of the rod, after the action of the impulse, then the equations of motion are 3mu = p.

[motion of C.G. *i.e.* B] ...(1)

$3mk^2\ \omega$ = Pa. [taking moments about B] ...(2)

and $3mk^2 = ma^2 + ma^2$

so that $k^2 = \frac{2a^2}{3}$

Now, K. E. = $\frac{1}{2}3mu^2 + \frac{1}{2}.3mk^2\omega 2.$

$$= \frac{5}{2} m \cdot \frac{4P^2}{2m5^2} = \frac{2P}{5m} = \frac{(K \cdot E)_2}{(K \cdot E)_1} = \frac{24}{5}.$$

Example 21:

An inelastic sphere of radius a rolls down a flight of perfectly rough steps, show that if the velocity of the centre on the first step exceeds $\sqrt{(ga)}$ *its velocity will be the same on every step, the step being such that, in its flight, the sphere never impinges on an edge.*

Solution:

Let E be the edge of the first step and let v be its velocity at E. Now the sphere has a tendency to turn about the edge E and let θ be the angle through which it turns in time t.

Now equations of motion of the sphere are given

by $ma\dot{\theta}^2 = ma\cos\theta - R$ ...(2)

(taking moments about E)

$$(2) \Rightarrow a^2\dot{\theta}^2 = -\frac{10ga}{7}\cos\theta + c$$

when $\theta = o, a\dot{\theta} = v, \Rightarrow c = v^2 + \frac{10ga}{7}$

$$\therefore a^2\dot{\theta}^2 = \frac{10ga}{7}(1-\cos\theta) + v^2 \text{ or } a\dot{\theta}^2 = \frac{10g}{7}(1-\cos\theta) + \frac{v^2}{a}]$$

$$\therefore (1) \text{ gives } \frac{10g}{7}(1-\cos\theta) + \frac{v^2}{a} = g\cos\theta - \frac{R}{m}$$

$$\text{or } R = m\left[g\cos\theta - \frac{10g}{7}(1-\cos\theta) + \frac{v^2}{a}\right] \qquad \therefore (R)_{\theta=0} = m\left(g - \frac{v^2}{a}\right)$$

obviously, $(R)_{\theta=0}$ will not remain positive, if

$v^2a > g$ *i.e.* if $v > \sqrt{(ga)}$

which implies that the sphere leaves the step at once if the velocity $v > \sqrt{(ga)}$. After travelling the distance on the first step, the sphere strikes the second step say at the point K. But the step is inelastic so the sphere will not rebound and will roll on the second step with the velocity v.

When it has rolled a distance x on the second step, let S be the force of friction, sufficient for pure rolling, then equations of motion are given by

$m\ddot{x} = S$ ...(3) and $m\frac{2a^2}{5}\ddot{\theta} = -Sa$ ...(4)

But the motion is of pure rolling, so we have

$$\dot{x} = a\dot{\theta} \Rightarrow \ddot{x} = a\ddot{\theta} \qquad ...(5)$$

from above, we have

$$\ddot{x} = 0 \Rightarrow \dot{x} = \text{constant} = v. = 0 \Rightarrow x = x = \text{constant} = v.$$

which implies that the sphere rolls on the second step with the uniform velocity v. Hence velocity at every step is the same.

Example 22:

Two uniform rods. AB, AC are freely joined at A and laid on a smooth horizontal table so that the angle BAC is a right angle. The rod AB is struck by a blow P at B in a direction perpendicular to AB, show that the initial velocity of A is $\frac{2P}{4m'+m'}$ *where, m, m′ are the masses of AB, AC respectively.*

Solution:

The initial motion of A is to be perpendicular to AB, hence the action at A (say X) must be along AC.

Let u_1 be the linear velocity of G_1 and ω_1 the angular velocity of AB. Then since impulse on AC is along CA, the rod AC will only move in the direction CA say with linear velocity u_2. Now for equation of motion, we have

$$\left.\begin{aligned} mu_1 &= P + X \qquad ...(1) \\ m\frac{1}{3}a^3\omega_1 &= (P - X)a, \quad ...(2) \end{aligned}\right\} \text{motion of AB}$$

$$m'\, u_2 = X \text{ motion of Ac.} \qquad ...(3)$$

But the rods are connected at A, so the motion of A as deduced form the motion of AB and AC must be the same

$$\Rightarrow u_1 = a\omega_1 = -u_2 \qquad ...(4)$$

Now putting the values of u_1, ω_1 and u_2 [form (1), (2), (3)] in (4) we have

$$\frac{(P+X)}{m} - \frac{3(P-X)}{m} = -\frac{X}{m'} \Rightarrow X = \frac{2Pm'}{4m'+m}$$

Whence substitution in (3) provides us

$$u_2 = \frac{X}{m'} = \frac{2P}{4m'+m} \text{ i.e. velocity of the rod AC} = \frac{2P}{4m'+m}$$

or velocity of A = $\frac{2P}{4m'+m}$

Example 23:

AB, BC and CD are three equal uniform rods lying in a straight line on a smooth plane, and they are freely jointed at B and C. A blow is applied at the centre of BC in a direction perpendicular to BC. If ω be the initial angular velocity of AB or CD, and θ the angle they make with BC any time, show that the angular velocity is $\dfrac{\omega}{\sqrt{(a+\sin^2\theta)}}$

Solution:

Let P be the impulse of the blow applied at G_2, the centre of BC, and X the impulsive action at B or C.

Just after the impulse, let U be the velocity of the centre of BC; and V be the velocity of centre of AB and CD, the angular velocities being ω.

Now, we have; $mU = P = 2X$ ...(1) for the rod BC

$$mV = X \quad ...(2)$$

and $\left. m\dfrac{a^2}{3}\omega = Xa \right\}$ for the rod AB or CD

Also, AB and BC are connected at B, so the velocity of B as deduced from BC and AB, must be the same *i.e.* $U = V + a\omega$ ...(4)

$\therefore$ (2) and (3) $\Rightarrow V = \dfrac{a\omega}{3}$ ...(5)

and $U = \dfrac{4a\omega}{3}$ ...(6)

After the action of the impulse, the rods are set in motion and move in the horizontal plane under finite forces. The rod BC retains is horizontal position where as AB, and DC turn about B and C.

At any time t, let AB or DC make an angle θ with BC and further let u be the velocity of BC. Now co-ordinates of G_1 relative to BC (with B as origin) are $x = a\cos\theta$, $y = a$ sing θ

$$\Rightarrow \dot{X} = -a\sin\theta\dot{\theta},\ \dot{y} = a\sin\dot{\theta}.$$

$\Rightarrow$ actual velocity of G_1 along CB and the right angles to CB is given by

$$\left(-a\sin\theta\dot{\theta}\right) \text{ and } \left(u - a\cos\theta\dot{\theta}\right).$$

Now K. E. of the system

$$= \frac{1}{2}mu^2 + \frac{1}{2\cdot}2m\left[u - a\dot{\theta}\cos\theta\right]^2 + a^2\dot{\theta}^2\sin^2\theta + \frac{a^2}{3}\dot{\theta}^2$$

$$= \frac{1}{2}m\left[3u^2 - 4au\dot{\theta}\cos\theta + \frac{8a^2}{3}\dot{\theta}^2\right]$$

When $\theta = 0$, $u = U$; and $\theta = w$

$\therefore$ Initial K. E. of the system $= \frac{1}{2}m\left[3U^2 - 4aU\omega + \frac{8a^2}{3}\omega^2\right]$.

Now there is no displacement, in the points of application of (reactions and weights), at right angles to the horizontal plane, so no work is done by these forces, and thus implying that there should be no change in the K. E. of system. *i.e.*

$$= \frac{1}{2}m\left[3U^2 - 4bu\dot{\theta}\cos\theta + \frac{8a^2}{3}\dot{\theta}^2\right]$$

$$= \frac{1}{2}m\left(3U^2 - 4aU\omega + \frac{8a^2}{5}\omega^2\right)$$

or $$3u^2 - 4au\dot{\theta}\cos\theta + \frac{8a^2}{3}\dot{\theta}^2 = \left(8a^2\omega^2/3\right) \qquad ...(7)$$

[substituting the value of U]

Again after the action of the impulse, the motion is under finite forces hence there is no force on the system in the horizontal plane. The horizontal momentum of the system remains constant viz;

$$mu + 2m\left(u - a\theta\cos\dot{\theta}\right) = mU + 2m\left(U - a\omega\right)$$

or $3u - 2a\dot{\theta}\cos\theta = 3U - 2a\omega$

or $3u - 2a\dot{\theta}\cos\theta = 2a\omega$ [putting the value of U] ...(8)

$\therefore$ (7) and (8) $\Rightarrow \dot{\theta} = \dfrac{\omega}{\sqrt{\left(1+\sin^2\theta\right)}}$ (eliminating u)

Example 24:

Four equal rods, each of mass m and length 2a are freely joined at their ends so as to form a rhombus. The rhombus falls with a diagonal vertical, and is moving with velocity V when it hits a fixed horizontal inelastic plane. Find the motion of the rods immediately after the impact, and show that their angular velocities are equal to

$\dfrac{3V\sin\alpha}{2a\left(1+3\sin^2\alpha\right)}$, *where α is the angle each rod makes with the vertical.*

Show also that the impact destroys a fraction $\left(\dfrac{3\sin\alpha}{1+3\sin^2\alpha}\right)$ *of the kinetic energy just before the impact.*

Solution:

Let PQRS be the rhombus formed of four equal rods each of length 2a and mass m, fall with the diagonal RP vertical. By symmetry, the motion of the rod RS is the same as the motion of RQ while the motion of the rod PS is the same that of PQ. Hence we need only to consider the motion of PQ and QR alone.

Just after impact at P with the horizontal plane, the rod PQ turns about P, say with angular velocity ω_1, while the rod QR turns about Q, say with angular velocity ω_2, with the direction as shown in the figure.

Due to symmetry, the impulsive action at R will be only horizontal, let it be equal to H. Further let X_1, Y_1 be the horizontal and vertical impulse at Q in opposite directions on the rods PQ, QR respectively. As after impact, R moves in vertical direction, so horizontal velocity of the point R relative to Q = 0 ⇒ horizontal velocity of Q + horizontal velocity of R relative to Q = 0

$$\Rightarrow 2a\,\omega_1 \cos\alpha + 2a\,\omega_2 \cos a = 0$$

$$\Rightarrow \omega_2 = -\,\omega_2,$$

Now we have horizontal velocity of $G_1 = a\omega_1 \cos a$, and vertical velocity of $G_1 = a\,\omega_1 \sin\alpha$.

⇒ horizontal velocity of G_2 = horizontal velocity of Q + horizontal velocity of G_2 relative to Q.

$$= 2\,a\,\omega_1 \cos\alpha + a\,\omega_2 \cos\alpha = a\omega_1 \cos a \qquad (\because\ \omega_2 = -\,\omega_1)$$

and vertical velocity of $G_2 = 2a\,\omega_1 \sin\alpha - a\omega_2 \sin\alpha = 3a\,\omega_1 \sin\alpha$.

Now considering the combined motion of PQ and QR and taking moments about P, we get

$$\left\{m\frac{4a^2}{3}\omega_1 + m\left(\frac{a^2}{3}\omega 1 + a\omega_1 \cos\alpha.3a\cos\alpha + 3a\omega_1 \sin\alpha.a\sin\alpha\right)\right\}$$

$$-\,2m\,Va\sin\alpha = 4a\cos\alpha\,H$$

$$\Rightarrow 2\omega_1 = \frac{V\sin\alpha}{a} + \frac{2H\cos\alpha}{ma} \qquad \text{...(3)}$$

∴ (2) and (3) give,

$$\omega_1\left(\frac{4}{3} + 4\sin^2\alpha\right) = -\frac{V\sin\alpha}{a} \Rightarrow \omega_1 = \frac{3V\sin\alpha}{2a\left(1+3\sin^2\alpha\right)}$$

Now K.E. just before the impulse $= \frac{1}{2}4mV^2 = 2mV^2 = E_1$ (say) and K.E. just after the impulse = 2 [K.E. of PQ + K.E. of QR] = E_2 (say)

$$\text{or } E_2 = 2\left[\frac{1}{2}m\frac{4a^2}{3}\omega_1^2 + \frac{1}{2}m\left(\frac{a^2}{3}\omega_1^2 + a^2\omega_1^2\cos^2\alpha + 9a^2\omega_1^2\sin^2\alpha\right)\right]$$

$$= \frac{8}{3}ma^2\left(1+3\sin^2\alpha\right)\omega_1^2 (\because \omega_2 = -\omega_1)$$

$$= \frac{3\sin^2\alpha}{1+3\sin^2\alpha}2m\ V^2 = \frac{3\sin^2\alpha E_1}{1+\sin^2\alpha}$$

Example 25(a):

An equilateral triangle, formed by uniform rods freely hinged at their ends, is falling freely with one side horizontal and upper-mass. If the middle point of this side be suddenly stopped, show that the impulsive actions at the upper and lower hinges are in the ration $\sqrt{(13)}$:1.

Solution:

The middle point O of the rod QR is suddenly stopped, so the impulse is imparted at O. Hence an impulsive action between the two rods at Q is generated. A similar impulse is generated at R as there is symmetry about O. The action at P will be horizontal due to symmetry. Let it be X in opposite directions on the two rods PQ and PR.

Just before the blow there was only vertical velocity for every rod. As the system has been stopped, there are no linear velocities and no angualr velocities for any rod after the blow. Now considering the motion of the rod PQ, or (PR), we get

$$X_1 - X = 0 \qquad ...(1)$$

$\because$ there is no horizontal velocity of G before and after the impulse]

Taking moments about G, we get X_1 LQ – Y_1 GL + X_1 LM = 0

$\Rightarrow x_1\, a \sin 60° - y_1\, a \cos 60°$

$+ X\, a \sin 60° + X\, a \sin 60° = 0$

$$\Rightarrow X_1\frac{a\sqrt{3}}{2} + X\frac{a\sqrt{3}}{2} - Y_1\frac{a}{2} = 0 \qquad ...(2)$$

[$\because$ *there is no angular velocity of PQ before and after the impulse]*

These give $X_1 = X$

and $Y_1 = 2\sqrt{3}X.$

$$\Rightarrow \frac{\text{reaction at Q}}{\text{reaction at P}} = \frac{\sqrt{\left(X_1^{\ 2} + Y_1^{\ 2}\right)}}{X} = \frac{\sqrt{\sqrt{\left(X^2 + 12X^2\right)}}}{X} = \frac{\sqrt{(13)}}{1}.$$

Example 25(b):

A square lamina ABCD rest on a smooth horizontal plane. If the corner A is made to move with velocity u along the BA produced, then determine the initial angular velocity of the lamina.

Solution:

For a square let us choose b = a. The corner A is made to move with velocity u along BA therefore V = u and θ = π. Hence angular velocity u along BA is obtained by putting V = u, θ = π.

$$\therefore \frac{4}{3}\left(a^2 + a^2\right) \text{ w} = \text{u (a cos } \pi + \text{a sing } \pi) \text{ or } \frac{8}{3}a^2\omega = -\frac{3u}{8a}\cdot$$

Above relation gives the required angular velocity.

Example 25(c):

A sphere of mass m falls with velocity V on a perfectly rough inclined plane of mass M and angle a which rests on a smooth horizontal plane. Show that the vertical velocity of the centre of the sphere immediately after the impact is $\frac{5(M+m)V\sin^2\alpha}{7M+2m+5m\sin^2\alpha}$, *the bodies being all supposed perfectly inelastic.*

Solution:

Just before the impact velocity of the sphere is V in vertical direction $(\downarrow)$ there being no angular velocity then. Just after the impact let u and v be the velocities and ω the angular velocity of the sphere as shown in the figure.

Just after the impact, the inclined plane also begins to move in the horizontal direction. Now equations of motion are:

$$\left.\begin{aligned} m(-u) &= R\sin\alpha - S\cos\alpha; \quad (1) \\ m(v - V) &= -R\cos\alpha - S\sin\alpha \quad (2) \end{aligned}\right\}$$

(motion of the C.G. of the sphere

and $$m\frac{2a^2}{5}\omega = Sa \qquad \text{...(3)}$$

Also the motion of the inclined plane is given by

$MV' = R \sin \alpha - S \cos \alpha.$ ...(4)

Now horizontal velocity of the point K of the sphere

= horizontal velocity of the point K of the inclined plane.

$\therefore$ u + a ω cos α = V′

and vertical velocity of the point K of the sphere

= vertical velocity of the point K of the inclined plane.

$\therefore$ v − a ω sin α = 0 ⇒ v = aω sin α.

From (1) and (4) it is easy to see that MV′ = − mu, *i.e.* $V' = -\dfrac{mu}{M}$

$$\Rightarrow u + a\omega \cos a = -\frac{mu}{M} \Rightarrow u = -\frac{Ma\omega\cos\alpha}{M+m}.$$

Now multiplying (2) by sin a and (1) by cos a, and then adding these to (3), we obtain − mu cos α + m (v − V) sin α + m $\dfrac{2a\omega}{5} = 0.$

$$\Rightarrow \frac{Ma\omega\cos^2\alpha}{M+m} + (v - V)\sin\mu + \frac{2a\omega}{5} = 0\left(\because v = -\frac{Ma\omega\cos^2\alpha}{M+m}\right)$$

$$\Rightarrow \left[\frac{Ma\omega\cos^2\alpha}{M+m} + \frac{2}{5}\right]\frac{v}{\sin\alpha} + (v - V)\sin\alpha = 0[\because v = a\omega\sin\alpha]$$

$$\Rightarrow [5M\cos^2\alpha + 2(M+m) + (M+m)\sin^2\alpha]v$$

$$= 5\ (M + m)\ V\ \sin^2\alpha$$

$$\Rightarrow [7M + 2M + 5m\sin^2]\alpha$$

$$\Rightarrow v = \frac{5(M+m)V\sin^2\alpha}{7M + 2m\ \ 5m\sin^2\alpha}\cdot$$

Example 26:

A top consists of a thin uniform spherical shell of radius a and centre C and is free to move about a fixed point O of its surface. The radius OC makes an angle θ with upward down vertical Oz and ϕ is the angle between the plane zOC and a fixed vertical plane zOx.

Initially $\theta = \dfrac{1}{2}\pi, \dot{\phi} = \left(\dfrac{2n}{5}\right),$

where n = $\sqrt{\left(\dfrac{15g}{2a}\right)}$

Show that at time t,

$$\cos\theta = \tanh^2\left\{\left(\frac{3g}{10a}\right)\right\}.$$

Solution:

We have equations of motion of the top as

$A(\dot{\theta}^2 + \dot{\phi}^2 \sin^2\theta) + 2mgh\cos\theta = E$ and $A\dot{\phi}^2 \sin^2\theta + Cn\cos\theta = D$

where D,E are constants to be determined by initial conditions.

Initially $\theta = \frac{1}{2}\pi, \theta = 0, \dot{\phi} = \frac{2n}{5}$.

Putting these in the above equations, we get

$$E = A\left(\dot{\theta} + \frac{4n^2}{25}\right) = (4n^2/25)A \text{ and } D = A\frac{2n}{5} = \frac{2n}{5}A$$

Thus the above equations can be written as

$$A(\dot{\theta}^2 + \dot{\phi}^2 \sin^2\theta) + 2mgh\cos\theta = \left(\frac{4n^2}{25}\right)A \quad ...(1)$$

$$\text{and } A\dot{\phi}\sin^2\theta + Cn\cos\theta = (2n/5)A \quad ...(2)$$

Again C = moment of inertia about OC

= moment of inertia of a hollow sphere about a diameter

= $(2ma^2/3)$

A = moment of inertia about OA

= moment of inertia of a hollow sphere about a line at ⊥ OC

$= (2ma^2/3) + ma^2 = (5ma^2/3)$

and h = distance of C.G. from O = a

$$\text{or} \quad u = \tanh\frac{\lambda t}{2} \Rightarrow u^2 = \frac{\sinh^2\frac{1}{2}\lambda t}{\cosh^2\frac{1}{2}\lambda t} = \frac{\cosh\lambda t - 1}{\cosh\lambda t + 1} = \frac{1 - \operatorname{sech}\lambda t}{1 + \operatorname{sech}\lambda t}$$

$$\therefore 1 + u^2 = \frac{2}{1 + \operatorname{sech}\lambda t} \quad \textit{i.e.} \quad 2\cos\theta = \frac{2}{1 + \operatorname{sech}\lambda t}$$

or sec θ = 1 + sech λt, or sec θ = 1 + sech $\sqrt{(mgh/A)}\,t$.

Example 27(a):

A circular disc, of radius a, has a thin rod pushed through its centre perpendicular to its plane, the length of the rod being equal to the radius of the disc. Show that the system can not spin with the rod vertical unless the angular velocity is greater then $\sqrt{\left(\frac{20g}{a}\right)}$

Solution:

With usual notations, in this case

$$A=\left(\frac{Ma^2}{a}\right)+Ma^2=\left(\frac{5Ma^2}{4}\right), C=\left(\frac{Ma^2}{2}\right) \text{ and } h = a$$

For the top to execute steady motion with axis vertical, the necessary condition is that

$$C^2n^2 > 4AMgh \quad \sqrt{\left(\frac{4AMgh}{C^2}\right)}$$

$$\text{or } \; n > \sqrt{\left(\frac{20g}{a}\right)} \text{ putting for A, C and h.}$$

Example 27(b):

A tennis ball of hallow spherical space is given by underground cut, and hits the ground at the other side of the net at a distance c from it. If u and v its horizontal and vertical velocities and ω its angular velocity when it hits the perfect rough ground, show that the ball will return back towards the net if 2aω>3u.

Further show that it will rebound over the net if $c<\dfrac{2ev(2a\omega-3u)}{5g}$

will touch the net over head if $k<\dfrac{5c}{2}\cdot\dfrac{2ev(2a\omega-3u)-5gc}{(2a\omega-3u)^2}$, *where e is the coefficient of restitution and h is the height of the net.*

Solution:

After the impact let u_1 and v_1 be the velocities of the centre of gravity G of the ball and ω_1 its angular velocity.

Let F be the impulsive friction and R the impulsive normal reactions of motion if the ball are

$$m\,(u_1 + u) = F \quad ...(1) \quad m\,(v_1 + v) = R \qquad ...(2)$$

$$\text{and } m\frac{2}{3}a^2(\omega_1-\omega)=-Fa \qquad ...(3)$$

where $ev = v_1$,

Just after the impact there is no horizontal velocity of the point of contact,

$$\text{so} \quad u_1 - a\omega_1 = 0 \qquad ...(4)$$

From (1) and (3), we obtain

$$u_1+u+\frac{2}{3}a(\omega_1\ \omega)=0$$

$$\Rightarrow 5u_1 = 2a\omega - 3u \Rightarrow u_1 = \frac{2a\omega - 3u}{5}$$

Hence the ball will return back towards the net if u_1 is positive when $2a\omega - 3u>0$ or when $2a\omega>3u$.

Second part. After rebouding from the ground, the ball moves in parabolic path. Now considering the horizontal motion of the ball, we have $t = \frac{c}{u_1}$

Now t, must be less than the time of reaching the ground.

$$\therefore \frac{5c}{2a\omega - 2u} < \frac{2ev}{g}$$

$$\text{or } c < \frac{2ev(2a\omega - 3u)}{5g}.$$

Third part. Let the vertical height to which the ball rises be h then have

$$h' = v_1 t - \frac{1}{2}gt^2 = ev - \frac{1}{2}gt^2 \qquad (\because v_1 = ev).$$

$$= \frac{5}{2}e\frac{2ev\left(2a\omega - 3u\right) - 5gc}{\left(2a\omega - 3u\right)^2}$$

Example 27(c):

Three particles of equal masses are attached to the ends, A and C, and the middle point B of light rod ABC, and the system is at rest on a smooth table. The particle C is struck by a blow at right angles to the rod: show that the energy communicated when the system is free as 24 : 25.

Solution:

Let A, B, C be the three points of a rod, where the three particles, each of mass m are placed. Now let the end A be free, and an impulse P be applied at C, such that the rod begins to rotate about the point B. Then, if the velocity of mass at is u_1, the velocities of the masses A and C are u_1 $u_1 + a\omega_1$ respectively.

Case I. *When the end A is free*

We have

$$mu_1 + m\,(u_1 - a\omega_1) + (u_1 + a\omega_1)$$

$$= P \Rightarrow u_1 = \frac{P}{3m} \qquad ...(1)$$

Taking moments about C, we have

$mu_1 + m\ (u_1 - a\omega_1)\ 2a = 0 \Rightarrow 3u - 2a\omega_1 = 0.$ ...(2)

$$\Rightarrow a\omega_1 = \frac{P}{2m}[u \sin g(1)].$$

Velocity of the mass at C $= u_1 + aw_1 = \dfrac{P}{3m} + \dfrac{P}{2m} = \dfrac{5P}{6m}$

and the K.E. communicated to the rod $= \dfrac{1}{2}$ impulse at C $\times$ (velocity of C)

$$= \frac{1}{2}P \cdot \frac{5P}{6m} = \frac{1}{12}\frac{5P^2}{m} = E_1 \text{ (Say).}$$

Case II. *When A is fixed*

In this case the rod will begin to rotate about A, with an angular velocity ω and an impulsive thrust (= X say) will be generated at A.

Now the moment about A gives

$$m\ (2a)^2\ \omega + ma^2\omega = p.$$

$$2a \Rightarrow m5a\omega = 2P \Rightarrow a\omega = \frac{2P}{5m}\cdot$$

$\therefore$ Velocity communicated to C = $2a\omega$

($\because$ C describes a circle, about A, of radius 2a)

hence its velocity perpendicular to AC $= 2a\omega = \dfrac{4P}{5m}\cdot$

Hence the kinetic energy communicated in this case becomes

$$E_2 = \frac{1}{2}P \text{ (Velocity of C)}$$

$$= \frac{1}{2}P\frac{4P}{5m} = \frac{2P^2}{5m} = E_2 \text{ (say).}$$

$$\Rightarrow \frac{E_1}{E_2} = \frac{(5p^2/12m)}{(2P^2/5m)} = \frac{25}{24} \Rightarrow \frac{E_2}{E_1} = \frac{24}{25}\cdot$$

Example 27(d):

Three equal uniform rods AB, BC, CD are freely jointed and placed in a straight line on a smooth table. The rod AB is struck at its end A by a by a blow which is perpendicular to its length, find the resulting motion and show that the velocity of the centre of AB is 19 times that of CD and its angular velocity 11 times that of CD.

Solution:

Let each rod be of length 2a and mass m and P be the impulse of the blow at A. Hence there will be impulsive action at B and C in opposite directions, which also taken as X and Y respectively.

Just, before the blow, the three rods are at rest and after the blow let u_1, ω_1; u_2, ω_2; u_3, ω_3 be the velocities of C.G.'s and the angular velocities of the rods AB, BC and CA respectively. Then we have

For, AB, $mu_1 = P - X,\ m\frac{a^2}{3}\omega_1 = pa + Xa \cdot$ ...(A)

For, BC, $mu_2 = X - Y,\ m\frac{a^2}{3}\omega 3 = Xa + Ya \cdot$ ...(B)

For, CD, $mu_3 = Y,\ m\ \frac{a^2}{3}\omega_3 = Y\ a.$ ...(C)

Now, velocity of B should be the same as deduced from motion of AB and BC and similarly velocity of C should be the same as deduced from the motion of BC and CD respectively.

For B $u_1 - a\omega_1 = u_2 + a\omega_2$ ...(1)

For C $u_2 - a\omega_2 = u_3 + a\omega_3$ ...(2)

$$\Rightarrow u_1 - a\omega_1 = \frac{P - X}{m} - \frac{3(P + X}{m} = \frac{-2P - 4X}{m}$$

From (B), we have $u_2 + a\omega_2 = \frac{X - Y}{m} + \frac{3(X + Y)}{m} = \frac{4X + 2Y}{m}$

$\therefore$ (1) $\Rightarrow 2P - 4X = + 2\ Y$

or $4X + Y = -P.$ ...(3)

Again from (B), we obtain $u_2 - a\omega_2 = \frac{X - Y}{m} + \frac{3(X + Y)}{m} = \frac{-2X - 4Y}{m}$

and $u_3 + a\omega_3 = \frac{Y}{m} + \frac{3Y}{m} = \frac{4Y}{m}.$ [using C]

$\Rightarrow -2X - 4Y = 4Y$ or $X + 4Y = 0.$ [using (2)] ...(4)

Solving (3) and (4), we get $X = -\frac{4P}{15}$ and $Y = \frac{P}{15}$

$$\Rightarrow u_1 = \frac{1}{m}\left(P + \frac{4P}{15}\right) = \frac{19P}{15m}, \omega_1 = \frac{3}{ma}\left(P - \frac{4P}{15}\right) = \frac{11P}{5ma}$$

$$u_2 = \frac{1}{m}\left(-\frac{4P}{15} - \frac{P}{15}\right) = -\frac{P}{3m}, \omega_2 = \frac{3}{ma}\left(-\frac{4P}{15} - \frac{P}{15}\right) = \frac{-5P}{5ma}$$

$$u_3 = \frac{1}{15m}, \omega_3 = \frac{3}{ma}\left(\frac{P}{15}\right) = \frac{P}{5ma}$$

$$\Rightarrow u_1 : u_3 = \frac{19P}{15m} : \frac{P}{15m} = 19 : 1 \text{ i.e. } u_1 = 19u_3$$

i.e. velocity of centre of AB is 19 times that of CD.

Further, $\omega_1 : \omega_2 = \frac{11P}{5ma} : \frac{P}{5ma} = 11 : 1$ *i.e.* $\omega_1 = 11\omega_3$

i.e. the angular velocity of AB is 11 times that of CD.

Example 27(e):

A uniform flat rod, of length 2a, rests on a rough horizontal plane with its weight uniformly distributed. A horizontal force P large enough to produce motion is applied suddenly at the end perpendicular to the length of the rod. Show that initially point, where x is given by the positive root of the equation

$$x^3 - \left(\frac{1}{3} - \frac{2P}{\mu W}\right)a^2x - \frac{2}{3}\frac{P}{\mu W}a^3 = 0,$$

W being the weight of the rod and μ the coefficient of friction.

Solution:

Let the impulse P be given at the end A and let O′ be the point about which the rod begins to turn where GO′ = x. Just after the application of the impulse, let ω be the angular velocity of the rod. Hence velocity of the centre of gravity G just after the impuise is x ω as shown in the figure. Due to the impulse P, there is an impulsive friction at each point of the rod.

Now consider an element δy of the rod to the right and also to the left of the point O′, where O′ Q = y.

Friction on each of these elements $= \mu\frac{W}{2a}\delta y$

Now taking moments about the point O′, we have

$$* M\left(\frac{a^2}{3} + x^2\right)\omega = P(a + x) - \int_0^{a+x} \mu\frac{W}{2a}y\,dy - \int_0^{a-x} \mu\frac{W}{2a}y\,dy.$$

[The moments of impulsive frictions are negative because they tend to decrease ω].

$$= P(a + x) - \frac{\mu W}{2a}\left[\frac{y^2}{2}\right]_0^{a-x}$$

$$\text{i.e. } M\left(\frac{a^2}{3} + x^2\right)\omega = P(a + x) - \frac{\mu W}{2a}\left(a^2 + x^2\right).$$

Also moment of centre of gravity G, gives

Change in angular moments about O′ is equal to the sum of the moments of the impulses about O′

$$Mx\omega = P - \int_0^{a+x} \frac{\mu W}{2a} dy + \int_0^{a-x} \frac{\mu W}{2a} dy.$$

[The friction on the right hand of O is opposite to P, while on the left is in direction of P].

$$\Rightarrow Mx\omega = P - \frac{\mu W}{2a}(a+x) + \frac{\mu W}{2a}(a-x) \Rightarrow Mx\omega = P - \frac{\mu W}{2a}x \quad ...(2)$$

Whence eliminating ω between (1) and (2), we readily obtain

$$\left(\frac{a^2}{3}+x^2\right)\left(\frac{P}{x}-\frac{\mu W}{2a}\right) = P(a+x) - \frac{\mu W}{2a}\left(a^2+x^2\right)$$

$$\Rightarrow x^3 - \left(\frac{1}{3}-\frac{2P}{\mu W}\right)a^2x - \frac{2}{3}\frac{P}{\mu W}a^3 = 0.$$

Example 28:

A lamina in the form of an equilateral triangle ABC lies on a smooth horizontal plane. Suddenly it receives a blow at A in a direction parallel to BC, which causes A to move with velocity V. Determine the instantaneous velocity of B and C and describe the subsequent motion of the lamina.

Solution:

As the impulse is parallel to BC, the velocity of G_1, the centre of gravity of the triangular lamina, AVC must also be parallel to BC. Let this velocity be u, and ω be the angular velocity Δ of AVC

Equations of motion of the C.G. of the lamina are : mu = P and ...(1)

$$*\frac{m}{3}\left(\frac{a^2}{3}+\frac{a^2}{3}+\frac{a^2}{3}\right)\omega = P\frac{2}{3}a\sqrt{3} \quad ...(2)$$

(taking moments about G) $\left(\because \text{GD etc.} = a/\sqrt{3}\right)$.

∴ Velocity of the point A = (velocity of G + velocity of A relative to G)

or $V = u + \frac{2}{3}a\sqrt{3\omega}.$ Now (1) and (2)

$$\Rightarrow a\omega = 2\sqrt{3\cdot u\cdot}$$

$$\Rightarrow V = u + 4u = 5u$$

i.e. $u = \frac{1}{5}V$ and $a\omega = \frac{2\sqrt{3}}{5}V.$

* Triangle ABC is kinetically equivalent to three particles, each one third of the mass of the triangle, placed at the middle points of the sides.

$\therefore$ velocity of B = velocity of G + velocity of B relative to G

$$= u + \frac{2}{3}a\sqrt{3\omega}$$

$\Rightarrow$ velocity of B parallel to BC

= (velocity of B parallel to BC) + velocity of B relative to G parallel to BC),

$$= u - \frac{2}{3}a\sqrt{3\omega\cos 60^0} = -a\omega = -\frac{2\sqrt{3}}{5}V.$$

Finally, proceeding in the same manner, the velocities of C are

$-\frac{V}{5}$ and $\frac{2\sqrt{3}}{5}V$ along and perpendicular to BC.

Example 29:

A square plate, of side 2a, is falling with velocity u, a diagonal being vertical, when an inelastic string attached to the middle point of an upper edge becomes tight in vertical position. Show that the impulsive tension of the string is 4/7 Mu, where M is the mass of the plate.

Solution:

When the string becomes tight, a jerk experienced by the string resulting an impulsive tension in the string say, T.

Just after the impulse let u′ be the vertical velocity of G, and ω be the angular velocity of the square, while just before the jerk, the velocity of G is u and there no angular velocity. It is to note that there will be no velocity in the horizontal direction, as there is no horizontal impulse.

Equations of motion of the square plate ABCD are:

$$M(u' - u) = -T \quad ...(1) \quad \text{and} \quad M\frac{2a^2}{3}\omega = T\frac{a}{\sqrt{2}} \qquad ...(2)$$

Also, the velocity of K relative to G is $a\omega$, at right angles to GK. Hence its resolved part is $a\omega \cos 45^0$ *i.e.* $a\omega\frac{1}{\sqrt{2}}$ in vertical upward direction. But just after the impulse, the point K is reduced to rest.

$\Rightarrow$ vertical velocity of the point K = 0 $\Rightarrow u' - \dfrac{a\omega}{\sqrt{2}} = 0.$ ...(3)

$$\therefore \left\{u - \left(\frac{T}{m}\right)\right\} - \left(\frac{3T}{4M}\right) = \textit{i.e.}\ \ T = \frac{4}{7}Mu.$$

Example 30:

A light string is wound round the circumference of a uniform reel of radius a and radius of gyration k about its axis. The free end of the string being tied to a fixed point, the reel is lifted up and let fall so that, at the moment when the sting becomes tight, the velocity of the centre of reel is u and the string is vertical. Find the change in the motion and show that the impulsive tension is $mu\left(\dfrac{k^2}{a^2+k^2}\right)$

Solution:

When the string becomes tight, a jerk is experienced by the string resulting an impulsive tension in the string, say T.

Just after the jerk let v be the velocity of the centre of gravity G and ω the angular velocity, while just before the jerk the velocity of G is u and there is no angular velocity.

Equations of motion of the reel are,

$m\ (v - u) = -\ T$ ...(1)

and $mk^2\ \omega = Ta$ ...(2)

Just after the impulse, the velocity of the point contact K is zero

$\Rightarrow v - a\omega = 0$ ...(3)

$$\therefore \left(u - \frac{1}{m}T\right) - \frac{Ta^2}{mk^2} = 0 \ \Rightarrow T = mu\left(\frac{k^2}{a^2+k^2}\right)$$

Example 31:

A uniform inelastic rod falls without rotation, being inclined at an angle β to the horizon and hits a smooth fixed peg at a distance form its upper end equal to one third of its length. Show that the lower end begins to descend vertically.

Solution:

Let AB be the rod which strikes the fixed smooth peg at C; the inclination of the rod, say, β at that time. Let S be the impulse at the peg C perpendicular to the rod. Just before the impact, the rod was falling without rotation under gravity. Hence it must have then only the vertical velocity V. Just after impact, let u, v, be the horizontal and vertical velocities of the C.G. of the rod AB and let ω its angular velocity.

Now equations of motion of the rod AB are as follows:

$$\left.\begin{array}{ll} mu = S\sin\beta & (1) \\ m(v - V) = -S\cos\beta, & (2) \end{array}\right\} \text{motion of C.G.}$$

and $\quad m\frac{a^2}{3}\omega = S\cdot\frac{a}{3} \quad ...(3) \qquad$ (taking moments about C.G.)

Substituting the value of S from (3) in (1) and (2), we readily get

$u = -a\omega \sin\beta$, $v = v - a\omega \cos\beta$.

Now velocity of the end A relative to G = aω(perpendicular to AB)

and horizontal velocity at A = horizontal velocity of G + horizontal velocity of A relative to G

$u + a\omega \sin\beta = -a\omega \sin\beta + a\omega \sin\beta$ (substituting value of u)

and vertical velocity of A = vertical velocity of G

+ vertical velocity of A relative of G.

$= v + a\omega \cos\beta = (V - a\omega \cos\beta) + a\omega \cos\beta = V$.(substituting for v)

implies that after the impact the lower end A begins to descend vertically.

EXERCISES

1. A square lamina, ABCD rests on a smooth horizontal plane. If the corner A is made to move with velocity u along the line produced then determine the initial angular velocity of the lamina [– 3u/8a].
2. A billard ball is at rest on a horizontal table and is struck by a horizontal table and is struck by a horizontal blow, in a vertical plane passing through the centre of the ball; if the initial motion is one of pure rolling, find the height of the point struck above the table.
3. When the axis of a symmetrical top is stationary and the spin is large and equal to u, a blow J is applied perpendicular to the axis at a distance c from the fixed point. Prove that the maximum angular deviation of the axis is approximately

   $2\tan^{-1}(Jc/Cn)$

   C being the moment of inertia of the top about its axis of symmetry.
4. Two equal uniform rods AB and AC, are freely hinged at A and rest in a straight line on a smooth table. A blow is struck at B perpendicular

to the rods; show that the kinetic energy generated is 7/4 times what it would be if the rods were rigidly fastened together at A.

5. A top is a right cone of height 1 and vertical [angle, 90° and move about its vertex under gravity. Initially it started rotating with angular velocity $5\sqrt{(3g/l)}$, the axis being instantaneously at rest and inclined as 60° to the upward vertical. Show that in the subsequent motion the angle between the axis and vertical oscillations between 60°. and 120°.
6. A top is processing steadily with one end of its axis fixed and its axis horizontal. If a small weight W be suddenly attached to the other end of its axis without impulse, show that an oscillatory motion is set up in which the maximum dip of the axis below the horizontal is approximately.

   $2W\ a/(A^2\ \Omega + C^2n^2)$,

   Where Ω is the initial angular velocity of precession and a the length of the axis.
7. A top of vertex O, is set in motion so that the angular momentum Cn is equal to that about the vertical drawn upward through O. The inclination of the axis to the vertical was initially i, and every point on the axis was moving horizontally. Determine if the axis will become vertical in a finite time, and if it does, show that the velocity of the centre of gyration P at that instant is equal to

   $$2\sqrt{(gl)}\tan\frac{1}{4}i\left\{\sqrt{\left(p-\cos^2\frac{1}{4}i\right)}\right\}$$ where $4A^2\ gp = C^2\ n^2\ l$ and $OP = l$.
8. Three uniform rods, AB, BC, CD are hinged freely at their ends B and C, so to form three sides of a square and are laid on a smooth table; the end is struck by a horizontal blow P at right angles to AB. Show that the initial velocity of A is nineteen times of D and that the impulsive action at B and C are respectively (5/12) P and (1/12) P.
8. A ring of wire of radius c rests on the top of smooth fixed sphere of radius a and is set rotating about the vertical diameter of the sphere. If the ring is slightly displaced, prove that the motion is unstable if the angular velocity is less than

   $$\left[2g\frac{2a^2-c^2}{c^4}\sqrt{\{(a^2-c^2)\}}\right]^{1/2}$$

10. A symmetrical top is set in motion on a rough horizontal plane with an angular motion about its figure, the axis being inclined at angle i to the vertical. Show that between the great approach to end recess from the vertical, the centre of gravity describes an are

$$h \tan^{-1}\left(\frac{\sin i}{p - \cos i}\right),$$

where p and h have the usual meanings.

11. Prove that a top can be set spinning steadily with its axis of the figure horizontal; by an impulsive couple about an axis making with the vertical and angle

$$\tan^{-1}\left(\frac{C^2n^2}{Amgh}\right)$$

12. Two equal uniform rods, AB and AC, are freely jointed at A, and placed on a smooth table so as to be at right angles. The rod AC is struck by a blow at C in a direction perpendicular to itself, show that the resulting velocities of the middle points of AB and AC are in the ratio 2 : 1.